PROBLEMS and SOLUTIONS for STATICS+++

ROBERT J. BONENBERGER
JAMES W. DALLY

University of Maryland, College Park

College House Enterprises, LLC
Knoxville, Tennessee

The manuscript was prepared using Microsoft Word 11 point Times New Roman font. The Word files were converted to pdf files using Adobe Acrobat DC Pro.

College House Enterprises, LLC.
5713 Glen Cove Drive
Knoxville, TN 37919, U. S. A.
Phone and FAX (865) 558 6111
Email jwd@collegehousebooks.com
http://www.collegehousebooks.com

ISBN 978-1-935673-34-7

CONTENTS
Problems and Solutions for Statics+++

CHAPTER 1 Basic Concepts in Mechanics

1.1 Design a lever that will permit you to lift a weight of 1000 N a distance of 100 mm. In the design, prepare a dimensioned sketch of the lever, state the force that you apply to the lever, and show the analysis proving that you will be able to lift the weight and move it upward by the specified distance.

1.2 The *Sojourner*, a mechanical rover that explored Mars during a mission in 1997, had a mass of 10.5 kg. Determine the weight of *Sojourner* (in newtons) on (a) Earth, (b) Mars. Assume Mars has a mass of 0.64×10^{24} kg and a radius of 3390 km.

1.3 Weather satellites are often placed in geosynchronous orbit around Earth, so that they appear stationary in the sky when viewed from a point on the equator. The orbital altitude for such a satellite is 35,800 km. If a satellite weighs 500 N on the Earth's surface, calculate its weight in orbit. Assume Earth has a mass of 5.976×10^{24} kg and a radius of 6370 km.

1.4 An astronaut weighs 150 lb on the Earth's surface. When orbiting Earth in the space shuttle *Atlantis*, however, the astronaut's weight is measured as 130 lb. Determine the altitude of the shuttle's orbit. Assume Earth has a mass of 5.976×10^{24} kg and a radius of 6370 km.

1.5 Suppose that you are in a space ship traveling in a circular orbit about the Earth, as shown in the figure to the right. Using a spreadsheet, prepare a graph showing your weight as a function of your position relative to the surface of the Earth. Consider radii from $1.0(R_e)$, on the Earth's surface, to $15(R_e)$.

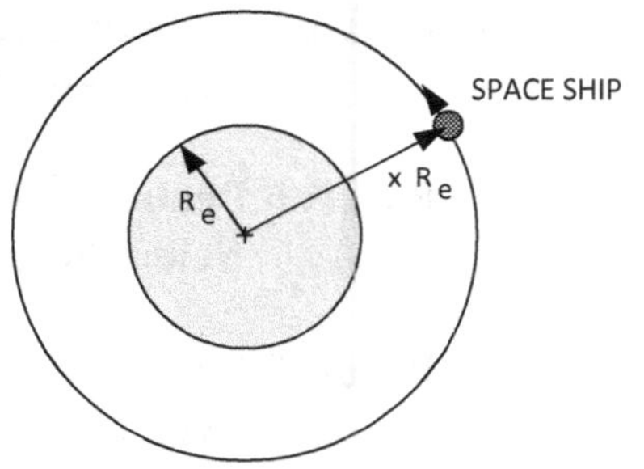

1.6 Determine the gravitational force exerted on the Earth by (a) the sun, (b) the moon. Also find the ratio of the forces (F_{sun}/F_{moon}). Assume masses of 5.976×10^{24} kg for the Earth, 1.990×10^{30} kg for the sun and 7.350×10^{22} kg for the moon. The mean distance (center-to-center) between the sun and the Earth is 149.6×10^6 km and between the moon and the Earth is 384×10^3 km.

1.7 Prove whether or not the blocks, shown in the figure (a) and (b) below, are in equilibrium.

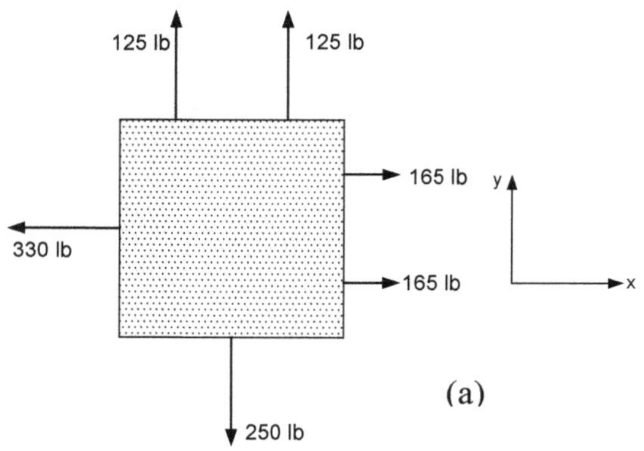

(a)

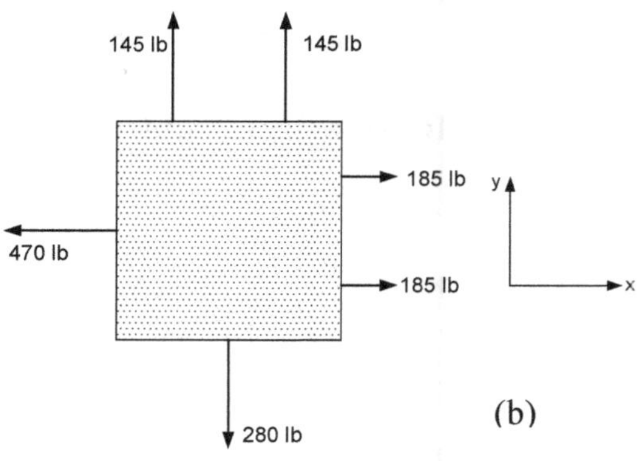

(b)

1.8 Prove whether or not the blocks, shown in the figures (a) and (b) below, are in equilibrium.

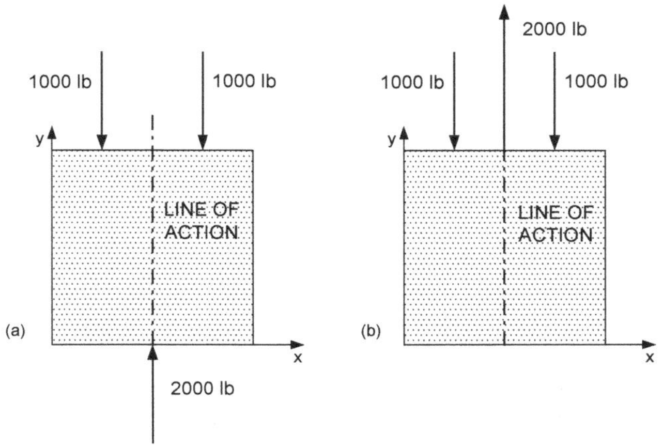

1.9 Draw a free body diagram of an automobile parked on a level street showing the reaction forces on the tires. Also show the active forces on the pavement. Assume that the auto weighs 3,200 lb. State the assumption made in distributing the weight of the auto among the four tires.

1.10 Determine the weight of a body with the following masses in SI units:

(a) 65 kg (b) 19 kg (c) 58 kg (d) 17.3 kg (e) 34.7 kg (f) 81.2 kg

1.11 Determine the weight of a body with the following masses in U. S. Customary units:

(a) 7.5 slug (b) 12 slug (c) 65 slug (d) 4.2 slug (e) 33.7 slug (f) 49.4 slug

1.12 If a pressure of 3,000 psi acts on a piston with a diameter of 1.75 in., determine the force required to maintain the piston in equilibrium.

1.13 If a pressure of 300 MPa acts on a piston with a diameter of 150 mm, determine the force required to maintain the piston in equilibrium.

1.14 If a force of 3,000 lb acts on a piston with a diameter of 5.0 in., determine the pressure required to maintain the piston in equilibrium.

1.15 If a force of 5,000 N acts on a piston with a diameter of 100 mm, determine the pressure required to maintain the piston in equilibrium.

1.16 If a piston is in equilibrium under a pressure of 1.5 ksi and a force of 10.5 kip, determine the diameter of the piston.

1.17 If a piston is in equilibrium under a pressure of 55 MPa and a force of 675 kN, determine the diameter of the piston.

1.18 List the four basic quantities and give their units in both the SI and U. S. Customary systems.

1.19 Determine the SI equivalent of the masses listed below.

(a) 7.5 slug (b) 12 slug (c) 65 slug (d) 4.2 slug (e) 33.7 slug (f) 49.4 slug

1.20 Determine the U. S. Customary equivalent of the masses listed below.

(a) 65 kg (b) 19 kg (c) 58 kg (d) 17.3 kg (e) 34.7 kg (f) 81.2 kg

1.21 Determine the SI equivalent of the following forces:

(a) 2,400 lb (b) 425 lb (c) 688 lb (d) 18 tons (e) 52 kip (f) 95 kip

1.22 Determine the U. S. Customary equivalents for the following forces:

(a) 2,234 N (b) 89.8 N (c) 13.2 kN (d) 96.9 mN (e) 142.3 μN (f) 44.8 MN

1.23 Determine the SI equivalent of the following lengths:

(a) 13 ft
(b) 153 in.
(c) 41.3 yards
(d) 16.4 miles
(e) 6 ft - 2 in.
(f) 1.8 miles - 662 yards

1.24 Determine the U. S. Customary equivalents for the following lengths:

(a) 7.4 m
(b) 500 mm
(c) 12.9 cm
(d) 13.2 km
(e) 18.8 mm
(f) 64.3 cm

1.25 Convert the following speeds into SI equivalent units:

(a) 60 mph
(b) 45 mph
(c) 30 mph
(d) 22 ft/s
(e) 90 ft/s
(f) 580 mph
(g) 16 ft/s
(h) 224 ft/s
(i) 187 mph

1.26 Convert the following speeds into U. S. Customary equivalent units:

(a) 120 km/h
(b) 61 km/h
(c) 88 km/h
(d) 123 m/s
(e) 510 m/s
(f) 17.5 km/s
(g) 312 m/h
(h) 16.7 m/s
(i) 43.5 km/s

1.27 Convert the following stresses into SI equivalent units:

(a) 12,000 psi
(b) 2,320 psi
(c) 1,980 psi
(d) 132.5 ksi
(e) 30×10^6 psi
(f) 10.5×10^6 ksi
(g) 194,200 psi
(h) 32 psi
(i) 96,230 psi

1.28 Convert the following stresses into U. S. Customary equivalent units:

(a) 1,000 MPa
(b) 121 MPa
(c) 86.4 MPa
(d) 1,400 kPa
(e) 144 kPa
(f) 9,642 kPa
(g) 207 GPa
(h) 16,430 MPa
(i) 42.3 kPa

1.29 Prepare a short written description of the differences among scalars, vectors, and tensors. Also prepare a sketch illustrating a quantity of each type.

1.30 Consider each quantity listed in Table 1.4 and classify it as a scalar, vector or tensor.

Table 1.4
Unit Conversion Factors

Quantity	U. S. Customary	SI Equivalent
Acceleration	ft/s^2 in/s^2	0.3048 m/s^2 0.0254 m/s^2
Area	ft^2 in^2	0.0929 m^2 645.2 mm^2
Distributed Load	lb/ft lb/in.	14.59 N/m 0.1751 N/mm
Energy	ft-lb	1.356J
Force	kip = 1000 lb lb	4.448 kN 4.448N
Impulse	lb-s	4.448 N-s
Length	ft in mi	0.3048 m 25.40 mm 1.609 km
Mass	lb mass slug ton mass	0.4536 kg 14.59 kg 907.2 kg
Moments or Torque	ft-lb in-lb	1.356 N-m 0.1130 N-m
Area Moment of Inertia	in^4	0.4162 x 10^6 mm^4
Power	ft-lb/s hp	1.356 W 745.7 W
Stress and Pressure	lb/ft^2 lb/in^2 (psi) ksi = 1000 psi	47.88 Pa 6.895 kPa 6.895 MPa
Velocity	ft/s in/s mi/h (mph)	0.3048 m/s 0.0254 m/s 0.4470 m/s
Volume	ft^3 in^3 gal	0.02832 m^3 16.39 cm^3 3.785 L
Work	ft-lb	1.356 J

Problem 1.2:

Sojourner rover:

$m = 10.5\ kg$

Mars:

$m_M = 0.64 \times 10^{24}\ kg$

$r_M = 3390\ km$

find: $\boxed{W = ?}$.

a) on Earth

b) on Mars

① $W = mg = (10.5)(9.807)$

② $\boxed{W = 103.0\ N}$. on Earth (a)

③ $W = G\dfrac{m\,m_M}{r_M^2} = (6.673 \times 10^{-11})\dfrac{(10.5)(0.64 \times 10^{24})}{(3390 \times 10^3)^2}$

④ $\boxed{W = 39.02\ N}$. on Mars (b)

Problem 1.3:

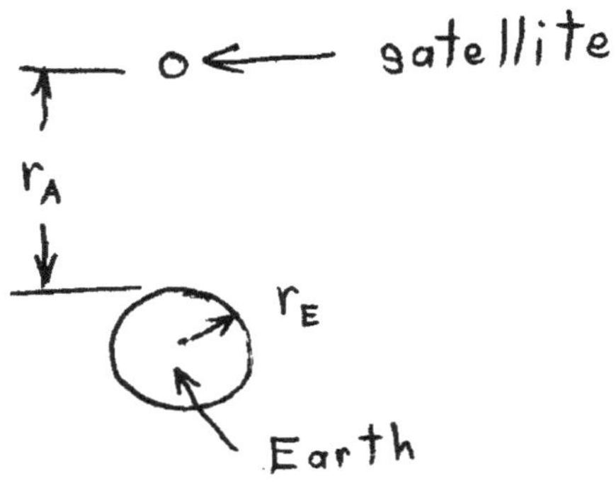

$$r_E = 6{,}370 \text{ km}$$
$$r_A = 35{,}800 \text{ km}$$
$$W_S = 500 \text{ N} \leftarrow \text{on Earth}$$
$$m_E = 5.976 \times 10^{24} \text{ kg}$$

find: $\boxed{W_S = ?}$. ← in orbit

① $W_S = m_S g \Rightarrow m_S = \dfrac{W_S}{g} = \dfrac{500 \ (\text{kg-m/s}^2)}{9.807 \ \text{m/s}^2}$

② $m_S = 50.98 \text{ kg}$

③ in orbit:

$$W_S = F = G \frac{m_S m_E}{r^2}$$

④ $r = r_A + r_E = 6{,}370 + 35{,}800 = 42{,}170 \text{ km}$

$$= 42.17 \times 10^6 \text{ m}$$

⑤ $W_S = \left(6.673 \times 10^{-11} \dfrac{\text{m}^3}{\text{kg-s}^2}\right) \dfrac{(50.98 \text{ kg})(5.976 \times 10^{24} \text{ kg})}{(42.17 \times 10^6 \text{ m})^2}$

⑥ $\boxed{W_S = 11.43 \text{ N}}$.

Problem 1.14:

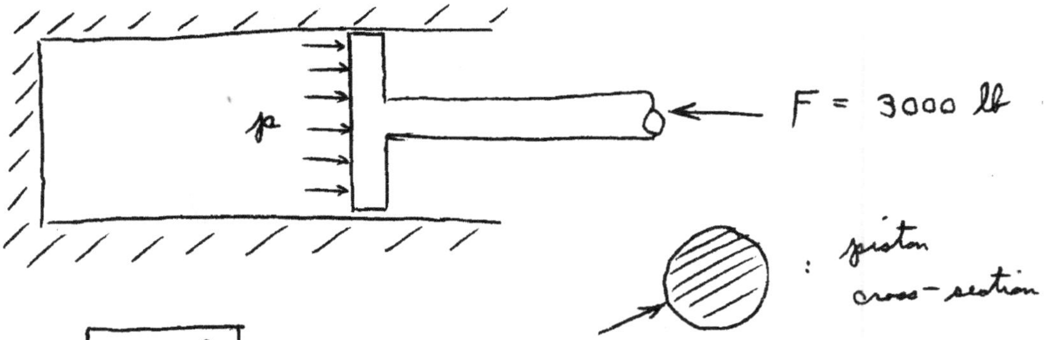

find : $\boxed{p = ?}$.

① $F = pA = p\left(\frac{\pi}{4} d^2\right)$

② $p = \dfrac{4F}{\pi d^2} = \dfrac{4\,(3000\text{ lb})}{\pi\,(5\text{ in})^2}$

③ $\boxed{p = 152.8\text{ psi}}$.

Problem 1.16:

piston :

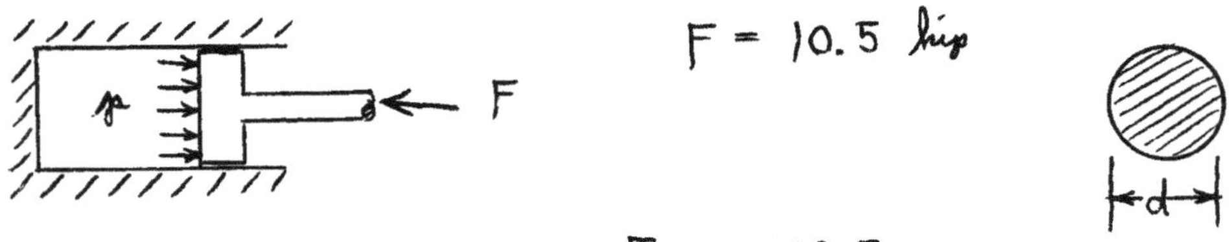

$p = 1.5$ ksi

$F = 10.5$ kip

* equilibrium

find : $\boxed{d = ?}$.

① $F = pA \Rightarrow A = \dfrac{F}{p} = \dfrac{10.5}{1.5} = 7\text{ in}^2$

② $A = \frac{\pi}{4} d^2 \Rightarrow d = \sqrt{\frac{4}{\pi} A} = \sqrt{\frac{4}{\pi}(7)}$

③ $\boxed{d = 2.985\text{ in}}$.

Problem 1.21(a,f):

a) $F = 2400\ lb$ find: $\boxed{F = ?\ (\text{in SI})}$.

f) $F = 95\ kip$

① $F = (2400\ lb)\left(\frac{4.448\ N}{lb}\right) = \boxed{10{,}680\ N}$. (a)

② $F = (95\ kip)\left(\frac{4.448\ kN}{kip}\right) = \boxed{422.6\ kN}$. (f)

Problem 1.22:

given: find: $\boxed{F = ?\ (\text{in FPS})}$.

a) $F = 2234\ N$ d) $F = 96.9\ mN$

b) $F = 89.8\ N$ e) $F = 142.3\ \mu N$

c) $F = 13.2\ kN$ f) $F = 44.8\ MN$

① $F = (2234\ N)\left(\frac{lb}{4.448\ N}\right) = \boxed{502.2\ lb}$. (a)

② $F = (89.8\ N)\left(\frac{lb}{4.448\ N}\right) = \boxed{20.19\ lb}$. (b)

③ $F = (13.2\ kN)\left(\frac{kip}{4.448\ kN}\right) = \boxed{2.968\ kip = 2{,}968\ lb}$. (c)

④ $F = (96.9 \times 10^{-3}\ N)\left(\frac{lb}{4.448\ N}\right) = \boxed{0.02179\ lb}$. (d)

⑤ $F = (142.3 \times 10^{-6}\ N)\left(\frac{lb}{4.448\ N}\right) = \boxed{3.199 \times 10^{-5}\ lb}$. (e)

⑥ $F = (44.8 \times 10^{6}\ N)\left(\frac{lb}{4.448\ N}\right) = \boxed{10.07 \times 10^{6}\ lb = 10{,}070\ kip}$. (f)

Problem 1.27:

given:

a) $\sigma = 12{,}000 \text{ psi}$

b) $\sigma = 2320 \text{ psi}$

c) $\sigma = 1980 \text{ psi}$

d) $\sigma = 132.5 \text{ ksi}$

e) $\sigma = 30 \times 10^6 \text{ psi}$

f) $\sigma = 10.5 \times 10^6 \text{ ksi}$

g) $\sigma = 194{,}200 \text{ psi}$

h) $\sigma = 32 \text{ psi}$

i) $\sigma = 96{,}230 \text{ psi}$

find: $\boxed{\sigma = ? \text{ (in SI)}}$.

① $\sigma = (12{,}000 \text{ psi})\left(\dfrac{6.895 \text{ kPa}}{\text{psi}}\right) = \boxed{82{,}740 \text{ kPa} = 82.74 \text{ MPa}}$ (a).

② $\sigma = (2320 \text{ psi})\left(\dfrac{6.895 \text{ kPa}}{\text{psi}}\right) = \boxed{16{,}000 \text{ kPa} = 16.00 \text{ MPa}}$. (b)

③ $\sigma = (1980 \text{ psi})\left(\dfrac{6.895 \text{ kPa}}{\text{psi}}\right) = \boxed{13{,}650 \text{ kPa} = 13.65 \text{ MPa}}$. (c)

④ $\sigma = (132.5 \text{ ksi})\left(\dfrac{6.895 \text{ MPa}}{\text{ksi}}\right) = \boxed{913.6 \text{ MPa}}$. (d)

⑤ $\sigma = (30 \times 10^6 \text{ psi})\left(\dfrac{6.895 \text{ kPa}}{\text{psi}}\right) = 206.9 \times 10^6 \text{ kPa} = 206.9 \text{ GPa}$ (e).

⑥ $\sigma = (10.5 \times 10^6 \text{ ksi})\left(\dfrac{6.895 \text{ MPa}}{\text{ksi}}\right) = 72.40 \times 10^6 \text{ MPa} = 72.40 \text{ TPa}$ (f).

⑦ $\sigma = (194{,}200 \text{ psi})\left(\dfrac{\text{ksi}}{1000 \text{ psi}}\right)\left(\dfrac{6.895 \text{ MPa}}{\text{ksi}}\right) = 1339 \text{ MPa} = 1.339 \text{ GPa}$ (g).

⑧ $\sigma = (32 \text{ psi})\left(\dfrac{6.895 \text{ kPa}}{\text{psi}}\right) = \boxed{220.6 \text{ kPa}}$. (h)

⑨ $\sigma = (96{,}230 \text{ psi})\left(\dfrac{\text{ksi}}{1000 \text{ psi}}\right)\left(\dfrac{6.895 \text{ MPa}}{\text{ksi}}\right) = \boxed{663.5 \text{ MPa}}$. (i)

Problem 1.28(a,d):

a) $\sigma = 1000 \text{ MPa}$

d) $\sigma = 1400 \text{ kPa}$

find: $\boxed{\sigma = ? \text{ (in FPS)}}$.

① $\sigma = (1000 \text{ MPa})\left(\dfrac{\text{ksi}}{6.895 \text{ MPa}}\right) = \boxed{145.0 \text{ ksi}}$. (a)

② $\sigma = (1400 \text{ kPa})\left(\dfrac{\text{psi}}{6.895 \text{ kPa}}\right) = \boxed{203.0 \text{ psi}}$. (d)

CHAPTER 2 Forces and Moments

2.1 Prepare a free body diagram (FBD) showing the following loads applied to a beam:

(a) concentrated load
(b) uniformly distributed load
(c) non-uniform distributed load
(d) two reaction loads

2.2 Suppose you are to analyze the rope used in a tug of war. Make a section cut of the rope at a location between the two teams and draw a diagram representing the stresses acting on the area exposed by the section cut. Also draw a diagram representing the internal force acting on the area exposed by the section cut.

2.3 Show a graphic representation of a force vector with a magnitude (F) and an orientation (θ) relative to the positive x-axis for:

(a) F = 7,500 lb and θ = 105°
(b) F = 1,400 N and θ = 210°
(c) F = 11.2 kN and θ = 48°
(d) F = 1.620 kip and θ = 155°

2.4 Show that the body, illustrated in figures (a) and (b) below, is in equilibrium before and after sliding the 4000 lb force along its line of action.

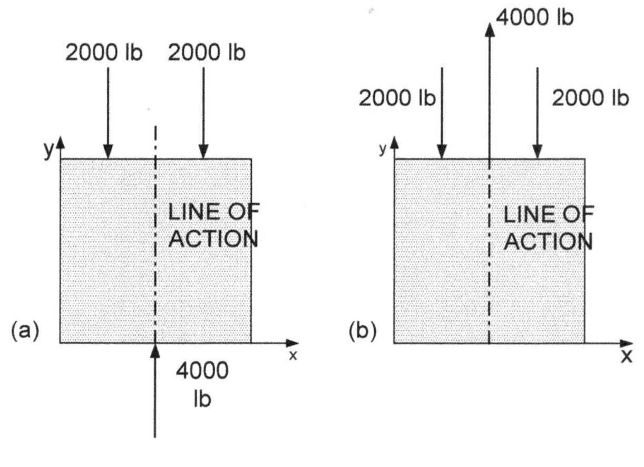

2.5 Draw a disk shaped body and apply a vertical force directed downward at its center. Also show a reactive force at its point of contact with a horizontal support. (a) Describe the condition for equilibrium of the disk. (b) Slide the applied force along its line of action to another location and describe the "new" condition for equilibrium.

2.6 Prepare a drawing showing the reactive forces on the short beam illustrated in the figure to the right.

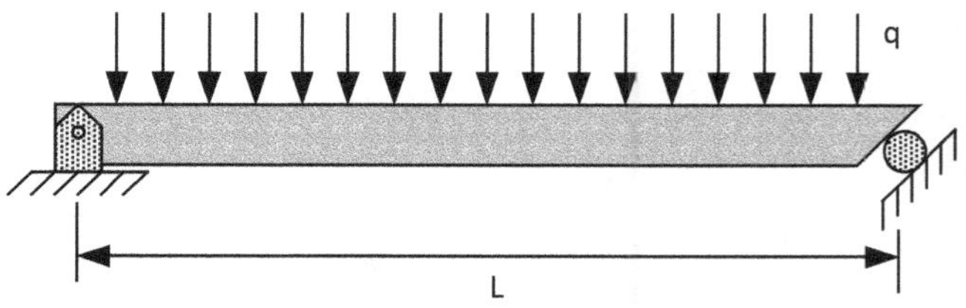

2.7 Prepare three drawings showing the reactive forces on each of the three cylinders in the stack shown in the figure to the right. Note that the cylinders are of the same diameter, and they are bonded together at the three contact points.

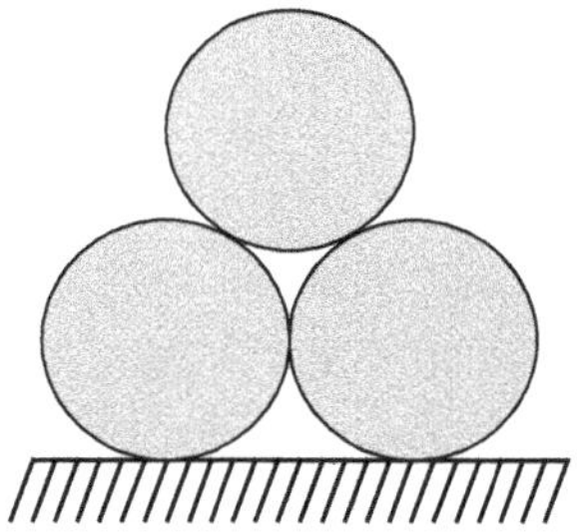

2.8 For the vectors **A** and **B,** illustrated in the figure shown to the left, determine the magnitude and direction of the following resultant vectors:

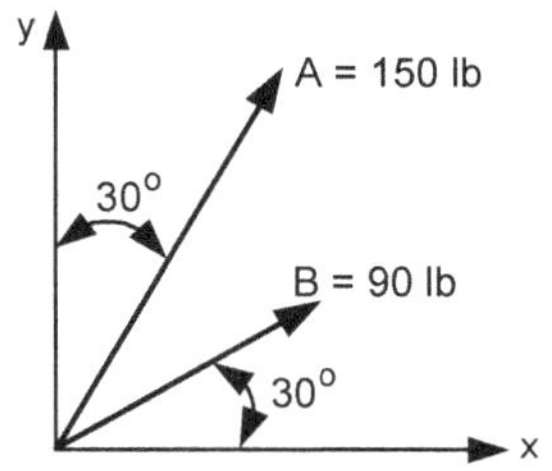

(a) $\mathbf{S_v} = \mathbf{A} + \mathbf{B}$
(b) $\mathbf{D_v} = \mathbf{A} - \mathbf{B}$
(c) $\mathbf{D_v} = \mathbf{B} - \mathbf{A}$.

2.9 For the vectors **A** and **B,** illustrated in the figure shown to the right, determine the magnitude and direction of the following resultant vectors:

(a) $\mathbf{S_v} = \mathbf{A} + \mathbf{B}$
(b) $\mathbf{D_v} = \mathbf{A} - \mathbf{B}$
(c) $\mathbf{D_v} = \mathbf{B} - \mathbf{A}$.

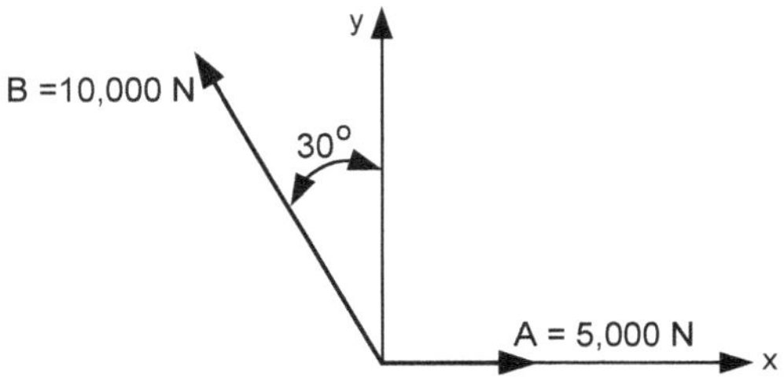

2.10 For the vectors **A** and **B,** illustrated in the figure shown to the left, determine the magnitude and direction of the following resultant vectors:

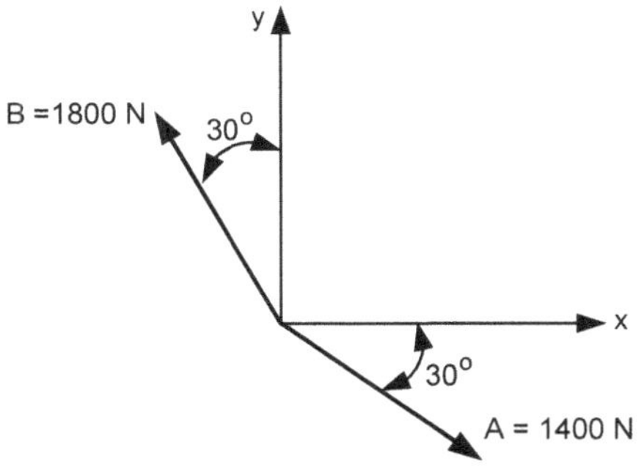

(a) $\mathbf{S_v} = \mathbf{A} + \mathbf{B}$
(b) $\mathbf{D_v} = \mathbf{A} - \mathbf{B}$
(c) $\mathbf{D_v} = \mathbf{B} - \mathbf{A}$.

2.11 Prepare a drawing similar to the one shown below illustrating the vector subtraction $\mathbf{D_v} = \mathbf{B} - \mathbf{A}$.

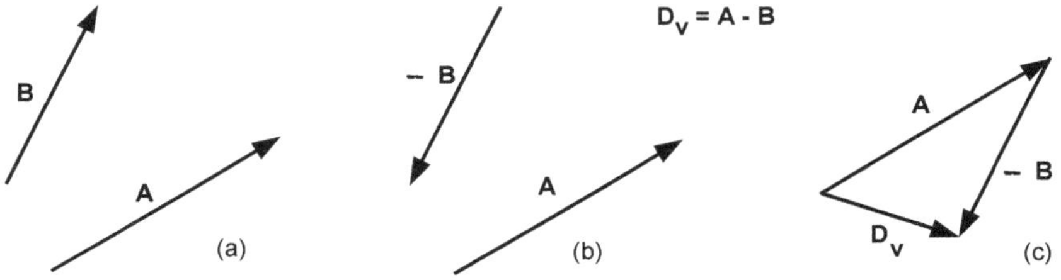

2.12 From the figure shown to the right, determine the Cartesian components of the force vectors: (a) **A** and (b) **B**.

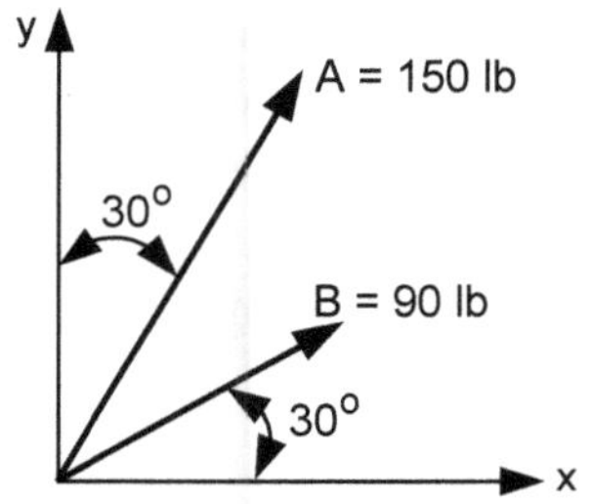

2.13 From the figure shown to the left, determine the Cartesian components of the force vectors:

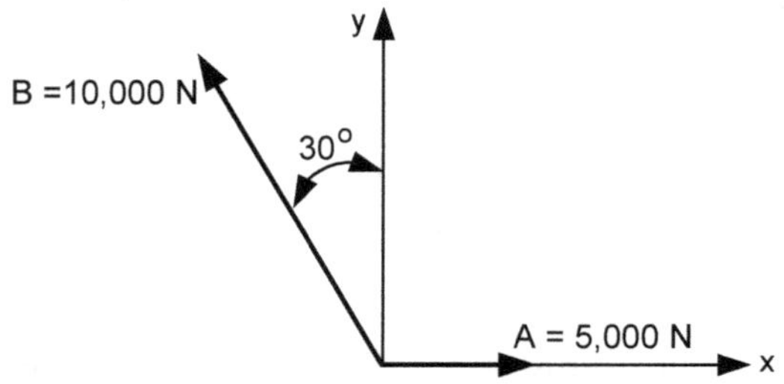

(a) **A** and (b) **B**.

2.14 From the figure shown to the right, determine the Cartesian components of the force vectors:

(a) **A** and (b) **B**.

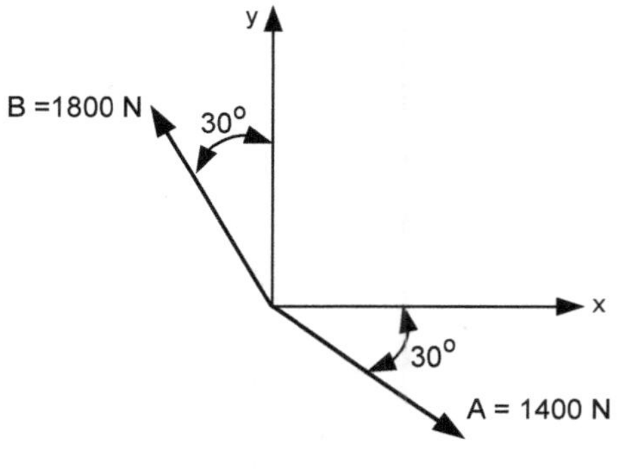

2.15 Using the Cartesian components of vectors **A** and **B,** in the figure shown to the left, determine the magnitude and direction of:

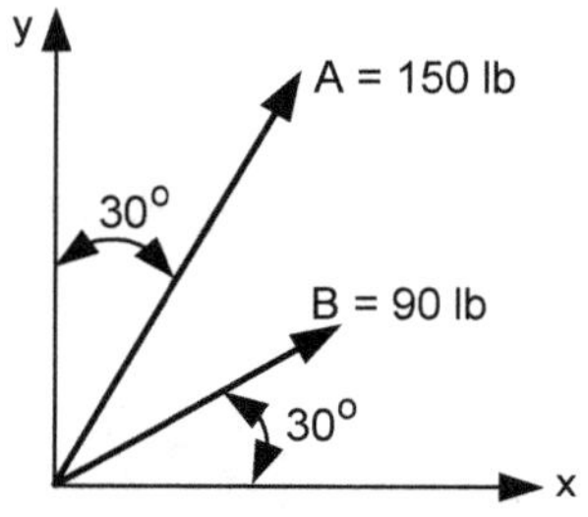

(a) $\mathbf{S_v} = \mathbf{A} + \mathbf{B}$
(b) $\mathbf{D_v} = \mathbf{A} - \mathbf{B}$
(c) $\mathbf{D_v} = \mathbf{B} - \mathbf{A}$.

2.16 Using the Cartesian components of vectors **A** and **B,** in the figure shown to the right, determine the magnitude and direction of:

(a) $\mathbf{S_v} = \mathbf{A} + \mathbf{B}$
(b) $\mathbf{D_v} = \mathbf{A} - \mathbf{B}$
(c) $\mathbf{D_v} = \mathbf{B} - \mathbf{A}$.

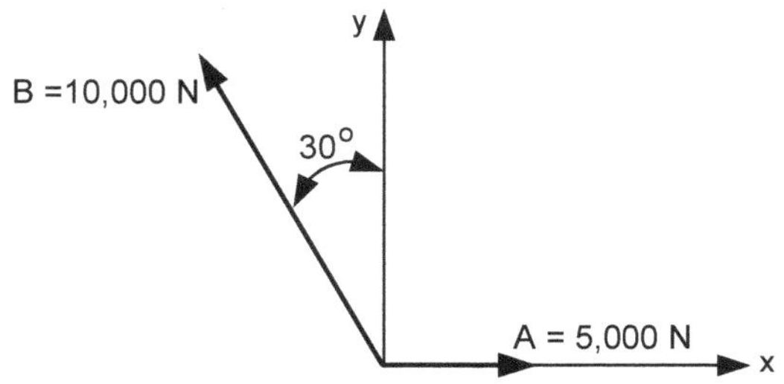

2.17 Using the Cartesian components of vectors **A** and **B,** in the figure shown to the left, determine the magnitude and direction of:

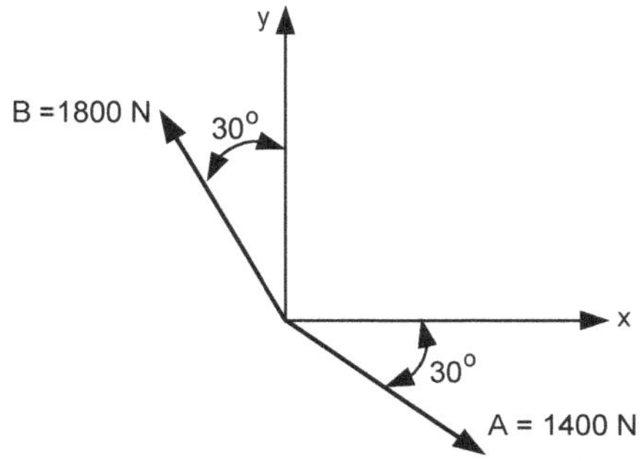

(a) $\mathbf{S_v} = \mathbf{A} + \mathbf{B}$
(b) $\mathbf{D_v} = \mathbf{A} - \mathbf{B}$
(c) $\mathbf{D_v} = \mathbf{B} - \mathbf{A}$.

2.18 Determine the force components along the x and y' axes due to the force of 18,440 lbs that is pulling on the eyebolt shown in the figure to the right.

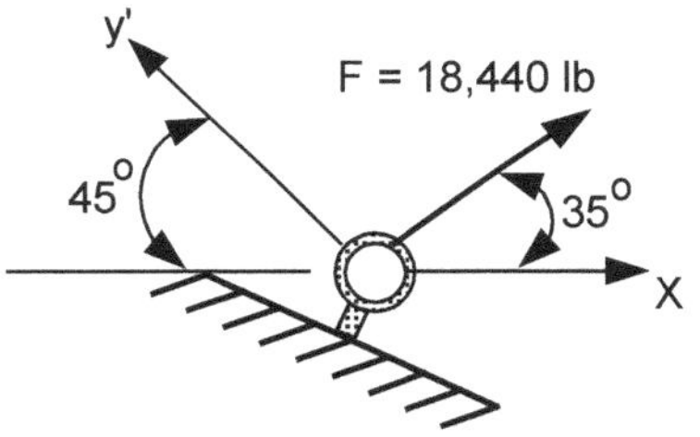

2.19 A linkage, shown in the figure to the left, is subjected to a force of 2,150 N. Determine the components of the force along the axes of links AB and AC.

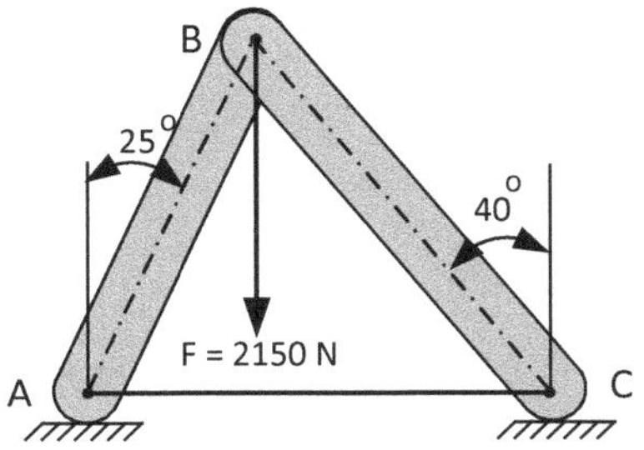

2.20 Determine the angle θ so that the link AB, shown in the figure to the right, has a component of force of 1,250 lb along its axis. The angle ϕ is fixed at 6.4 degrees.

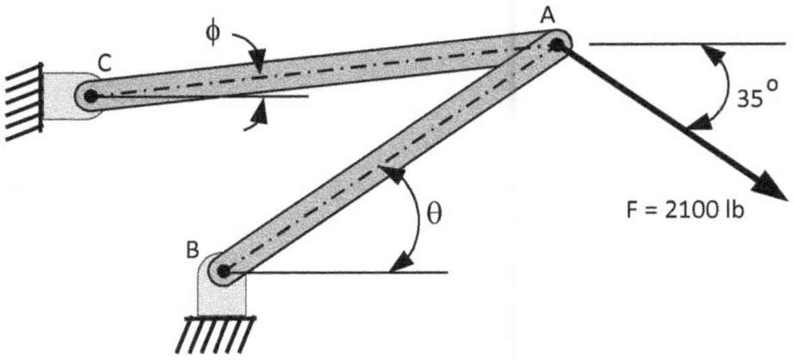

2.21 Determine the vector sum $\mathbf{S_v}$ of the three vectors shown in the figure to the right. Specify its direction relative to the x-axis.

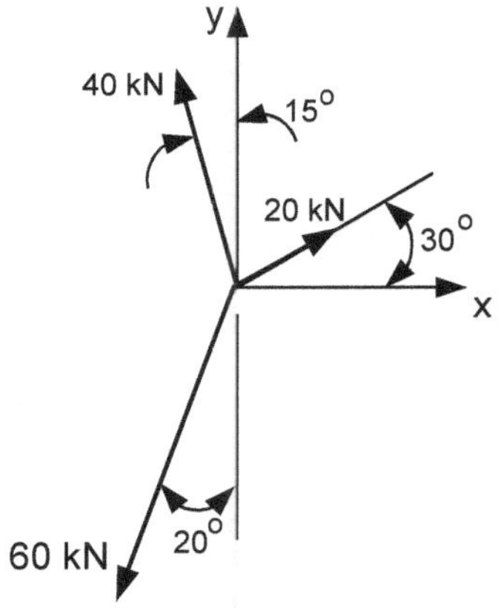

2.22 Determine the magnitude and the direction of the force **F**, shown in the figure to the right if the box is to be lifted from the floor without swinging. The weight of the box is 35 kN.

2.23 If θ is fixed at 55°, in the figure shown to the right, calculate the maximum weight of the box that can be lifted without it swinging when it clears the floor.

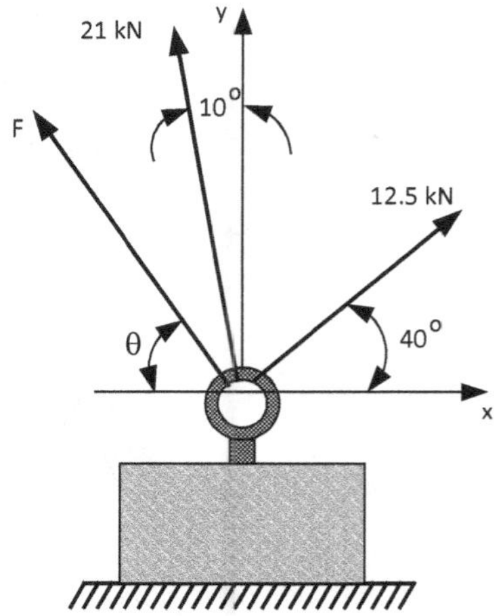

2.24 The gusset plate, illustrated in the figure below, is subjected to four forces due to the attached uniaxial structural members. If the vector sum of the four forces is zero, determine the forces F_1 and F_2.

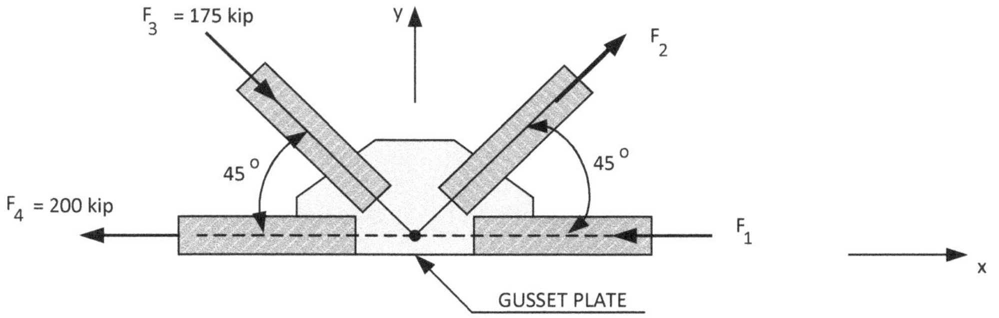

2.25 Three forces act on the bracket, illustrated in the figure to the right. Determine F_1 and θ if the magnitude of one of the Cartesian components of the vector sum S_{vx} = 16,200 lb. Note, $S_{vy} = 0$.

2.26 If the vector sum $\mathbf{S_v}$ of the three forces is 17,500 lb, and is oriented at 10° relative to the x-axis, determine F_1 and θ. Reference the figure to the right.

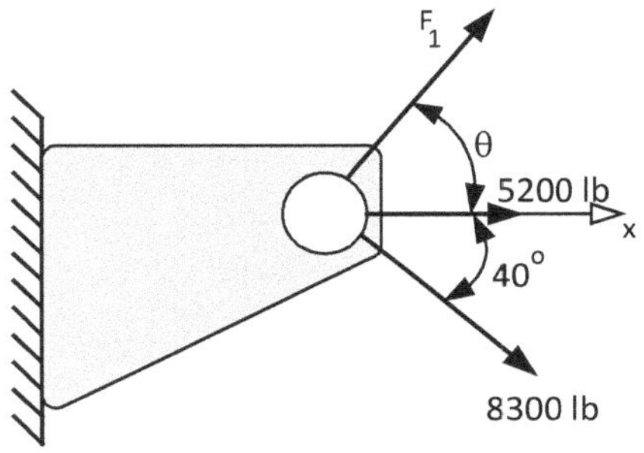

2.27 Define coplanar forces, and show a sketch illustrating four coplanar force vectors.

2.28 Define concurrent forces, and show a sketch illustrating four concurrent force vectors.

2.29 Prepare a sketch of three forces that are:

(a) coplanar, concurrent (b) non-coplanar, concurrent (c) coplanar, non-concurrent

2.30 Suppose you exert a force of 125 N on a single ended lug wrench with a 320-mm long handle to loosen a wheel bolt. What moment (torque) are you applying to the bolt?

2.31 You exert a force of 150 N with each hand on the handles of a double-ended lug wrench to loosen a wheel bolt. If each handle is 225 mm long, what moment (torque) are you applying to the bolt?

2.32 Determine the moment about point O due to the force shown in the figure to the right.

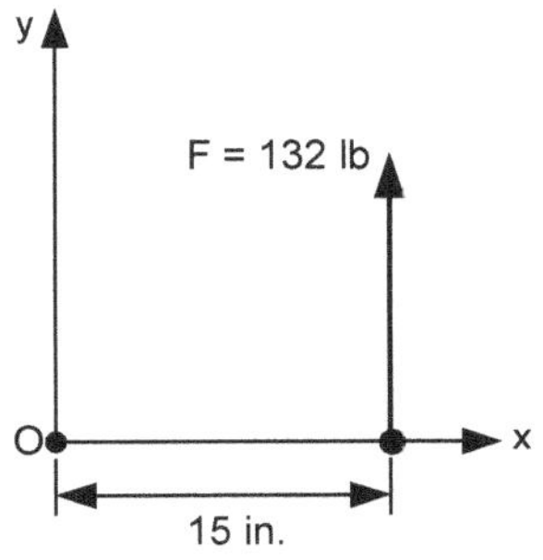

2.33 Determine the moment about point O due to the force shown in the figure to the left.

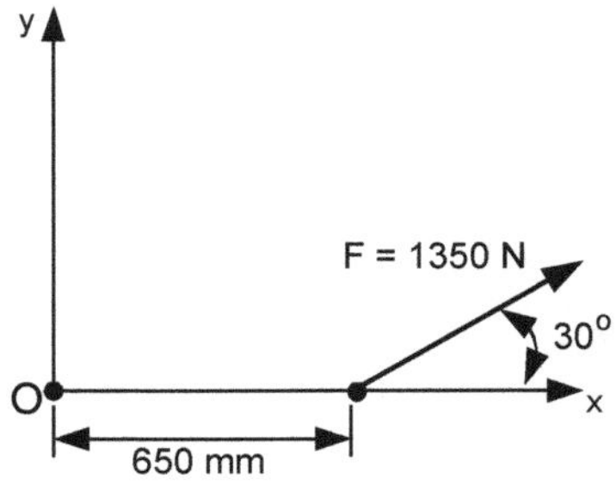

2.34 Determine the total moment about point O due to the two forces shown in the figure to the right.

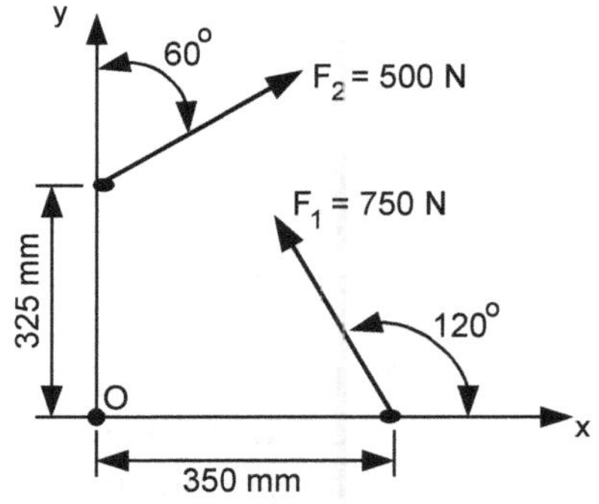

2.35 Express the force **F** or moment **M** in vector format if its Cartesian components are:

(a) $F_x = 2{,}100$ N, $F_y = -2{,}300$ N, and $F_z = 1{,}250$ N
(b) $F_x = 78$ lb, $F_y = -185$ lb, and $F_z = 124$ lb
(c) $M_x = 300$ N-m, $M_y = 400$ N-m, and $M_z = 500$ N-m
(d) $M_x = 740$ in-lb, $M_y = 425$ in-lb, and $M_z = -520$ in-lb

2.36 Determine the magnitude and direction, relative to the positive x-axis, for each of the vectors listed:

(a) $F_x = 1{,}400$ N, $F_y = 900$ N and $F_z = 0$
(b) $F_x = 355$ lb, $F_y = 520$ lb and $F_z = 0$
(c) $M_x = 1{,}800$ N-m, $M_y = 1{,}350$ N-m and $M_z = 0$
(d) $M_x = 655$ ft-lb, $M_y = -335$ ft-lb and $M_z = 0$

2.37 A position vector **r** is constructed in space between the origin (0, 0, 0) and point P. Write **r** as a Cartesian vector and determine the magnitude (r) and coordinate direction angles (α, β, γ) if:

(a) P = (7, 4, 6) in.
(b) P = (110, 65, 70) mm
(c) P = (0.120, 0.155, 0.170) m
(d) P = (–10, 3, –8) ft

2.38 Determine the coordinate direction angles (α, β, γ) for the following forces:

(a) $\mathbf{F} = (4\,\mathbf{i} + 5\,\mathbf{j} - 6\,\mathbf{k})$ N
(b) $\mathbf{F} = (-3\,\mathbf{i} + 2\,\mathbf{j} - 6\,\mathbf{k})$ lb
(c) $\mathbf{F} = (-3\,\mathbf{i} - 3\,\mathbf{j} + 7\,\mathbf{k})$ kN
(d) $\mathbf{F} = (5\,\mathbf{i} - 2\,\mathbf{j} - 8\,\mathbf{k})$ kip

2.39 A space force **F** has a known magnitude (F) and two known coordinate direction angles (α, β). Write a vector equation for **F**, specify the three components (F_x, F_y, F_z) and prepare a drawing showing the force vector in a three-dimensional Cartesian coordinate system for:

(a) F = 1,200 N, $\alpha = 60°$ and $\beta = 30°$
(b) F = 1,500 lb, $\alpha = 75°$ and $\beta = 50°$
(c) F = 17 kN, $\alpha = 135°$ and $\beta = 60°$
(d) F = 25 kip, $\alpha = 50°$ and $\beta = 140°$

2.40 Write a vector equation for the summation of two space forces $\mathbf{F_1}$ and $\mathbf{F_2}$. The force F_1 is 22.5 kN in magnitude with coordinate direction angles of $\alpha_1 = 60°$, $\beta_1 = 60°$ and $\gamma_1 = 135°$. The force F_2 is 30.0 kN in magnitude with coordinate direction angles of $\alpha_2 = 60°$, $\beta_2 = 30°$ and $\gamma_2 = 90°$. Prepare a drawing showing this resultant force vector in a three-dimensional Cartesian coordinate system.

2.41 Determine the angle between the following pairs of space forces:

(a) $\mathbf{F_1} = (3\,\mathbf{i} + 6\,\mathbf{j} - 3\,\mathbf{k})$ N and $\mathbf{F_2} = (-2\,\mathbf{i} + 5\,\mathbf{j} + 7\,\mathbf{k})$ N
(b) $\mathbf{F_1} = (5\,\mathbf{i} - 1\,\mathbf{j} + 4\,\mathbf{k})$ lb and $\mathbf{F_2} = (8\,\mathbf{i} - 2\,\mathbf{j} + 3\,\mathbf{k})$ lb
(c) $\mathbf{F_1} = (-2\,\mathbf{i} + 6\,\mathbf{j} - 3\,\mathbf{k})$ kN and $\mathbf{F_2} = (-3\,\mathbf{i} + 4\,\mathbf{j} - 4\,\mathbf{k})$ kN

2.42 Determine the magnitude of a single force component produced by two space forces ($\mathbf{F_1}$, $\mathbf{F_2}$) that is directed along a line of action which lies in the x–y plane and makes an angle of θ with the positive x-axis:

(a) $\mathbf{F_1} = (3\,\mathbf{i} + 6\,\mathbf{j} - 3\,\mathbf{k})$ lb; $\mathbf{F_2} = (-2\,\mathbf{i} + 3\,\mathbf{j} + 5\,\mathbf{k}.)$ lb; and $\theta = 30°$
(b) $\mathbf{F_1} = (4\,\mathbf{i} + 3\,\mathbf{j} - 2\,\mathbf{k})$ kN; $\mathbf{F_2} = (3\,\mathbf{i} - 5\,\mathbf{j} + 6\,\mathbf{k})$ kN; and $\theta = 40°$
(c) $\mathbf{F_1} = (2\,\mathbf{i} - 4\,\mathbf{j} - 2\,\mathbf{k})$ kip; $\mathbf{F_2} = (-4\,\mathbf{i} + 3\,\mathbf{j} - 5\,\mathbf{k})$ kip; and $\theta = 120°$

2.43 A force vector **F**, with components F_x, F_y and F_z is applied to a structure at point Q. Determine the moment of **F** about the origin O of the coordinate system when:

(a) $F_x = 4{,}000$ N, $F_y = 2{,}500$ N, $F_z = 1{,}500$ N and Q = (2.3, 3.4, 4.6) m
(b) $F_x = -2{,}000$ lb, $F_y = 2{,}500$ lb, $F_z = 1{,}200$ lb and Q = (3.4, –2.5, 4.6) ft
(c) $F_x = 2.3$ kN, $F_y = 5.5$ kN, $F_z = 3.3$ kN and Q = (1.4, 3.9, 2.6) m

2.44 If a structure is loaded at point Q with a force **F**, determine the moment $\mathbf{M_O}$ about the origin O. Prepare a drawing showing the moment vector $\mathbf{M_O}$ in a three-dimensional Cartesian coordinate system. Values for **F** and Q are:

(a) $\mathbf{F} = (125\,\mathbf{i} + 112\,\mathbf{j} + 45\,\mathbf{k})$ lb and Q = (2, –3, 15) ft
(b) $\mathbf{F} = (3.8\,\mathbf{i} + 3.3\,\mathbf{j} - 4.5\,\mathbf{k})$ kN and Q = (–2.2, 2.6, –1.8) m
(c) $\mathbf{F} = (55\,\mathbf{i} - 89\,\mathbf{j} + 42\,\mathbf{k})$ lb and Q = (–3, 4, 8) ft

2.45 The cell phone tower illustrated in the figure to the right is supported by three cables that are maintained with tension forces $F_A = F_B = F_C = 800$ lb. The cables are anchored into the ground plane at locations A, B and C. Because the structural strength and rigidity of the tower is along its axis, we design the cable support system to exhibit a vector sum $\mathbf{S_v}$ that is directed along the axis of the tower from point D to the tower support point O. If the tower height H = 120 ft and the anchor locations are given in the figure, determine the position for the anchor at point B to achieve the design objective.

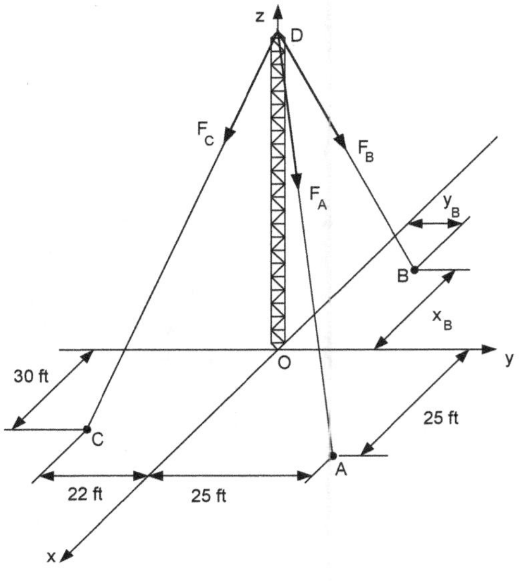

2.46 The cell phone tower, illustrated in the figure to the right, is supported by three cables that are maintained with tension forces $F_A = 800$ lb and $F_C = 1{,}000$ lb. The cables are anchored into the ground plane at locations A, B and C. Because the structural strength and rigidity of the tower is along its axis, we design the cable support system to exhibit a vector sum $\mathbf{S_v}$ that is directed along the axis of the tower from point D to the tower support point O. If the tower height H = 120 ft and the anchor locations are given in the figure, determine the force F_B that must be maintained in cable BD to achieve the design objective.

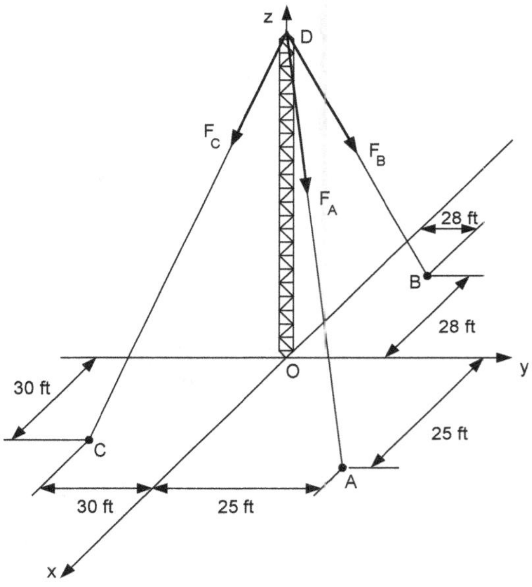

2.47 The boom of a crane extends from its base at point O to its tip at point Q as indicated in the figure below. A three-dimensional coordinate system has been established in this illustration with the point O at the origin and the point Q defined with coordinates (– 1, 3, 8). A force with a magnitude of 21 kN is applied by the boom onto a cable that extends from the tip of the boom to point P. Point P is located on the x–y plane with coordinates shown in the figure below. If the coordinates are expressed in meters, determine the following quantities:

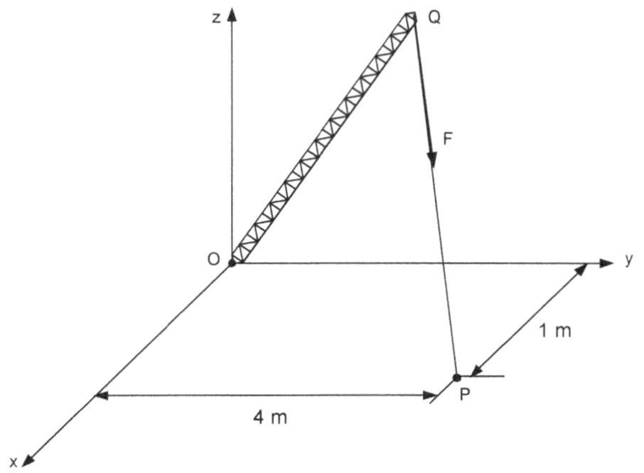

1. The force **F** (in vector form).
2. The force components F_x, F_y, and F_z.
3. The moment $\mathbf{M_O}$ (in vector form).
4. The moment components M_x, M_y, and M_z.
5. The magnitude of the moment (M_O).
6. The unit vector giving the direction of $\mathbf{M_O}$.
7. The angle between **F** and boom OQ.
8. The projection of **F** along boom OQ.

2.48 The boom of a crane extends from its base at point O to its tip at point Q as indicated in the figure below. A three-dimensional coordinate system has been established in this illustration with the point O at the origin and the point Q defined with coordinates (3, 2, 18). A force with a magnitude of 7,500 lb is applied by the boom onto a cable that extends from the tip of the boom to point P. Point P is located on the x–y plane with coordinates shown in the figure below. If the coordinates are expressed in feet, determine the following quantities:

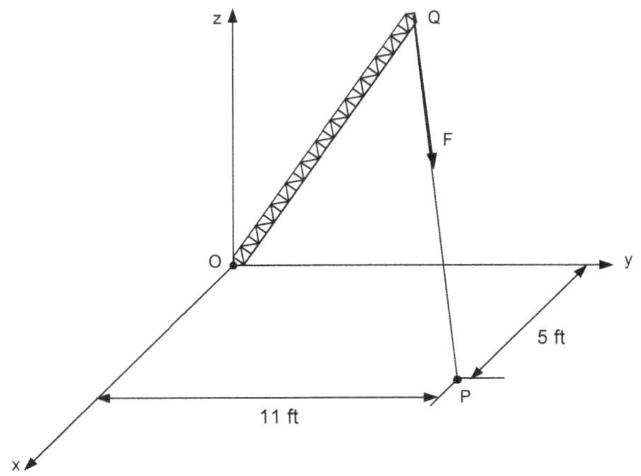

1. The force **F** (in vector form).
2. The force components F_x, F_y, and F_z.
3. The moment $\mathbf{M_O}$ (in vector form).
4. The moment components M_x, M_y, and M_z.
5. The magnitude of the moment (M_O).
6. The unit vector giving the direction of $\mathbf{M_O}$.
7. The angle between **F** and boom OQ.
8. The projection of **F** along boom OQ.

Problem 2.16:

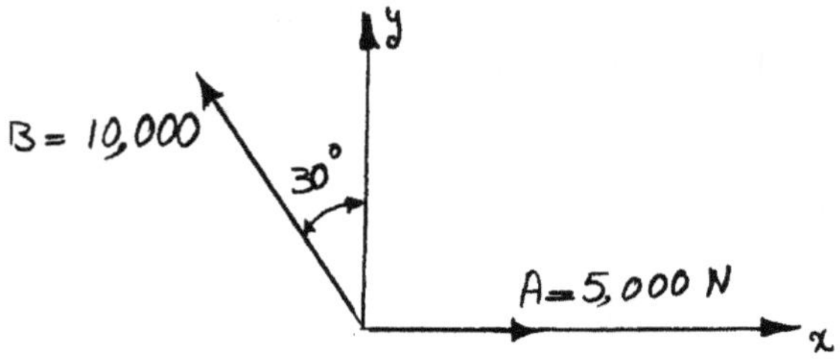

$A_x = 5{,}000$ N

$A_y = 0$ N

$B_x = B\cos\theta = 10{,}000\cos(120°) = -5{,}000$ N

$B_y = B\sin\theta = 10{,}000\sin(120°) = 8660.3$ N

a) $S_v = A + B$

$S_{vx} = A_x + B_x = 5000 - 5000 = 0$ N

$S_{vy} = A_y + B_y = 0 + 8660.3 = 8660.3$ N

$S_v = \sqrt{S_{vx}^2 + S_{vy}^2} = \sqrt{0^2 + 8660.3^2} = 8660.3 \text{ N} \Rightarrow \boxed{S_v = 8660.3 \text{ N}}$

$\theta = \sin^{-1}\left(\frac{S_{vy}}{S_v}\right) = \sin^{-1}(1) \Rightarrow \boxed{\theta = 90°}$

b) $D_v = A - B$

$D_{vx} = A_x - B_x = 5{,}000 - (-5{,}000) = 10{,}000$ N

$D_{vy} = A_y - B_y = 0 - 8660.3 = -8660.3$ N

$D_v = \sqrt{D_{vx}^2 + D_{vy}^2} = \sqrt{(10{,}000)^2 + (-8660.3)^2} = 13228.8 \text{ N} \rightarrow \boxed{D_v = 13228.8 \text{ N}}$

$\theta = \sin^{-1}\left(\frac{D_{vy}}{D_v}\right) = \sin^{-1}\left(\frac{-8660.3}{13228.8}\right) = -40.9° \rightarrow \boxed{\theta = -40.9°}$

or $\boxed{\theta = 319.1°}$

c) $D_v = B - A$

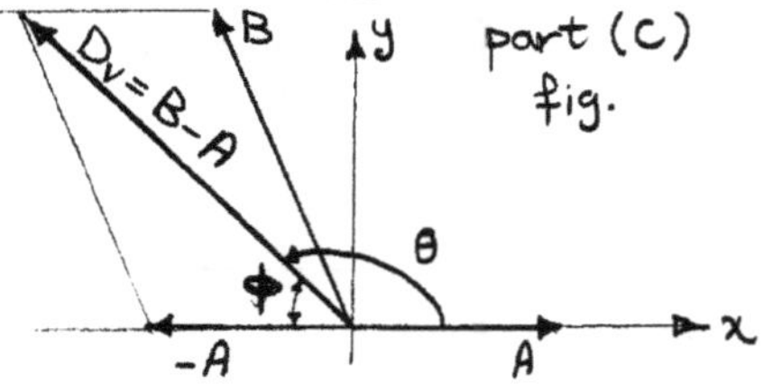

part (c) fig.

$D_{vx} = B_x - A_x = -10{,}000$ N

$D_{vy} = B_y - A_y = 8660.3$ N

$\boxed{D_v = 13228.8 \text{ N}}$

$\phi = \sin^{-1}\left(\frac{D_{vy}}{D_v}\right) = \sin^{-1}\left(\frac{8660.3}{13228.8}\right) = 40.9° \rightarrow \theta = 180° - 40.9°$

$\boxed{\theta = 139.1°}$

Problem 2.17:

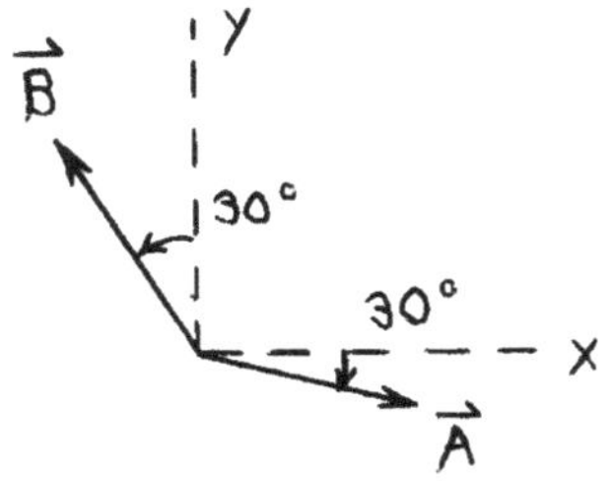

$A = 1400\ N$

$B = 1800\ N$

find: a) $\boxed{S_v,\ \theta_{S_v} = ?}$

b) $\boxed{D_v,\ \theta_{D_v} = ?}$

c) $\boxed{D_v,\ \theta_{D_v} = ?}$

① $A_x = (1400)\cos(30°) = 1212\ N$

② $A_y = -(1400)\sin(30°) = -700\ N$

③ $B_x = -(1800)\sin(30°) = -900\ N$

④ $B_y = (1800)\cos(30°) = 1559\ N$

⑤ $\vec{S}_v = \vec{A} + \vec{B} = (A_x + B_x)\vec{i} + (A_y + B_y)\vec{j}$

⑥ $\vec{S}_v = (1212 - 900)\vec{i} + (-700 + 1559)\vec{j}$

⑦ $\vec{S}_v = (312\vec{i} + 859\vec{j})\ N$

⑧ $S_v = \sqrt{(312)^2 + (859)^2} \Rightarrow \boxed{S_v = 913.9\ N}$.

a) $\theta_{S_v} = \tan^{-1}\left(\frac{859}{312}\right) \Rightarrow \boxed{\theta_{S_v} = 70.04°}$.

⑨ $\vec{D}_v = \vec{A} - \vec{B} = (A_x - B_x)\vec{i} + (A_y - B_y)\vec{j}$

⑩ $\vec{D}_v = (1212 + 900)\vec{i} + (-700 - 1559)\vec{j}$

⑪ $\vec{D}_v = (2112\vec{i} - 2259\vec{j})\ N$

⑫ $D_v = \sqrt{(2112)^2 + (-2259)^2} \Rightarrow \boxed{D_v = 3093\ N}$.

b) $\theta_{D_v} = \tan^{-1}\left(\frac{-2259}{2112}\right) \Rightarrow \boxed{\theta_{D_v} = -46.93°}$.

⑬ $\vec{D}_v = \vec{B} - \vec{A} = -(\vec{A} - \vec{B}) = (-2112\vec{i} + 2259\vec{j})\ N$

⑭ from previously: $\boxed{D_v = 3093\ N}$.

$\theta_{D_v} = -46.93° + 180°$

c) $\boxed{\theta_{D_v} = 133.1°}$.

Problem 2.18:

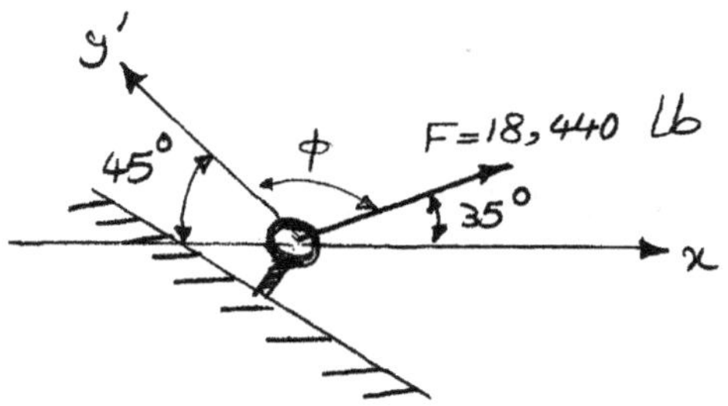

$F_x = ?$ $F_{y'} = ?$

$F_x = F\cos 35° = (18,440\ lb)\cos(35°) = 15105.2\ lb \rightarrow \boxed{F_x = 15105.2\ lb}$

$\phi = 180° - 35° - 45° = 100°$

$F_{y'} = F\cos\phi = F\cos 100° = (18440)\cos(100°) = -3202.1\ lb \rightarrow \boxed{F_{y'} = -3202.1\ lb}$

Problem 2.19:

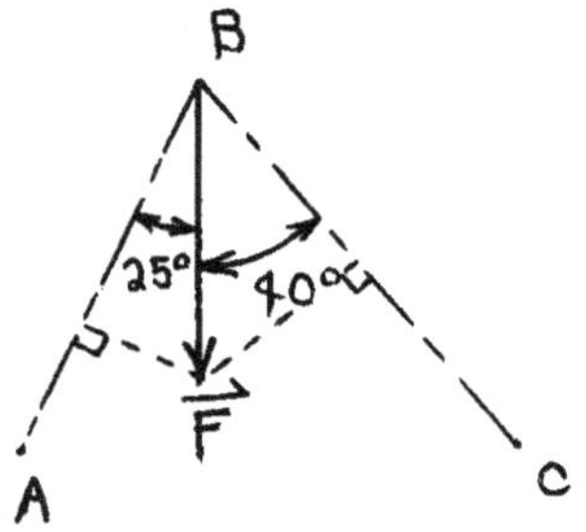

$F = 2150\ N$

find: $\boxed{F_{AB},\ F_{CB} = ?}$.

① $F_{AB} = F\cos\theta_{AB} = (2150)\cos(25°) \Rightarrow \boxed{F_{AB} = 1949\ N}$.

② $F_{CB} = F\cos\theta_{CB} = (2150)\cos(40°) \Rightarrow \boxed{F_{CB} = 1647\ N}$.

Problem 2.21:

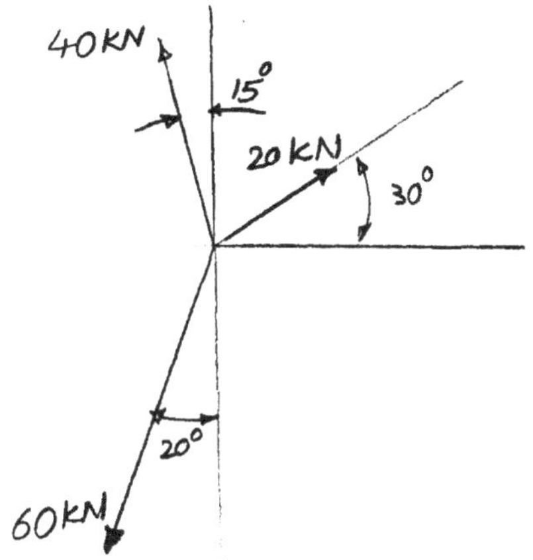

$$\left\{ \begin{array}{l} S_V = ? \\ \theta = ? \end{array} \right.$$

$$S_{V_x} = F_{1x} + F_{2x} + F_{3x} = 20 \cos(30^\circ) + 40 \cos(90^\circ + 15^\circ) + 60 \cos(270^\circ - 20^\circ)$$

$$\therefore \boxed{S_{V_x} = -13.553 \text{ kN}}$$

$$S_{V_y} = F_{1y} + F_{2y} + F_{3y} = 20 \sin(30^\circ) + 40 \sin(105^\circ) + 60 \sin(250^\circ)$$

$$\therefore \boxed{S_{V_y} = -7.744 \text{ kN}} \Rightarrow \boxed{\vec{S}_V = (-13.553\,\vec{i} - 7.744\,\vec{j}) \text{ kN}}$$

$$S = \sqrt{S_{V_x}^2 + S_{V_y}^2} \Rightarrow \boxed{S_V = 15.610 \text{ kN}}$$

$$\phi = \sin^{-1}\left(\frac{7.744}{15.610}\right) = 29.7^\circ \longrightarrow \theta = 180^\circ + \phi = 209.7^\circ$$

$$\therefore \boxed{\theta = 209.7^\circ}$$ relative to positive x-axis

Problem 2.22:

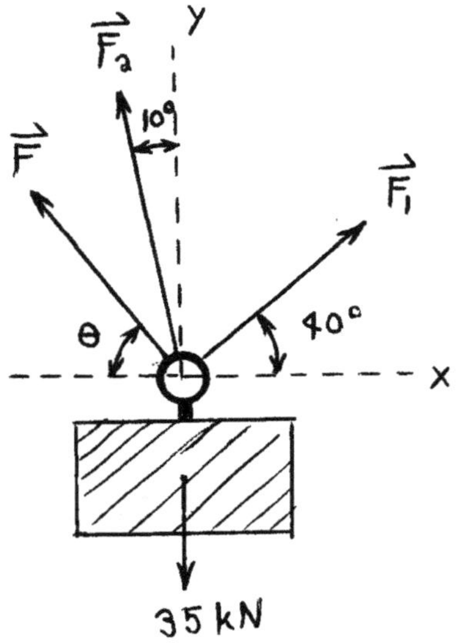

$F_1 = 12.5\ kN \qquad F_2 = 21\ kN$

* box does not swing when lifted

find: $\boxed{F, \theta = ?}$.

① $F_{1x} = (12.5)\cos(40°) = 9.576\ kN$

② $F_{1y} = (12.5)\sin(40°) = 8.035\ kN$

③ $F_{2x} = -(21)\sin(10°) = -3.647\ kN$

④ $F_{2y} = (21)\cos(10°) = 20.68\ kN$

⑤ $F_x = -F\cos\theta\ ; \quad F_y = F\sin\theta$

⑥ to prevent swinging while lifting:

$$S_{vx} = 0 = F_{1x} + F_{2x} + F_x = 9.576 - 3.647 - F\cos\theta$$

$$S_{vy} = 35 = F_{1y} + F_{2y} + F_y = 8.035 + 20.68 + F\sin\theta$$

⑦ $F\cos\theta = 5.929$
$F\sin\theta = 6.285$ } solve simultaneously

⑧ $$\frac{F\sin\theta}{F\cos\theta} = \tan\theta = \frac{6.285}{5.929} = 1.0600$$

⑨ $\boxed{\theta = 46.67°}$.

⑩ $F = \dfrac{5.929}{\cos(46.67°)} \Rightarrow \boxed{F = 8.640\ kN}$.

Problem 2.25:

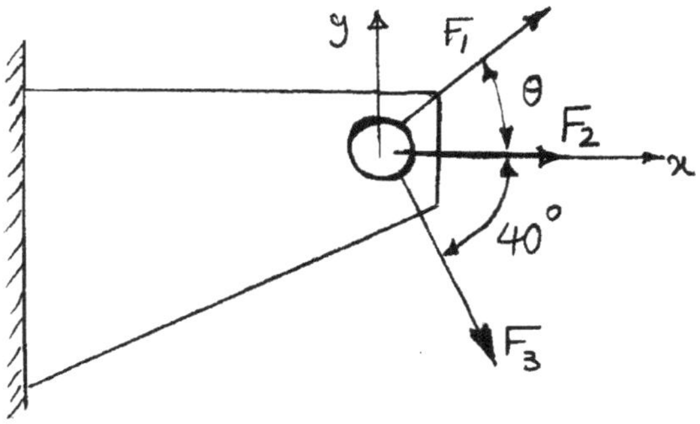

Correction to the problem: $S_{V_y} = 0$

$$\begin{cases} F_2 = 5{,}200 \text{ lb} \\ F_3 = 8{,}300 \text{ lb} \\ S_{V_x} = 16{,}200 \text{ lb} \end{cases} \qquad \begin{cases} F_1 = ? \\ \theta = ? \end{cases}$$

$$S_{V_x} = F_{1x} + F_{2x} + F_{3x} = F_1 \cos\theta + F_2 + F_3 \cos 40^\circ$$

$$16{,}200 = F_1 \cos\theta + 5{,}200 + 8{,}300 \cos 40^\circ$$

$$\therefore \boxed{F_1 \cos\theta = 4{,}641.8} \quad ①$$

$$S_{V_y} = F_{1y} + F_{2y} + F_{3y} = F_1 \sin\theta + 0 - F_3 \sin 40^\circ$$

$$0 = F_1 \sin\theta - 8{,}300 \sin 40^\circ$$

$$\therefore \boxed{F_1 \sin\theta = 5335.1} \quad ②$$

$$\frac{②}{①} = \frac{F_1 \sin\theta}{F_1 \cos\theta} = \frac{5335.1}{4641.8} \Rightarrow \tan\theta = 1.15 \rightarrow \boxed{\theta = 49^\circ}$$

$$① \rightarrow \boxed{F_1 = 7071.8 \text{ lb}}$$

Problem 2.26:

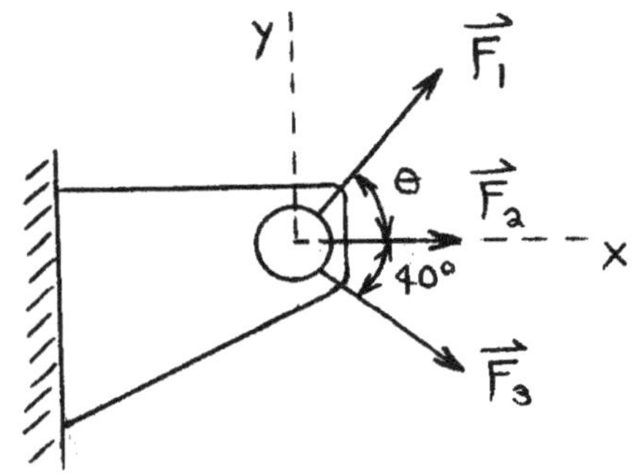

$F_2 = 5,200$ lb $F_3 = 8,300$ lb

$\theta_{S_V} = 10°$ $S_V = 17,500$ lb

find: $F_1, \theta = ?$

① $F_{2x} = 5,200$ lb; $F_{2y} = 0$

② $F_{3x} = (8,300)\cos(40°) = 6,358$ lb

③ $F_{3y} = -(8,300)\sin(40°) = -5,335$ lb

④ $F_{1x} = F_1 \cos\theta$; $F_{1y} = F_1 \sin\theta$

⑤ $S_{Vx} = (17,500)\cos(10°) = 17,234$ lb

⑥ $S_{Vy} = (17,500)\sin(10°) = 3,039$ lb

⑦ $S_{Vx} = F_{1x} + F_{2x} + F_{3x}$

⑧ $17,234 = F_1 \cos\theta + 5,200 + 6,358$

⑨ $F_1 \cos\theta = 5,676$

⑩ $S_{Vy} = F_{1y} + F_{2y} + F_{3y}$

⑪ $3,039 = F_1 \sin\theta + 0 - 5,335$

⑫ $F_1 \sin\theta = 8,374$

⑬ $\dfrac{F_1 \sin\theta}{F_1 \cos\theta} = \tan\theta = \dfrac{8,374}{5,676} = 1.4753$

⑭ $\boxed{\theta = 55.87°}$

⑮ $F_1 = \dfrac{5,676}{\cos(55.87°)} \Rightarrow \boxed{F_1 = 10,116 \text{ lb}}$

Problem 2.33:

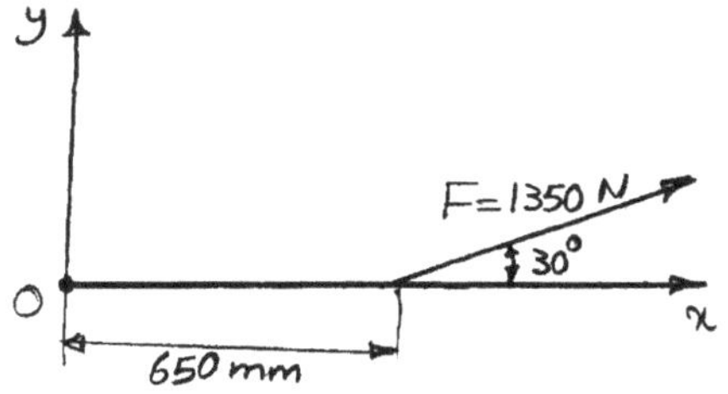

$M_O = ?$

$\overset{+}{\curvearrowleft} M_O = F_y \times d = (F \sin 30°) \times d = (1350 \times \frac{1}{2}\ N) \times (0.650\ m)$

$\therefore \boxed{\overset{+}{\curvearrowleft} M_O = 438.75\ N.m}$

Problem 2.34:

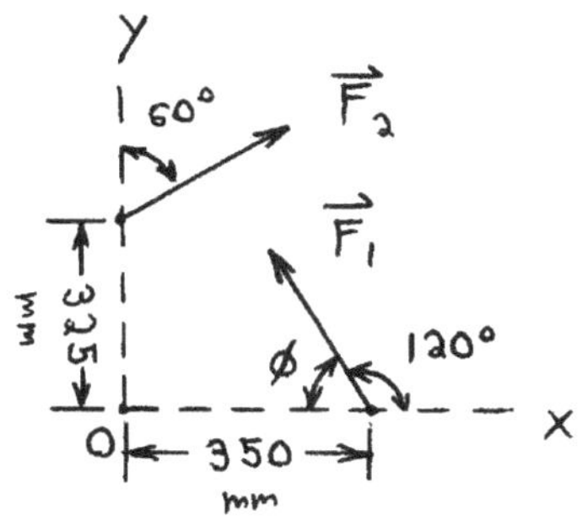

$F_1 = 750\ N \qquad F_2 = 500\ N$

find: $\boxed{M_o = ?}$

① $\phi = 180° - 120° = 60°$

② $F_{x_1} = -(750)\cos(60°) = -375\ N$

③ $F_{y_1} = (750)\ SIN(60°) = 649.5\ N$

④ $F_{x_2} = (500)\ SIN(60°) = 433.0\ N$

⑤ $F_{y_2} = (500)\cos(60°) = 250\ N$

⑥ $d_{\perp_1} = 350\ mm$, in x-direction

$d_{\perp_2} = 325\ mm$, in y-direction

⑦ $M_o = F_{y_1} d_{\perp_1} - F_{x_2} d_{\perp_2} = (649.5)(350) - (433)(325)$

⑧ $\boxed{M_o = 86{,}600\ N\text{-}mm = 86.6\ N\text{-}m\ (CCW)}$

* note: F_{x_1} + F_{y_2} do <u>not</u> contribute to M_o because their lines of action pass through pt. O

Problem 2.37(a):

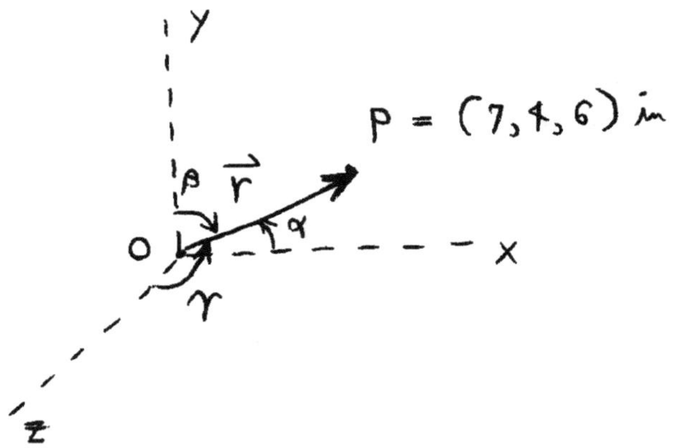

find: $\boxed{\vec{r},\ r,\ \alpha,\ \beta,\ \gamma = ?}$.

① $\vec{r} = (x_p - x_o)\vec{i} + (y_p - y_o)\vec{j} + (z_p - z_o)\vec{k}$

② $\vec{r} = (7-0)\vec{i} + (4-0)\vec{j} + (6-0)\vec{k}$

③ $\boxed{\vec{r} = (7\vec{i} + 4\vec{j} + 6\vec{k})\ in}$.

④ $r = \sqrt{(7)^2 + (4)^2 + (6)^2} = \sqrt{101}$

⑤ $\boxed{r = 10.05\ in}$.

⑥ $\cos\alpha = \frac{r_x}{r} = \frac{7}{10.05} \Rightarrow \boxed{\alpha = 45.85°}$.

⑦ $\cos\beta = \frac{r_y}{r} = \frac{4}{10.05} \Rightarrow \boxed{\beta = 66.55°}$.

⑧ $\cos\gamma = \frac{r_z}{r} = \frac{6}{10.05} \Rightarrow \boxed{\gamma = 53.34°}$.

Problem 2.37(b):

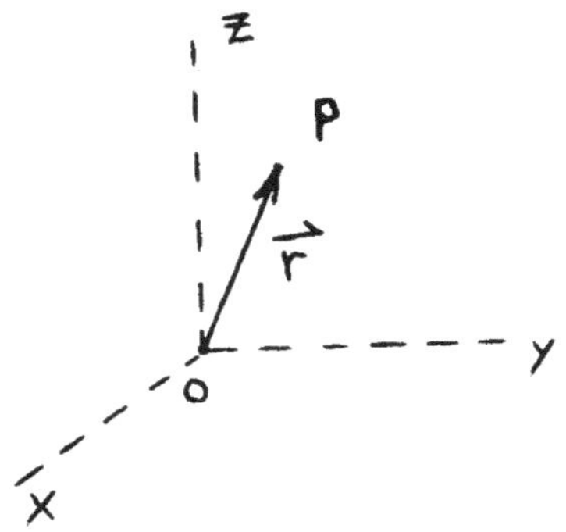

$P = (110, 65, 70)$ mm $\quad O = (0,0,0)$

find: $\boxed{\vec{r} = ?}$ (Cartesian vector)

$\boxed{r, \alpha, \beta, \gamma = ?}$

① $\vec{r} = (x_P - x_O)\vec{i} + (y_P - y_O)\vec{j} + (z_P - z_O)\vec{k}$

② $\vec{r} = (110 - 0)\vec{i} + (65 - 0)\vec{j} + (70 - 0)\vec{k}$

③ $\boxed{\vec{r} = (110\vec{i} + 65\vec{j} + 70\vec{k})\text{ mm}}$

④ $r = \sqrt{(110)^2 + (65)^2 + (70)^2} \Rightarrow \boxed{r = 145.7\text{ mm}}$

⑤ $\cos\alpha = \dfrac{r_x}{r} = \dfrac{110}{145.7} \Rightarrow \boxed{\alpha = 40.98^\circ}$

⑥ $\cos\beta = \dfrac{r_y}{r} = \dfrac{65}{145.7} \Rightarrow \boxed{\beta = 63.50^\circ}$

⑦ $\cos\gamma = \dfrac{r_z}{r} = \dfrac{70}{145.7} \Rightarrow \boxed{\gamma = 61.29^\circ}$

Problem 2.38(a):

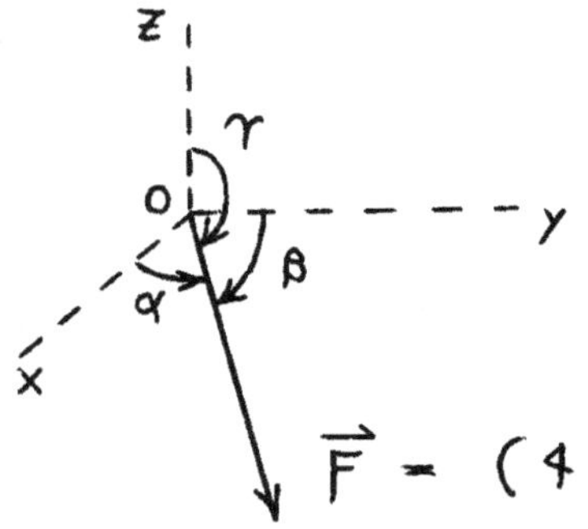

$\vec{F} = (4\vec{i} + 5\vec{j} - 6\vec{k})$ N

find: $\boxed{\alpha, \beta, \gamma = ?}$.

① $F = \sqrt{F_x^2 + F_y^2 + F_z^2} = \sqrt{(4)^2 + (5)^2 + (-6)^2} = \sqrt{77}$

② $F = 8.775$ N

③ $\cos\alpha = \dfrac{F_x}{F} = \dfrac{4}{8.775} = 0.45584 \Rightarrow \boxed{\alpha = 62.88°}$.

④ $\cos\beta = \dfrac{F_y}{F} = \dfrac{5}{8.775} = 0.56980 \Rightarrow \boxed{\beta = 55.26°}$.

⑤ $\cos\gamma = \dfrac{F_z}{F} = \dfrac{-6}{8.775} = -0.68376 \Rightarrow \boxed{\gamma = 133.1°}$.

Problem 2.39(b):

$F = 1{,}500$ lb, $\alpha = 75^\circ$, and $\beta = 50^\circ$

$F_x = ?$ $F_y = ?$ $F_z = ?$

$$\cos\alpha = \frac{F_x}{F} \Rightarrow \cos 75^\circ = \frac{F_x}{1500} \rightarrow \boxed{F_x = 388.23 \text{ lb}}$$

$$\cos\beta = \frac{F_y}{F} \Rightarrow \cos 50^\circ = \frac{F_y}{1{,}500} \rightarrow \boxed{F_y = 964.18 \text{ lb}}$$

$$\cos^2\alpha + \cos^2\beta + \cos^2\gamma = 1$$

$$\cos^2 75^\circ + \cos^2 50^\circ + \cos^2\gamma = 1 \rightarrow \cos^2\gamma = 0.52$$

$$\cos\gamma = \pm 0.72 \rightarrow \begin{cases} \gamma = \underline{43.86^\circ} \\ \gamma = 180 - 43.86^\circ = \underline{136.14^\circ} \end{cases}$$

$$F_z = F\cos\gamma = (1500)(\pm 0.72) = \pm 1081.50 \text{ lb} \rightarrow \boxed{F_z = \pm 1081.50 \text{ lb}}$$

or

$$F = \sqrt{F_x^2 + F_y^2 + F_z^2} \rightarrow 1500 = \sqrt{388.23^2 + 964.18^2 + F_z^2} \rightarrow \boxed{F_z = \pm 1081.50 \text{ lb}}$$

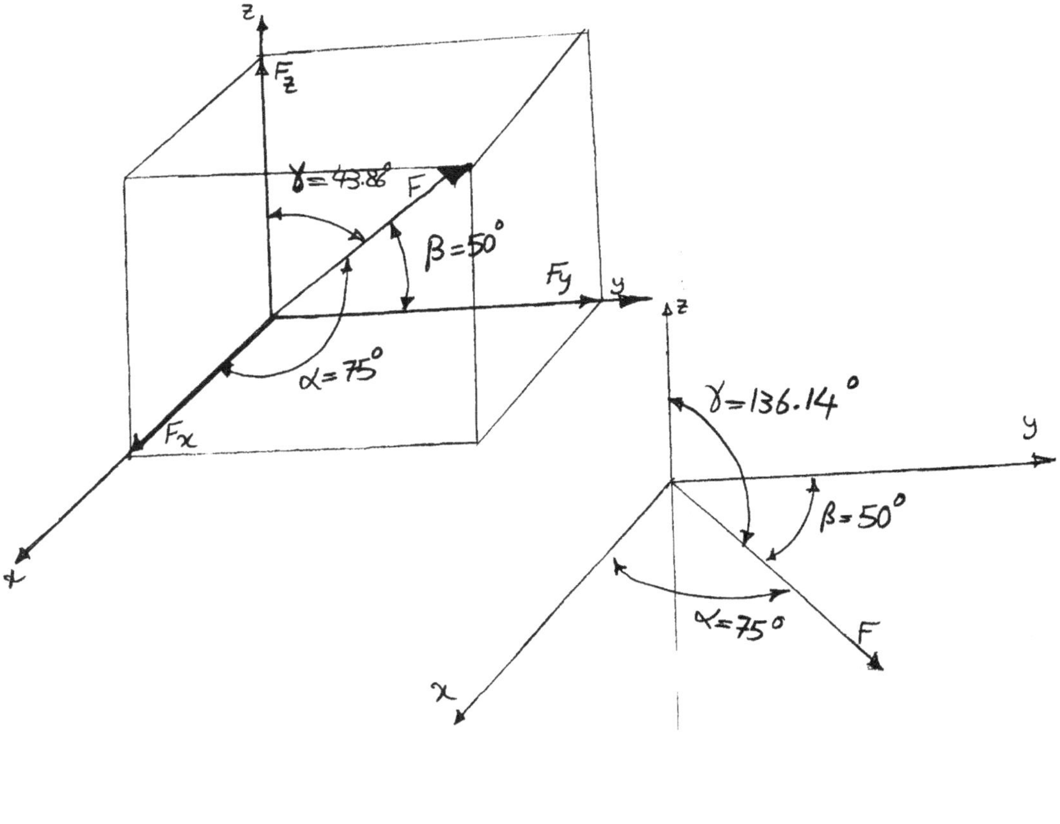

Problem 2.39(d):

3D vector:

$F = 25$ kip

$\alpha = 50°$; $\beta = 140°$

find: $\vec{F}$, F_x, F_y, F_z = ?

draw: $\vec{F}$ in 3D space

① $\cos^2\alpha + \cos^2\beta + \cos^2\gamma = 1$

② $\cos^2\gamma = 1 - \cos^2(50°) - \cos^2(140°) = 0$

③ $\cos\gamma = 0 \Rightarrow \gamma = 90°$

④ $F_x = F\cos\alpha = (25)\cos(50°) \Rightarrow$

⑤ $F_y = F\cos\beta = (25)\cos(140°) \Rightarrow$

⑥ $F_z = F\cos\gamma = (25)\cos(90°) \Rightarrow$

$F_x = 16.07$ kip

$F_y = -19.15$ kip

$F_z = 0$

⑦ $\vec{F} = F_x\vec{i} + F_y\vec{j} + F_z\vec{k}$

⑧ $\vec{F} = (16.07\vec{i} - 19.15\vec{j})$ kip

* vector lies in x-y plane

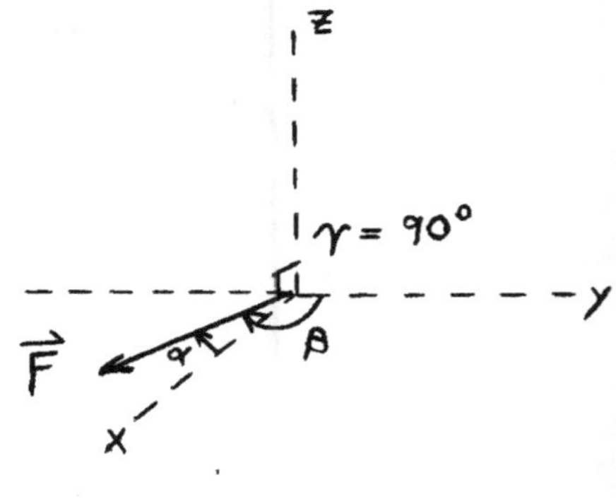

Problem 2.40:

$F_1 = 22.5 kN$ $\quad F_2 = 30 KN$

$\alpha_1 = 60°$ $\quad \alpha_2 = 60°$

$\beta_1 = 60°$ $\quad \beta_2 = 30°$

$\gamma_1 = 135°$ $\quad \gamma_2 = 90°$

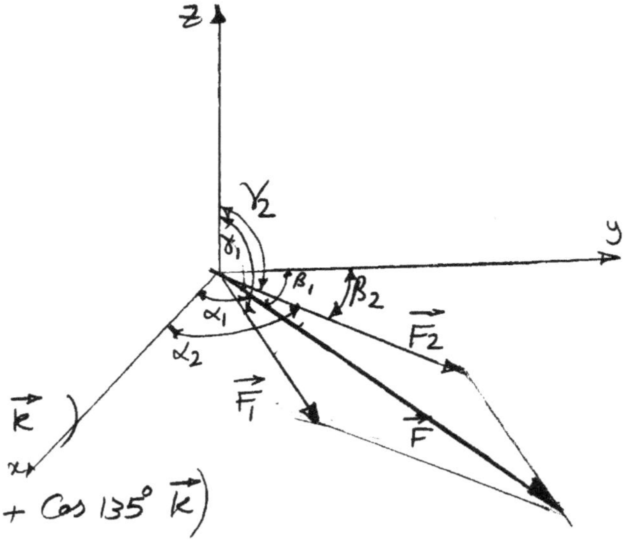

$\vec{F} = \vec{F_1} + \vec{F_2} = ?$

$\vec{F_1} = F_{1x}\vec{i} + F_{1y}\vec{j} + F_{1z}\vec{k}$

$\vec{F_1} = F_1(\cos\alpha_1\vec{i} + \cos\beta_1\vec{j} + \cos\gamma_1\vec{k})$

$\vec{F_1} = 22.5(kN)\left(\cos 60°\vec{i} + \cos 60°\vec{j} + \cos 135°\vec{k}\right)$

$\vec{F_1} = 11.25\vec{i} + 11.25\vec{j} - 15.9\vec{k}$

$\vec{F_2} = F_{2x}\vec{i} + F_{2y}\vec{j} + F_{2z}\vec{k}$

$\vec{F_2} = F_2(\cos\alpha_2\vec{i} + \cos\beta_2\vec{j} + \cos\gamma_2\vec{k})$

$\vec{F_2} = 30\left(\cos 60°\vec{i} + \cos 30°\vec{j} + \cos 90°\vec{k}\right)$

$\vec{F_2} = 15\vec{i} + 25.98\vec{j}$

$\vec{F} = \vec{F_1} + \vec{F_2} = (F_{1x} + F_{2x})\vec{i} + (F_{1y} + F_{2y})\vec{j} + (F_{1z} + F_{2z})\vec{k}$

$\boxed{\vec{F} = 26.25\vec{i} + 37.23\vec{j} - 15.9\vec{k}}$ (KN)

$F = 48.25 kN$

Problem 2.41(a):

$\vec{F_1} = (3i + 6j - 3k)\ N$

$\vec{F_2} = (-2i + 5j + 7k)\ N$

$\vec{F_1} \cdot \vec{F_2} = -6 + 30 - 21 = 3$

$F_1 = \sqrt{9 + 36 + 9} = \sqrt{54}$

$F_2 = \sqrt{4 + 25 + 49} = \sqrt{78}$

$$\theta = \cos^{-1}\left[\frac{(\vec{F_1} \cdot \vec{F_2})}{(F_1 F_2)}\right] = \cos^{-1}\left(\frac{3}{\sqrt{54}\sqrt{78}}\right) = 87.35°$$

$\boxed{\theta = 87.35°}$

Problem 2.41(b):

$\vec{F_1} = (5\vec{i} - \vec{j} + 4\vec{k})\ lb$ find : $\boxed{\theta = ?}$.

$\vec{F_2} = (8\vec{i} - 2\vec{j} + 3\vec{k})\ lb$

① $F_1 = \sqrt{(5)^2 + (-1)^2 + (4)^2} = 6.481\ lb$

② $F_2 = \sqrt{(8)^2 + (-2)^2 + (3)^2} = 8.775\ lb$

③ $\theta = \cos^{-1}\left(\frac{\vec{F_1} \cdot \vec{F_2}}{F_1 F_2}\right) = \cos^{-1}\left(\frac{(5)(8) + (-1)(-2) + (4)(3)}{(6.481)(8.775)}\right)$

④ $\boxed{\theta = 18.28°}$.

Problem 2.42(a):

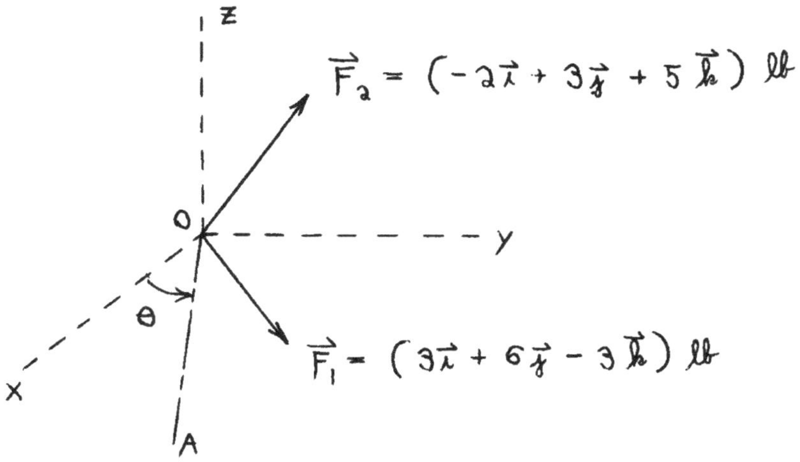

$\theta = 30°$

* OA lies in x-y plane

find: $\boxed{F_{OA} = ?}$

① $\vec{u}_{OA} = (\cos\theta)\vec{i} + (\sin\theta)\vec{j} = (\cos(30°))\vec{i} + (\sin(30°))\vec{j}$

② $\vec{u}_{OA} = 0.8660\,\vec{i} + 0.5\,\vec{j}$

③ $F_{OA_1} = \vec{F}_1 \cdot \vec{u}_{OA} = (3)(0.866) + (6)(0.5) + (-3)(0)$

④ $F_{OA_1} = 5.598$ lb

⑤ $F_{OA_2} = \vec{F}_2 \cdot \vec{u}_{OA} = (-2)(0.866) + (3)(0.5) + (5)(0)$

⑥ $F_{OA_2} = -0.232$ lb

⑦ $F_{OA} = F_{OA_1} + F_{OA_2} = 5.598 - 0.232$

⑧ $\boxed{F_{OA} = 5.366 \text{ lb}}$

Problem 2.43(a):

$\vec{F} = (4000\,\vec{i} + 2500\,\vec{j} + 1500\,\vec{k})\,N$ @ $Q = (2.3, 3.4, 4.6)\,m$

$O = (0, 0, 0)$

find: $\boxed{\vec{M}_o = ?}$.

① $\vec{r}_{OQ} = (2.3 - 0)\vec{i} + (3.4 - 0)\vec{j} + (4.6 - 0)\vec{k}$

② $\vec{r}_{OQ} = (2.3\,\vec{i} + 3.4\,\vec{j} + 4.6\,\vec{k})\,m$

③ $$\vec{M}_o = \vec{r}_{OQ} \times \vec{F} = \begin{vmatrix} \vec{i} & \vec{j} & \vec{k} \\ 2.3 & 3.4 & 4.6 \\ 4000 & 2500 & 1500 \end{vmatrix} \text{ N-m}$$

④ $$\begin{aligned}\vec{M}_o = &((3.4)(1500) - (4.6)(2500))\,\vec{i} \\ &+ ((4.6)(4000) - (2.3)(1500))\,\vec{j} \\ &+ ((2.3)(2500) - (3.4)(4000))\,\vec{k}\end{aligned}$$

⑤ $$\boxed{\begin{aligned}\vec{M}_o &= (-6{,}400\,\vec{i} + 14{,}950\,\vec{j} - 7{,}850\,\vec{k})\text{ N-m} \\ &= (-6.4\,\vec{i} + 14.95\,\vec{j} - 7.85\,\vec{k})\text{ kN-m}\end{aligned}}$$.

Problem 2.43(b):

$\vec{F} = (-2000i + 2500j + 1200k)$ lb

$Q = (3.4, -2.5, 4.6)$ ft

Position vector $\vec{r}$

$\vec{r} = r_x i + r_y j + r_z k = (3.4i - 2.5j + 4.6k)$ ft

Using the determinant format

$$\vec{M}_o = \begin{vmatrix} i & j & k \\ r_x & r_y & r_z \\ F_x & F_y & F_z \end{vmatrix} = \begin{vmatrix} i & j & k \\ 3.4 & -2.5 & 4.6 \\ -2000 & 2500 & 1200 \end{vmatrix} = \vec{r} \times \vec{F}$$

$$= [(-2.5 \times 1200 - 4.6 \times 2500)i + (-4.6 \times 2000 - 3.4 \times 1200)j + (3.4 \times 2500 - 2000 \times 2.5)k]$$

$$\boxed{\vec{M}_o = (-14500i - 13280j + 3500k) \text{ ft-lb}}$$

Problem 2.44(b):

$\vec{F} = (3.8\vec{i} + 3.3\vec{j} - 4.5\vec{k})$ kN @ $Q = (-2.2, 2.6, -1.8)$ m

$O = (0, 0, 0)$

find : $\boxed{\vec{M}_o = ?}$.

draw : $\vec{M}_o$ in 3D space

① $\vec{r}_{OQ} = (-2.2 - 0)\vec{i} + (2.6 - 0)\vec{j} + (-1.8 - 0)\vec{k}$

② $\vec{r}_{OQ} = (-2.2\vec{i} + 2.6\vec{j} - 1.8\vec{k})$ m

③
$$\vec{M}_o = \vec{r}_{OQ} \times \vec{F} = \begin{vmatrix} \vec{i} & \vec{j} & \vec{k} \\ -2.2 & 2.6 & -1.8 \\ 3.8 & 3.3 & -4.5 \end{vmatrix} \text{ kN-m}$$

④
$$\vec{M}_o = ((2.6)(-4.5) - (3.3)(-1.8))\vec{i} + ((-1.8)(3.8) - (-4.5)(-2.2))\vec{j} + ((-2.2)(3.3) - (3.8)(2.6))\vec{k}$$

⑤ $\boxed{\vec{M}_o = (-5.76\vec{i} - 16.74\vec{j} - 17.14\vec{k}) \text{ kN-m}}$.

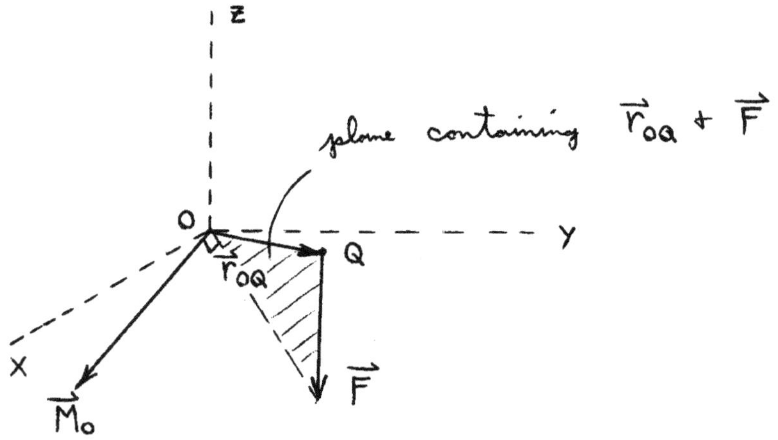

* note : $\vec{M}_o$ is ⊥ to plane containing $\vec{r}_{OQ}$ + $\vec{F}$

Problem 2.45:

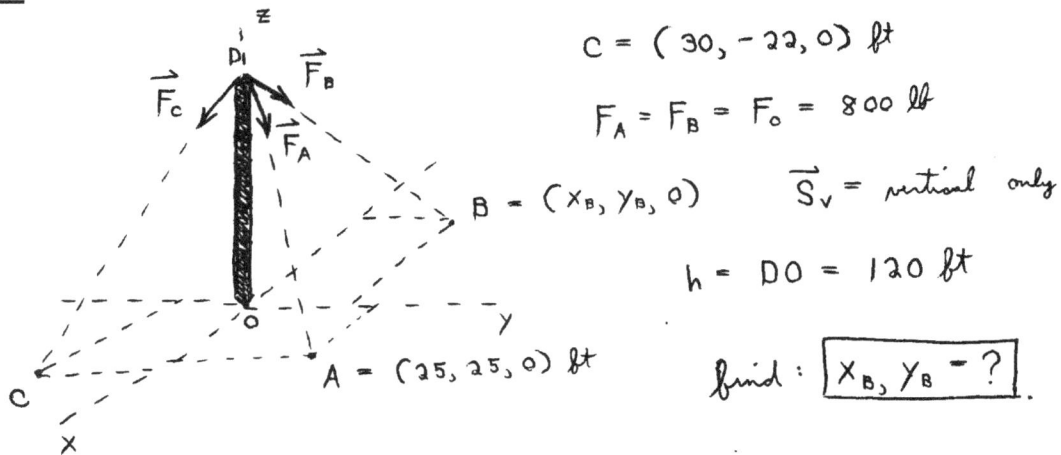

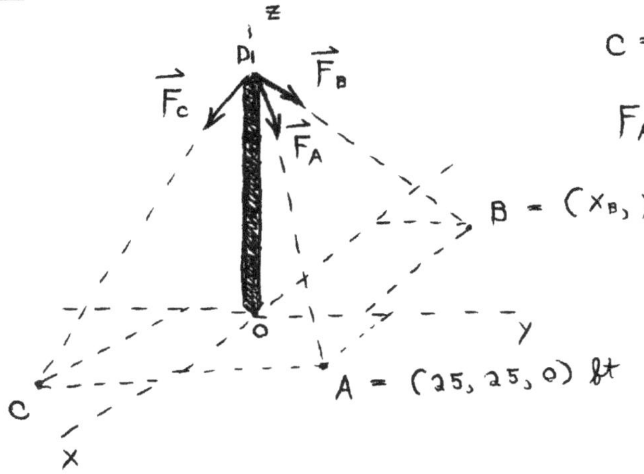

$C = (30, -22, 0)$ ft

$F_A = F_B = F_C = 800$ lb

$\vec{S}_v =$ vertical only

$h = DO = 120$ ft

find: $x_B, y_B = ?$

① $\vec{S}_v =$ vertical only $\Rightarrow S_{vx} = 0$ + $S_{vy} = 0$

② $\vec{r}_{DA} = (25-0)\vec{i} + (25-0)\vec{j} + (0-120)\vec{k}$

③ $\vec{r}_{DA} = (25\vec{i} + 25\vec{j} - 120\vec{k})$ ft

④ $r_{DA} = \sqrt{(25)^2 + (25)^2 + (-120)^2} = 125.1$ ft

⑤ $\vec{F}_A = F_A \dfrac{\vec{r}_{DA}}{r_{DA}} = (800)\dfrac{(25\vec{i} + 25\vec{j} - 120\vec{k})}{(125.1)}$

⑥ $\vec{F}_A = (159.9\vec{i} + 159.9\vec{j} - 767.4\vec{k})$ lb

⑦ $\vec{r}_{DC} = (30-0)\vec{i} + (-22-0)\vec{j} + (0-120)\vec{k}$

⑧ $\vec{r}_{DC} = (30\vec{i} - 22\vec{j} - 120\vec{k})$ ft

⑨ $r_{DC} = \sqrt{(30)^2 + (-22)^2 + (-120)^2} = 125.6$ ft

⑩ $\vec{F}_C = F_C \dfrac{\vec{r}_{DC}}{r_{DC}} = (800)\dfrac{(30\vec{i} - 22\vec{j} - 120\vec{k})}{(125.6)}$

⑪ $\vec{F}_C = (191.1\vec{i} - 140.1\vec{j} - 764.3\vec{k})$ lb

⑫ $S_{vx} = 0 = F_{Ax} + F_{Bx} + F_{Cx} = 159.9 + F_{Bx} + 191.1$

⑬ $F_{Bx} = -351$ lb

⑭ $S_{vy} = 0 = F_{Ay} + F_{By} + F_{Cy} = 159.9 + F_{By} - 140.1$

⑮ $F_{By} = -19.8$ lb

<u>Problem 2.45</u>: (con't)

(16) $F_{Bz} = \sqrt{F_B^2 - F_{Bx}^2 - F_{By}^2} = \sqrt{(800)^2 - (-351)^2 - (-19.8)^2}$

(17) $F_{Bz} = -718.6$ lb

(18) $\cos\gamma = \dfrac{F_{Bz}}{F_B} = \dfrac{r_{DBz}}{r_{DB}} \Rightarrow \dfrac{-718.6}{800} = \dfrac{-120}{r_{DB}}$

(19) $r_{DB} = 133.6$ ft

(20) $\cos\alpha = \dfrac{F_{Bx}}{F_B} = \dfrac{r_{DBx}}{r_{DB}} \Rightarrow \dfrac{-351}{800} = \dfrac{r_{DBx}}{133.6}$

(21) $r_{DBx} = -58.62 \text{ ft} = x_B - 0$

(22) $\boxed{x_B = -58.62 \text{ ft}}$

(23) $\cos\beta = \dfrac{F_{By}}{F_B} = \dfrac{r_{DBy}}{r_{DB}} \Rightarrow \dfrac{-19.8}{800} = \dfrac{r_{DBy}}{133.6}$

(24) $r_{DBy} = -3.304 \text{ ft} = y_B - 0$

(25) $\boxed{y_B = -3.304 \text{ ft}}$

Problem 2.47:

unit vector in $\vec{F}$ direction

$$\vec{u}_F = \frac{[(x_P - x_Q)i + (y_P - y_Q)j + (z_P - z_Q)k]}{[(x_P - x_Q)^2 + (y_P - y_Q)^2 + (z_P - z_Q)^2]^{1/2}}$$

$$= \frac{(1+1)i + (4-3)j - 8k}{\sqrt{4+1+64}}$$

$$= 0.2408i + 0.1204j - 0.9631k$$

force $\vec{F}$

$$\boxed{\vec{F} = F\vec{u}_F = (5.057i + 2.528j - 20.23k)\ kN}$$

force components

$$\boxed{F_x = 5.057\ kN\ ;\ F_y = 2.528\ kN\ ;\ F_z = -20.23\ kN}$$

position vector $\vec{r} = (-1, 3, 8)$

moment $\vec{M}_o$

$$\vec{M}_o = \vec{r} \times \vec{F} = \begin{vmatrix} i & j & k \\ -1 & 3 & 8 \\ 5.057 & 2.528 & -20.23 \end{vmatrix}$$

$$= [(-3 \times 20.23 - 8 \times 2.528)i + (8 \times 5.057 - 20.23)j + (-2.528 - 3 \times 5.057)k]$$

$$\boxed{\vec{M}_o = (-80.91i + 20.23j - 17.70k)\ kN\text{-}m}$$

moment components

$$\boxed{M_x = -80.91\ kN\text{-}m\ ;\ M_y = 20.23\ kN\text{-}m;\ M_z = -17.70\ kN\text{-}m}$$

magnitude of moment $\vec{M}_o$

$$M_o = [(-80.91)^2 + 20.23^2 + (-17.70)^2]^{1/2} = 85.26\ kN\text{-}m$$

$$\boxed{M_o = 85.26\ kN\text{-}m}$$

Problem 2.47: (con't)

unit vector giving the direction of $\vec{M_o}$

$$\vec{u}_{Mo} = (1/M_o)[\vec{M_o}] = (1/85{,}26)(-80{,}91i + 20{,}23j - 17{,}70k)$$

$$\boxed{\vec{u}_{Mo} = -0{,}949i + 0{,}237j - 0{,}208k}$$

The angle between $\vec{F}$ and boom OQ

→ angle between $\vec{F}$ and a vector QO

$$\theta = \cos^{-1}\left[\frac{\vec{F}\cdot\vec{QO}}{(F)(QO)}\right] = \cos^{-1}\left[\frac{(5{,}057i + 2{,}528j - 20{,}23k)\cdot(i - 3j - 8k)}{(5{,}057^2 + 2{,}528^2 + 20{,}23^2)^{1/2}\,(1+9+64)^{1/2}}\right]$$

$$\boxed{\theta = 28{,}15°}$$

Projection of $\vec{F}$ along boom OQ

$$F_{QO} = F\cos\theta = 21 \times \cos 28{,}15° = 18.52 \text{ kN}$$

$$\boxed{F_{QO} = 18{,}52 \text{ kN}} \text{ (compression)}$$

Alternate Method:

$$F_{OQ} = \vec{F}\cdot\vec{u}_{OQ} = \vec{F}\cdot\frac{\vec{r}_{OQ}}{r_{OQ}}$$

$$F_{OQ} = \frac{(5.056)(-1) + (2.528)(3) + (-20.22)(8)}{8.602}$$

$$\boxed{F_{OQ} = -18.51 \text{ kN}}$$

Problem 2.48:

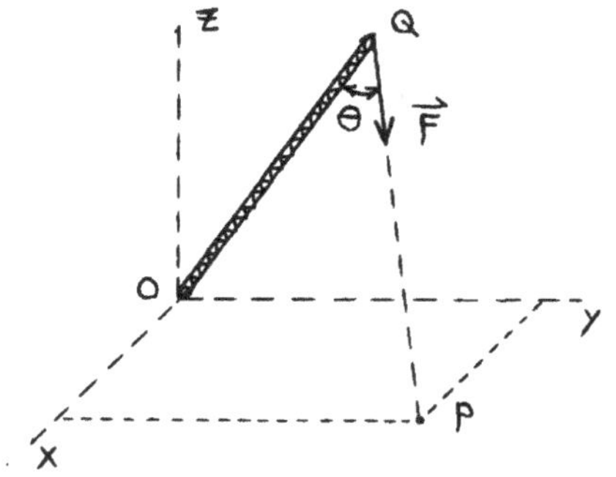

$F = 7500$ lb

$O = (0, 0, 0)$

$P = (5, 11, 0)$ ft

$Q = (3, 2, 18)$ ft

find: 1) $\vec{F} = ?$
2) $F_x, F_y, F_z = ?$
3) $\vec{M}_o = ?$
4) $M_x, M_y, M_z = ?$
5) $M_o = ?$
6) $\vec{u}_{M_o} = ?$
7) $\theta = ?$
8) $F_{OQ} = ?$

① $\vec{r}_{QP} = (5-3)\vec{i} + (11-2)\vec{j} + (0-18)\vec{k}$

② $\vec{r}_{QP} = (2\vec{i} + 9\vec{j} - 18\vec{k})$ ft

③ $r_{QP} = \sqrt{(2)^2 + (9)^2 + (-18)^2} = 20.22$ ft

④ $\vec{F} = F\vec{u}_{QP} = F\dfrac{\vec{r}_{QP}}{r_{QP}} = (7500)\dfrac{(2\vec{i} + 9\vec{j} - 18\vec{k})}{(20.22)}$

⑤ $\boxed{\vec{F} = (741.8\vec{i} + 3{,}338\vec{j} - 6{,}677\vec{k})\text{ lb}}$

⑥ $\boxed{F_x = 741.8\text{ lb};\quad F_y = 3{,}338\text{ lb};\quad F_z = -6{,}677\text{ lb}}$

⑦ $\vec{r}_{OQ} = (3-0)\vec{i} + (2-0)\vec{j} + (18-0)\vec{k} = (3\vec{i} + 2\vec{j} + 18\vec{k})$ ft

⑧ $\vec{M}_o = \vec{r}_{OQ} \times \vec{F} = \begin{vmatrix} \vec{i} & \vec{j} & \vec{k} \\ 3 & 2 & 18 \\ 741.8 & 3{,}338 & -6{,}677 \end{vmatrix}$ ft-lb

⑨ $\vec{M}_o = ((2)(-6{,}677) - (18)(3{,}338))\vec{i} + ((18)(741.8) - (3)(-6{,}677))\vec{j} + ((3)(3{,}338) - (2)(741.8))\vec{k}$

⑩ $\boxed{\vec{M}_o = (-73{,}438\vec{i} + 33{,}383\vec{j} + 8{,}530\vec{k})\text{ ft-lb}}$

⑪ $\boxed{M_x = -73{,}438\text{ ft-lb};\quad M_y = 33{,}383\text{ ft-lb};\quad M_z = 8{,}530\text{ ft-lb}}$

⑫ $M_o = \sqrt{(-73{,}438)^2 + (33{,}383)^2 + (8{,}530)^2}$

⑬ $\boxed{M_o = 81{,}119\text{ ft-lb}}$

⑭ $\vec{u}_{M_o} = \dfrac{\vec{M}_o}{M_o} = \dfrac{(-73{,}438\vec{i} + 33{,}383\vec{j} + 8{,}530\vec{k})}{(81{,}119)}$

⑮ $\boxed{\vec{u}_{M_o} = (-0.9053\vec{i} + 0.4115\vec{j} + 0.1052\vec{k})}$

Problem 2.48: (con't.)

⑯ $\vec{r}_{QO} = -\vec{r}_{OQ} = (-3\vec{i} - 2\vec{j} - 18\vec{k})\ ft$

⑰ $r_{QO} = \sqrt{(-3)^2 + (-2)^2 + (-18)^2} = 18.36\ ft = r_{OQ}$

⑱ $\theta = \cos^{-1}\left(\dfrac{\vec{r}_{QO} \cdot \vec{r}_{QP}}{r_{QO}\, r_{QP}}\right) = \cos^{-1}\left(\dfrac{(-3)(2) + (-2)(9) + (-18)(-18)}{(18.36)(20.22)}\right)$

⑲ $\boxed{\theta = 36.09^\circ}$.

⑳ $F_{OQ} = \vec{F} \cdot \vec{u}_{OQ} = \dfrac{\vec{F} \cdot \vec{r}_{OQ}}{r_{OQ}} = \dfrac{(741.8)(3) + (3{,}338)(2) + (18)(-6{,}677)}{(18.36)}$

㉑ $\boxed{F_{OQ} = -6{,}061\ lb}$.

CHAPTER 3 Free Body Diagrams and Equilibrium

3.1 An auto weighing 17,500 N is traveling on a straight highway down a hill with a grade of 3% at a constant velocity of 85 km/h. Construct a FBD of the auto showing all of the forces acting on it. A highway with a grade of 3% elevates 3 m for every 100 m in the horizontal direction.

3.2 A sports utility vehicle weighing 21,500 N is traveling on a straight highway up a hill with a grade of 2% at a constant velocity of 95 km/h. Construct a FBD of the vehicle showing all of the forces acting on it. A highway with a grade of 2% elevates 2 m for every 100 m in the horizontal direction.

3.3 An auto weighing 3,400 lb is traveling on a straight highway up a hill with a 2% grade at a constant velocity of 65 MPH. Construct a FBD of the auto showing all of the forces acting on it. A highway with a grade of 2% elevates 2 ft for every 100 ft in the horizontal direction.

3.4 A sports utility vehicle weighing 4,500 lb is traveling on a straight highway down a hill with a 3.5% grade at a constant velocity of 70 MPH. Construct a FBD of the auto showing all of the forces acting on it.

3.5 Prepare a sketch illustrating the force system described below and write the relevant equilibrium relations for it.

(a) Coplanar and concurrent
(b) Non-coplanar and concurrent
(c) Coplanar and non-concurrent
(d) Non-coplanar and non-concurrent

3.6 Construct a FBD for the mass and two-cable assembly shown in the figure to the right. Also determine the tension forces in cables AC and BC.

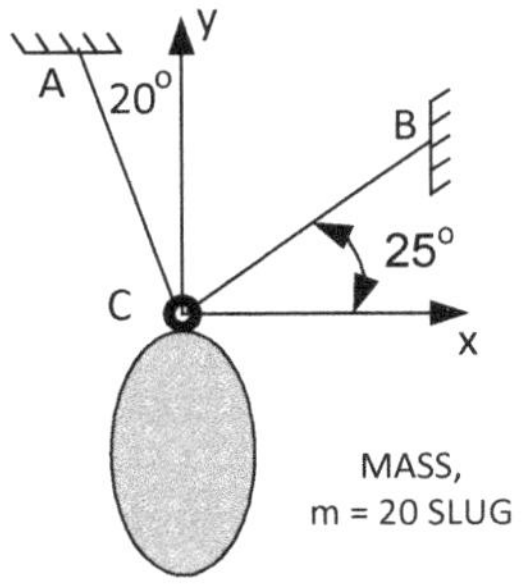

3.7 Construct a FBD for the mass and two-cable assembly shown in the figure to the left. Also determine the tension forces in cables AC and BC.

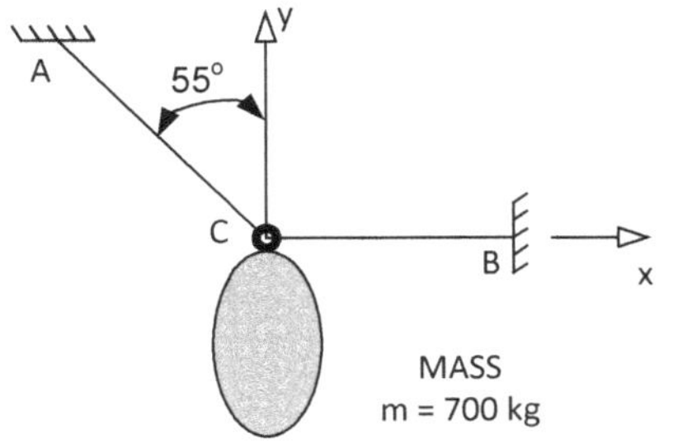

3.8 Construct a FBD for the mass and three-cable assembly shown in the figure to the right. Determine the tension forces in cables AC, BC and CD. Assume cable CD supports 30% of the mass.

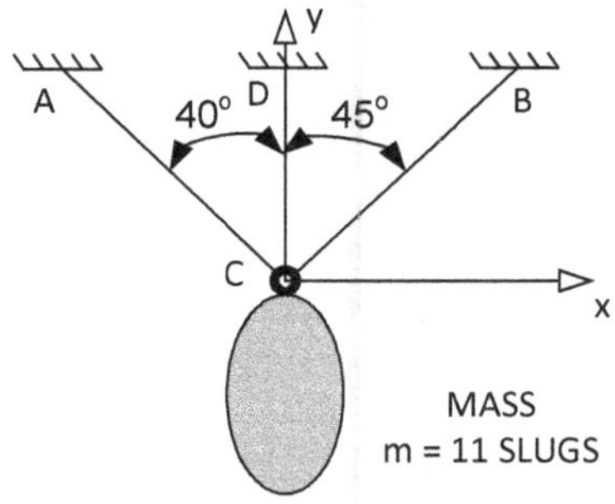

3.9 Determine the forces in cables AC and BC that support the signal lights at a traffic intersection as illustrated in the figure to the left. The signal lights have a mass m = 40 kg. The dimensions for the cable arrangement are given in meters.

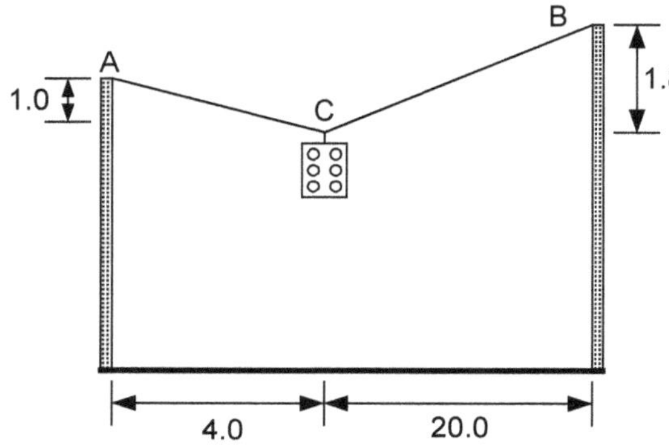

3.10 A container A that holds a 1.0 ton weight ton is lifted from the floor with a cable and pulley system shown in the figure to the right. Determine the equation for the force F required to lift the container as a function of α. Evaluate this equation to determine the force F if α = 20°. Comment on the effectiveness of this cable pulley system.

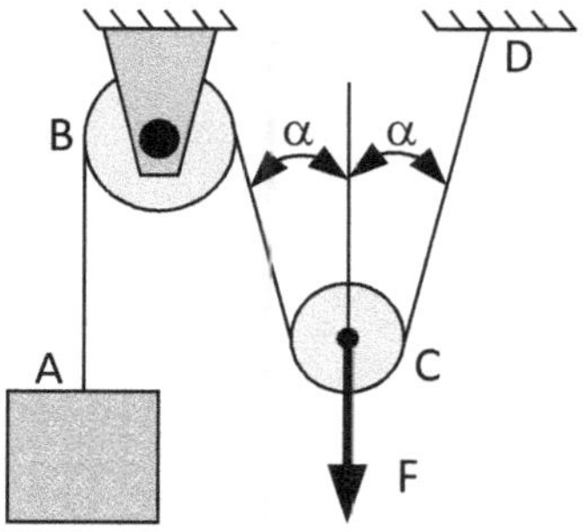

3.11 A container holding a mass m = 850 kg is lifted from the floor with a cable and pulley system shown in the figure to the right. Determine the force F required to lift the container as a function of α. Evaluate this equation to determine the force F if α = 17°. Comment on the effectiveness of this cable pulley system.

3.12 Using the pulley and cable arrangement shown in the figure to the right, design a lifting arrangement that requires only a force of 1,000 lb to lift a container weighing 1,800 lb. Assume that the pulleys are frictionless.

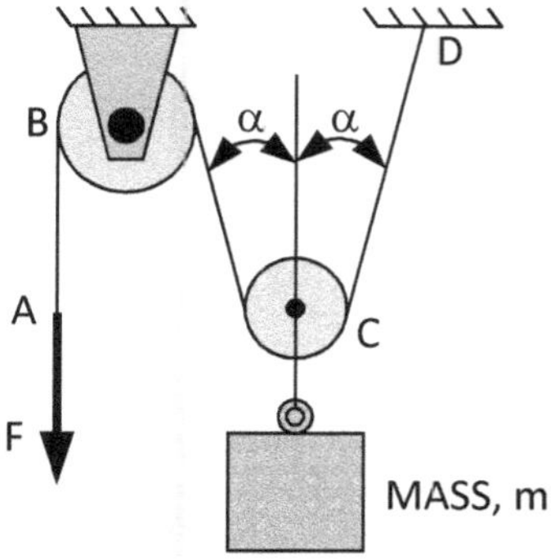

3.13 Using pulleys and cables, design a lifting arrangement that requires only a force of 500 lb to lift a one-ton container. Assume that the pulleys are frictionless.

3.14 For the pulleys and cable arrangement shown in Problem 3.11, determine the force required to lift a 4.5 kN weight as a function of the angle α the cables make with pulley C. Let the angle α vary from zero to 90°. It is suggested that you use a spreadsheet for these calculations and prepare a graph showing the results.

3.15 A differential hoist is illustrated in the figure to the right. At the top, two pulleys with radii r_1 and r_2 are fastened together so they turn as a single unit. A continuous cable passes around the smaller pulley, with a radius r_2, and then around the lower pulley with a radius r_3, and finally around the larger pulley with a radius r_1. We assume that the pulleys are frictionless relative to the shafts passing through their hubs. We also assume that the cables do not slip in the grooves of the pulleys. If the diameter of the lower pulley is $D = 2r_3 = r_1 + r_2$, determine the equation for the applied force F in terms of the load W, which is maintained in equilibrium.

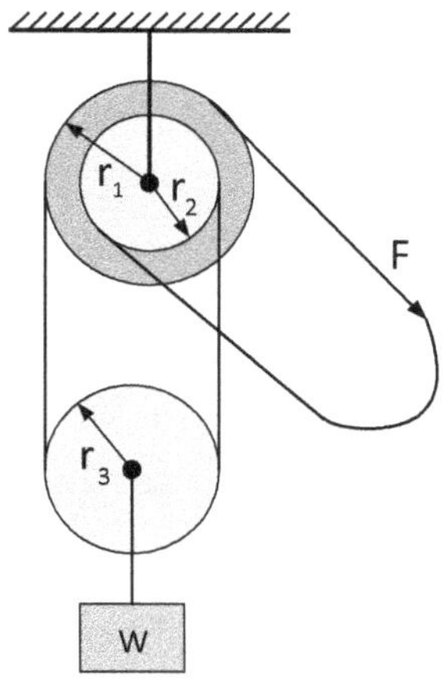

3.16 An artist has designed a bell using a long piece of pipe as the vibrating sound source. The pipe is to be supported using the wire ring arrangement shown in the figure to the left. Determine the force in each of the wires. The pipe has a mass of 43 kg.

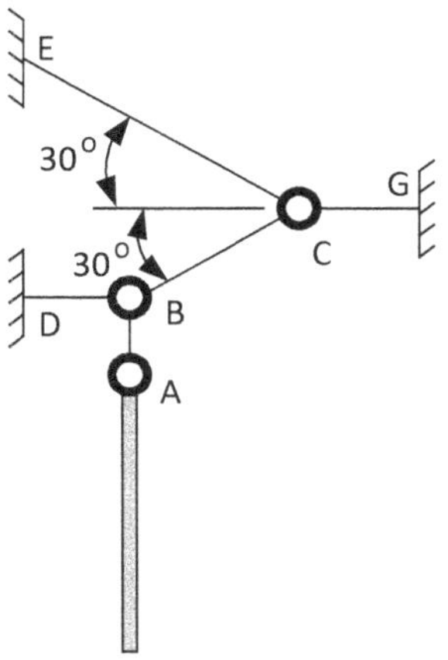

3.17 The pulley and cable, shown in the figure to the right, support a mass m = 12 slug suspended at point A. The cable B-A-C is 65 ft long and sags prior to the addition of the pulley and the container.

(a) If the pulley diameter is small relative to the span of 50 ft between the walls at B and C, determine the height h associated with the equilibrium position of the pulley and container.

(b) If the pulley and container are released from point C, describe the motion, which you would observe as the system reaches equilibrium.

(c) If the pulley and container are released from point B, describe the motion, which you would observe as the system reaches equilibrium.

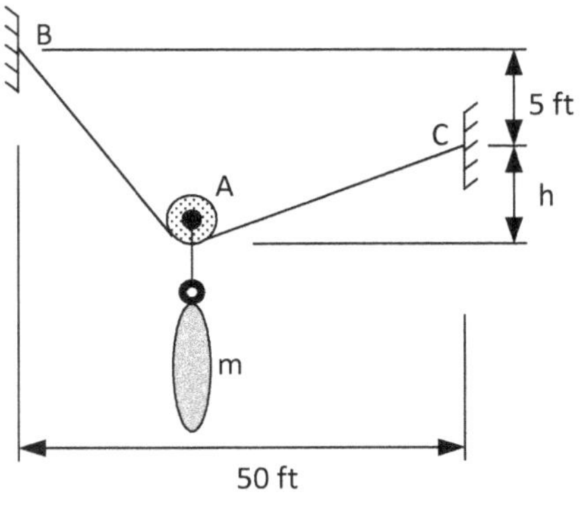

3.18 Two masses m_B and m_C are supported by the cable arrangement shown in the figure to the right. Determine the mass m_C and the forces in the cables AB, BC and CD if the mass m_B = 16.0 kg.

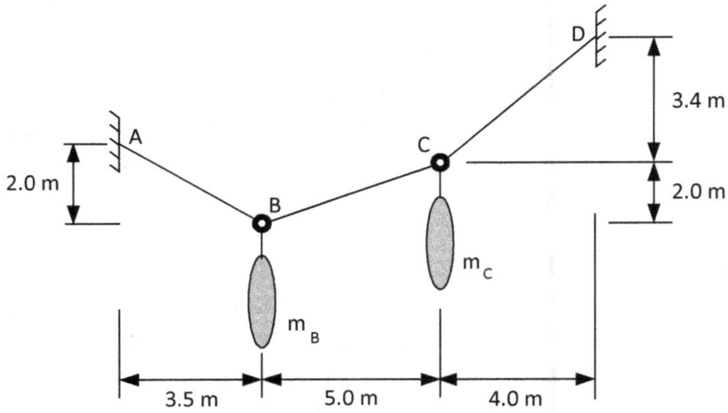

3.19 A bungee cord is stretched across a river as shown in the figure below. A stuntman weighing 140 lb is to cross the river by tightrope walking on the bungee cord. If the cord has a linear spring constant k = 60 lb/in., determine if the stuntman will remain dry.

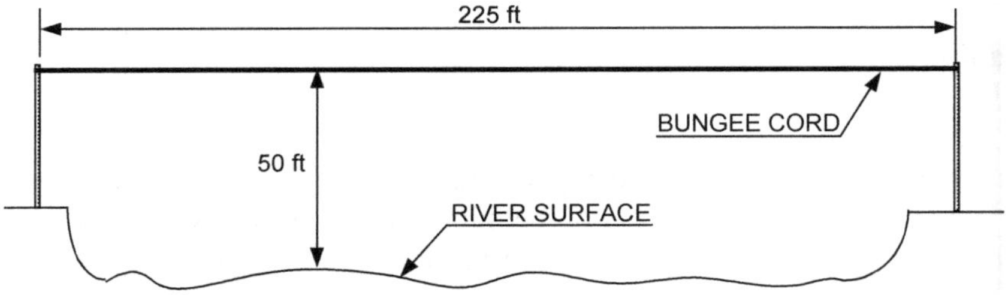

3.20 For the conditions described in Problem 3.19, determine the spring constant of the bungee cord if the stuntman is to barely wet the soles of his shoes.

3.21 Three springs are connected in series as shown in the figure to the right. Determine the extension of the entire arrangement if a force of 23.4 lb is applied to the last spring. Also determine the effective spring rate for the three springs in this series arrangement.

3.22 Three springs are connected in series as shown in the figure to the left. Determine the extension of the entire arrangement if a force of 70 N is applied to the last spring. Also determine the effective spring rate for the three springs in the series arrangement.

3.23 For the parallel spring arrangement determine the deflection of the lower bar shown in the figure to the right. The spring rates k_A = 12 N/mm. and k_B = 23 N/mm. The force F = 1145 N. Also determine the effective spring rate for the three springs in parallel.

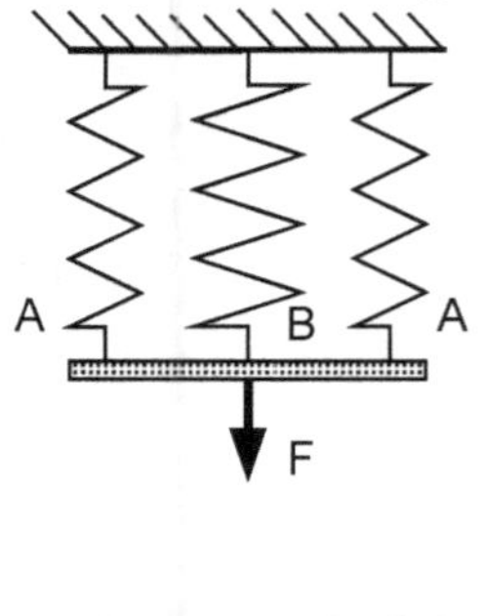

3.24 Three springs are used to support the rigid bar with a mass m as illustrated in the figure to the right. If $k_1 = 20$ lb/in., $k_2 = 28$ lb/in., $k_3 = 36$ lb/in. and m = 1.3 slug, determine the amount each spring elongates. The bar is constrained and remains parallel to the top surface to which the springs are attached.

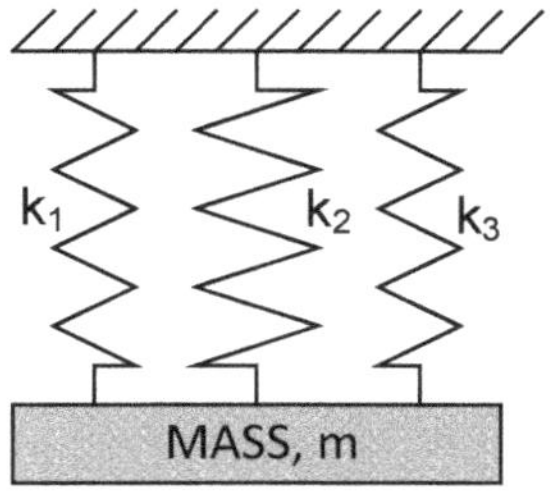

3.25 A cable and a spring are connected together at point C as shown in the figure to the left. Determine the equilibrium position of point C if a force F = 400 N is applied. The undeformed length L of both the cable and the spring before the application of the load is shown in the figure together with the spring rate k.

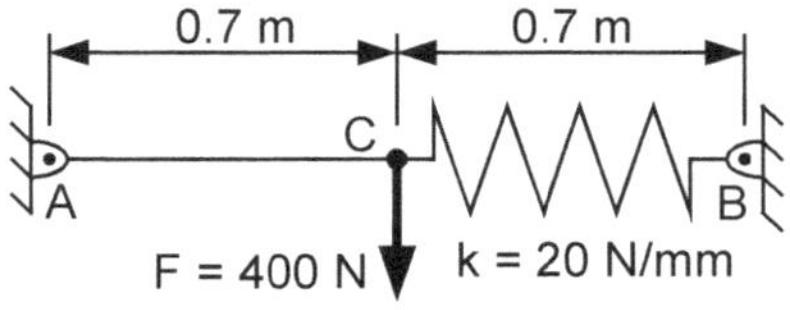

3.26 For the beam, shown in the figure to the right, solve for the reactions at the pin and clevis and the roller.

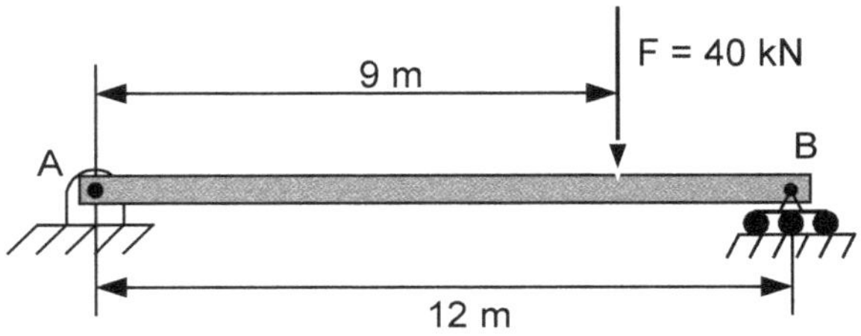

3.27 For the beam, shown in the figure to the left, solve for the reactions at the pin and clevis and the roller.

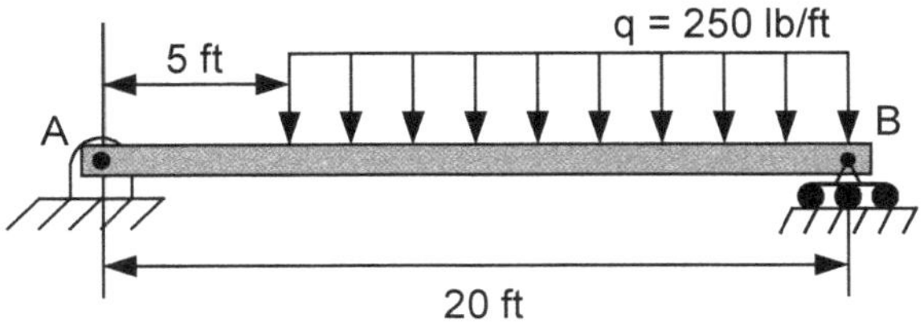

3.28 For the beam, shown in the figure to the right, solve for the reactions at the pin and clevis and the roller.

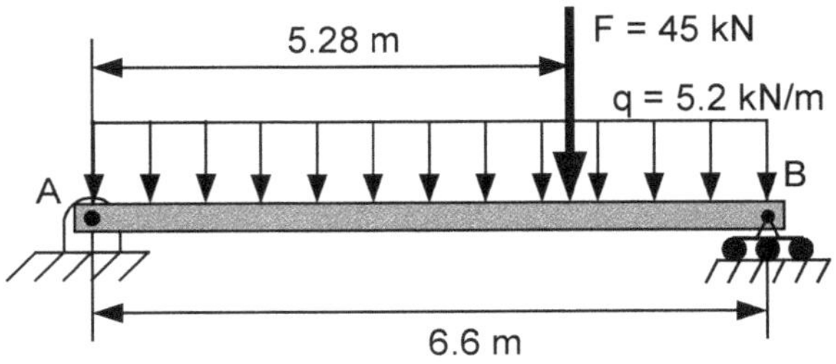

3.29 For the beam, shown in the figure to the right, solve for the reactions at the built-in end of the cantilever beam.

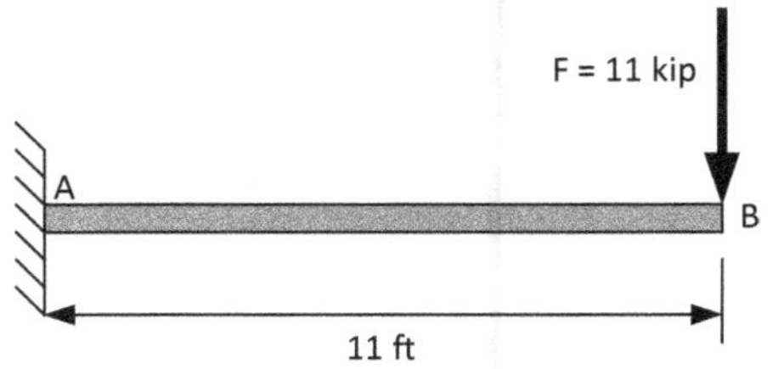

3.30 For the beam, shown in the figure to the left, solve for the reactions at the built-in end of the cantilever beam.

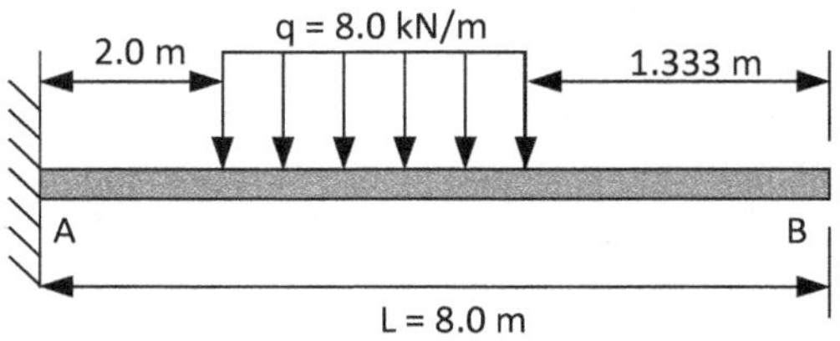

3.31 For the beam, shown in the figure below, solve for the reactions at the two supports.

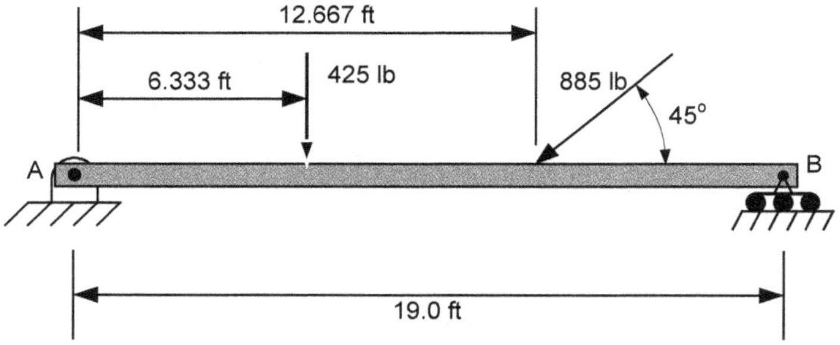

3.32 For the beam, shown in the figure below, determine the reactions at the two supports. The dimensions of the beam and the mass are given in the table below. Note that the arm CD is rigidly attached to the beam. Also, the radii of the two pulleys are the same and equal to 0.25 m.

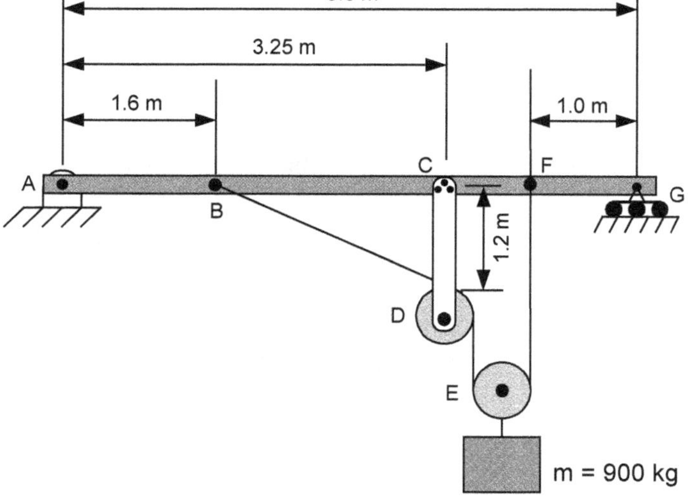

3.33 For the truss, shown in the figure to the right, determine the reactions at the two supports.

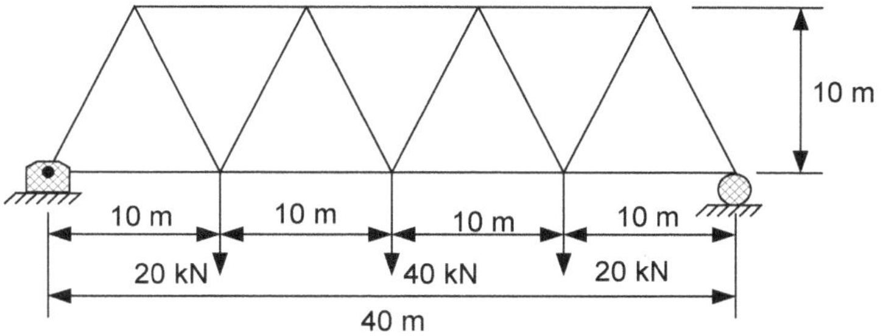

3.34 A long slender pole is to be erected by a winch and cable arrangement that is mounted on a truck, as illustrated in the figure to the left. If the pole weighs 2,000 lb and is 40 ft long, determine the reactions at point O and the force in the cable. Consider angles θ varying from 20° to 75°. A pivot is attached to the base of the pole so that it may be rotated without a resisting moment (i.e. acts like a pin).

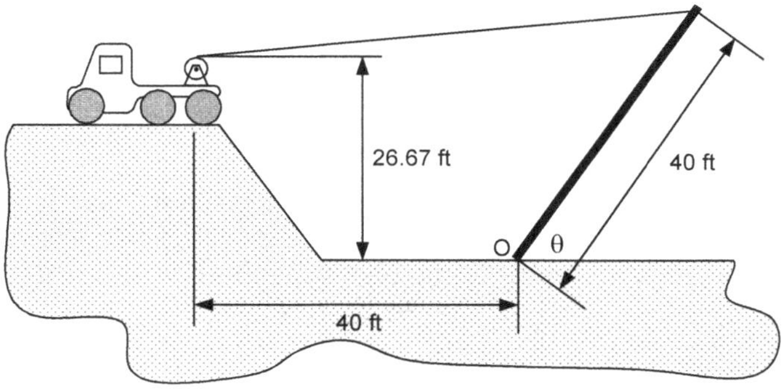

3.35 For the crane, described in the figure to the right, determine the maximum load W_L that can be lifted before the crane tips. Note, the crane weights 30 kN.

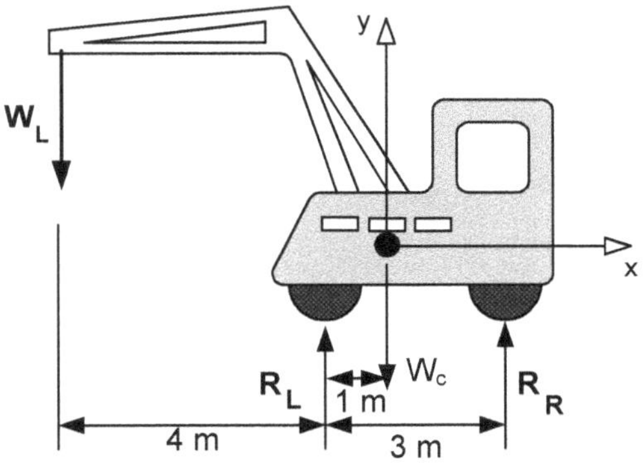

3.36 For the crane, illustrated in the figure to the left, determine the maximum load that can be lifted before the crane tips. Note, the crane weights 3.5 ton.

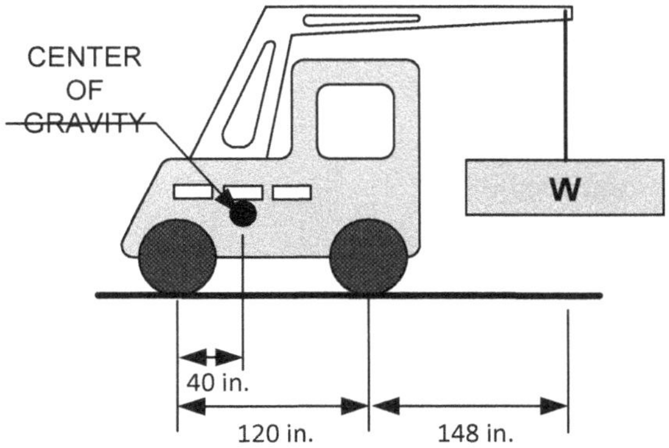

3.37 Determine the reaction forces and moments at the support for the stadium structure shown in the figure to the right.

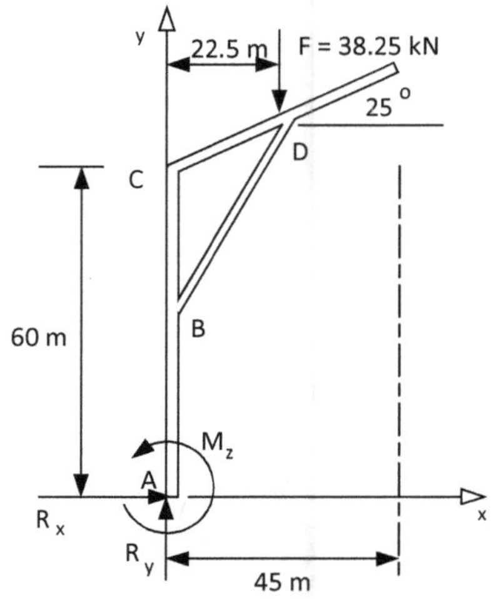

3.38 Determine the reaction forces at pins A and B that support the gusset plate shown in the figure to the left. Assume that the vertical components of the reaction forces at the pins are equal (i.e. $R_{Ay} = R_{By}$)

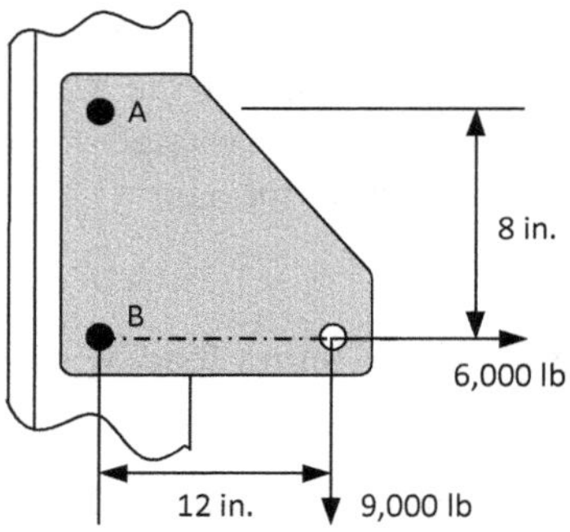

3.39 For the tension bar, shown in the figure to the right, determine the internal tension force P at sections A-A and B-B.

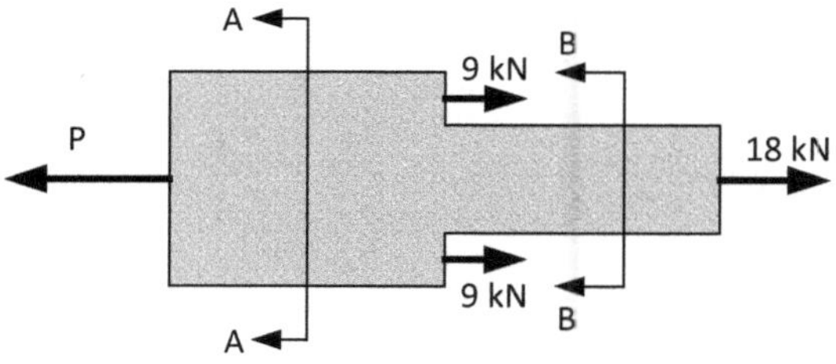

3.40 For the beam, shown in the figure to the right, determine the internal shear force V and the internal moment M at the position x = L/2.

3.41 For the beam, shown in the figure to the right, determine the internal shear force V and the internal moment M as a function of position for 0 < x < L. Prepare a graph of M and V as a function of x.

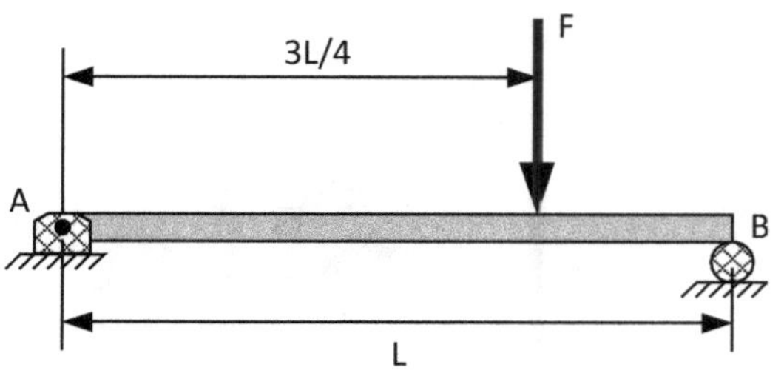

3.42 For the beam, shown in the figure to the right, determine the internal shear force V and the internal moment M at position x = L/3. Present your results in terms of q and L, which are known quantities.

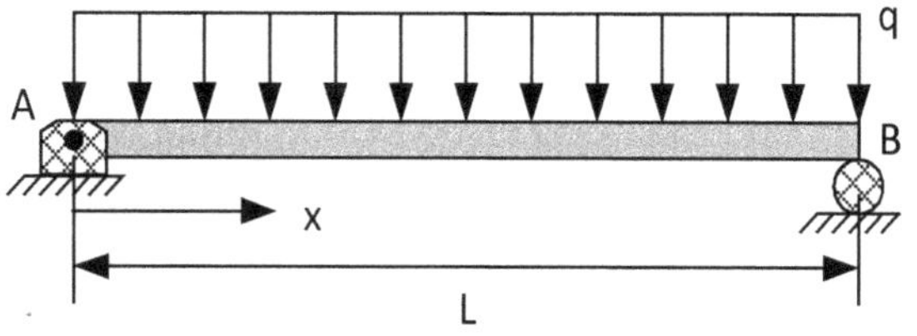

3.43 For the beam, shown, in the figure to the right, determine the internal shear force V and the internal moment M as a function of position x. Prepare a graph of M and V as a function of x.

3.44 For the beam, shown in the figure to the right, determine the internal shear force V and the internal moment M at position x = 3L/4. Present your results in terms of F and L, which are known quantities.

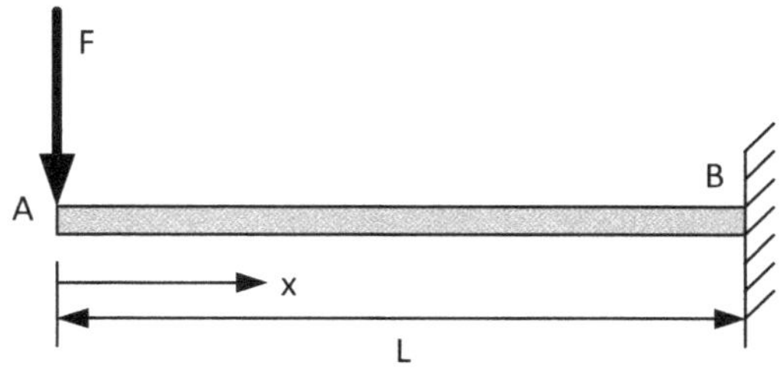

3.45 For the beam, shown in the figure to the right, determine the internal shear force V and the internal moment M as a function of position x. Prepare a graph of M and V as a function of x.

3.46 For the beam, shown in the figure to the lower right, determine the internal shear force V and the internal moment M at the position x = 12 ft.

3.47 For the beam, shown in the figure to the right, determine the internal shear force V and the internal moment M as a function of position x. Prepare a graph of M and V as a function of x.

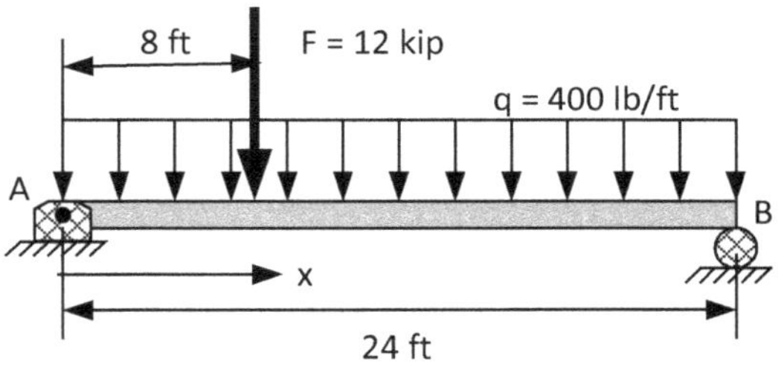

3.48 For the C-clamp, shown in the figure below, determine the internal forces P and V and the internal moment M_z at section A-A. The force F applied by the screw is 250 lb and the dimension D, which defines the distance from the screw to the centerline of the C-clamp is 1.5 in.

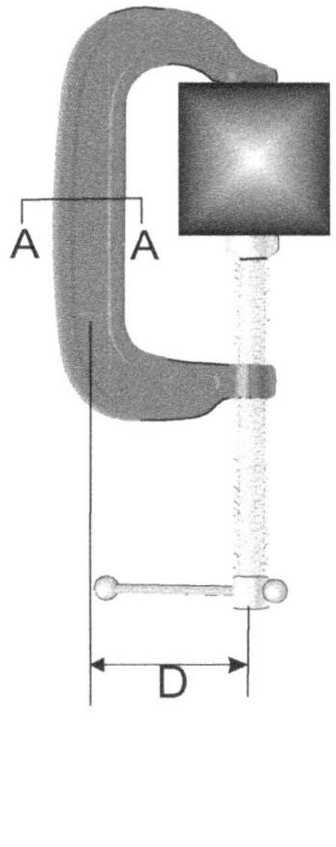

3.49 A new factory for heavy machinery has an assembly line with component parts stored above floor level. The components used in assembly are in elevated bins (#1, #2, and #3) as shown in the figure to the right. The elevated bins are suspended from a frame ABCD. Prepare a FBD of the left portion of the frame from Section A-A to the roller support at point A. Determine the internal forces and moment at section A-A as a function of the loads W_1, W_2 and W_3. Locate Section A-A at position $x = 3L/8$ and consider the loads W_1, W_2 and W_3 as known quantities. Assume that the centers of each bin are located at the following points: $x_1 = L/4$, $x_2 = L/2$ and $x_3 = 3L/4$.

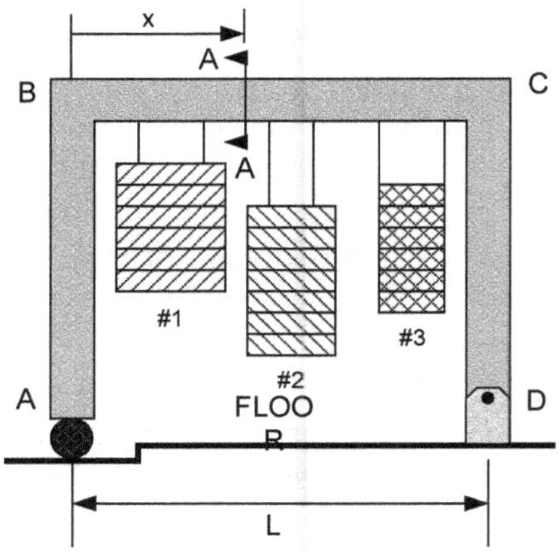

Problem 3.6:

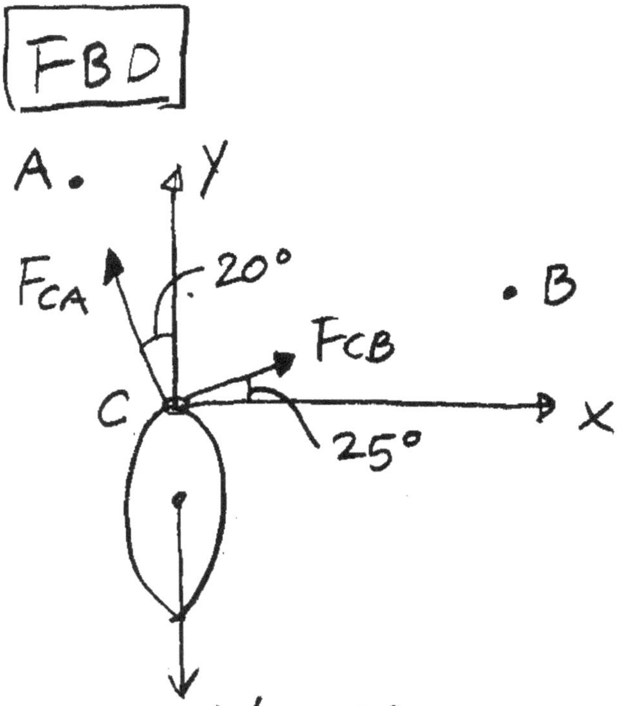

$$W = mg = 20 \times 32.17 = 643.4\ lb$$

$$\Sigma F_x = F_{CB}\cos 25° - F_{CA}\sin 20° = 0$$

$$\Sigma F_y = F_{CB}\sin 25° + F_{CA}\cos 20° - W = 0$$

$$F_{CB} = 220.9\ lb$$

$$F_{CA} = 585.3\ lb$$

Problem 3.7:

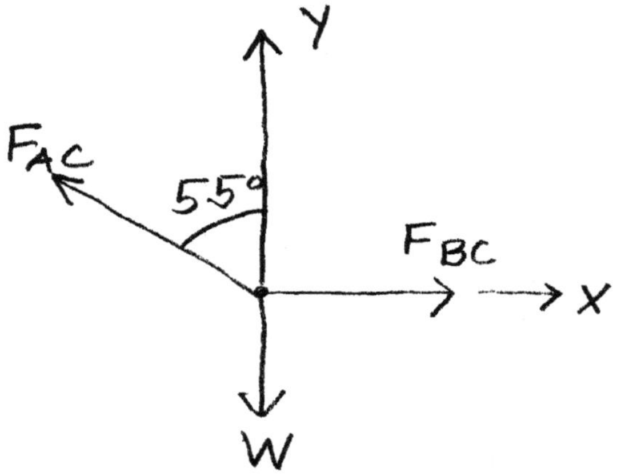

$$W = mg$$
$$= (700\ kg)(9.81)$$
$$= 6867\ N$$

$$\Sigma F_x = F_{BC} - F_{AC} \sin 55 = 0$$
$$F_{BC} = F_{AC} \sin 55$$
$$F_{BC} = 0.81915\ F_{AC} \quad (i)$$

$$\Sigma F_y = F_{AC} \cos 55 - W = 0$$
$$F_{AC} \cos 55 = W$$
$$F_{AC} = W / \cos 55$$
$$= 6867 / 0.57358$$
$$= 11{,}972$$

$$\boxed{F_{AC} = 11{,}970\ N}$$

Plug into Equation (i)

$$F_{BC} = 11{,}972\ (0.81915)$$
$$= 9806.9$$

$$\boxed{F_{BC} = 9807\ N}$$

Problem 3.9:

FBD:

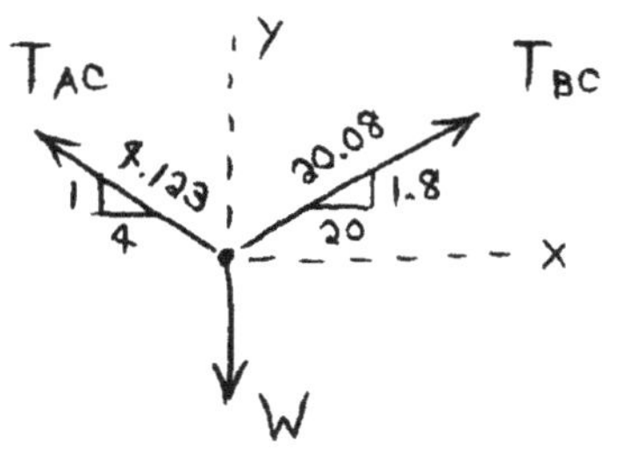

$m = 40$ kg

find: $\boxed{T_{AC},\ T_{BC} = ?}$.

① $W = mg = (40)(9.807) = 392.3$ N

② $\overset{+}{\rightarrow} \sum F_x = 0: \quad T_{BC}\dfrac{20}{20.08} - T_{AC}\dfrac{4}{4.123} = 0$

③ $+\uparrow \sum F_y = 0: \quad T_{BC}\dfrac{1.8}{20.08} + T_{AC}\dfrac{1}{4.123} - 392.3 = 0$

④ $T_{BC} = T_{AC}(0.9740)$

⑤ subing ④ in ③:

$T_{AC}(0.9740)(0.08964) + T_{AC}(0.2425) = 392.3$

⑥ $\boxed{T_{AC} = 1{,}189 \text{ N}}$.

⑦ $T_{BC} = (1{,}189)(0.9740)$

⑧ $\boxed{T_{BC} = 1{,}158 \text{ N}}$.

Alternate Method:

FBD

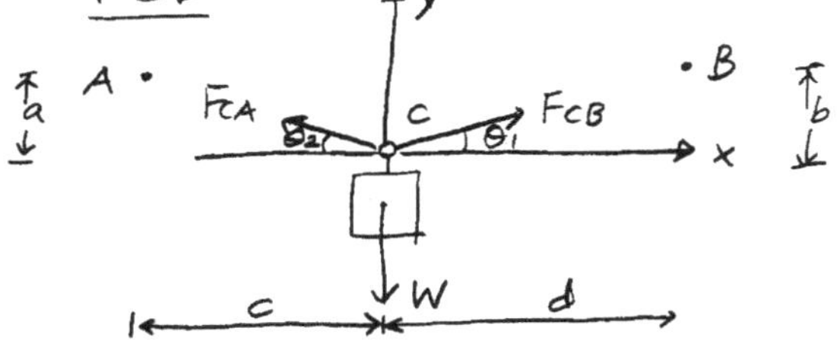

$W = mg = 40 \times 9.807 = 392.28$N

$\theta_1 = \tan^{-1}\left(\frac{b}{d}\right) = 5.14°$

$\theta_2 = \tan^{-1}\left(\frac{a}{c}\right) = 14.04°$

$\sum F_x = F_{CB}\cos\theta_1 - F_{CA}\cos\theta_2 = 0$

$\sum F_y = F_{CB}\sin\theta_1 + F_{CA}\sin\theta_2 - W = 0$

$\boxed{F_{CB} = 1158.35\text{N} \quad ; \quad F_{CA} = 1189.22\text{N}}$

Problem 3.10:

FBD's:

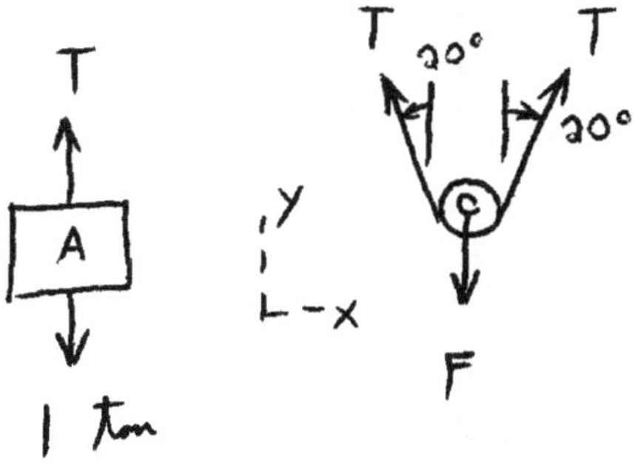

find: $\boxed{F = ?}$ + comment on effectiveness

* consider weight A:

① $\uparrow + \sum F_y = 0: \quad T - 1 \text{ ton} = 0 \quad \Rightarrow \quad T = 1 \text{ ton}$

* consider pulley C:

② $\uparrow + \sum F_y = 0: \quad 2(T \cos(20°)) - F = 0$

③ $F = 2(1)\cos(20°)$

④ $\boxed{F = 1.879 \text{ ton}}$.

∴ system is not effective because applied force F is much greater than wt. of container (1 ton)

Problem 3.11:

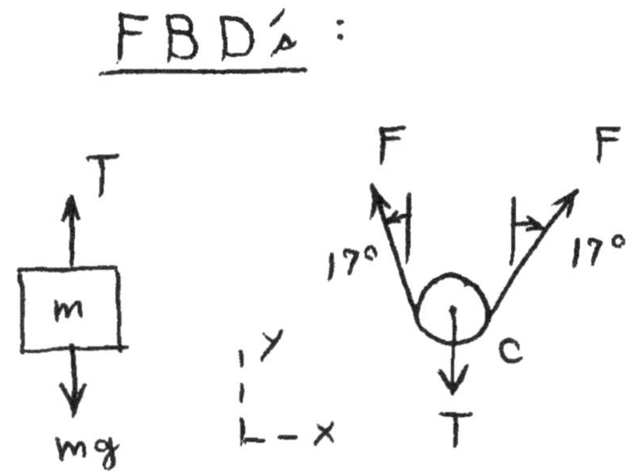

$m = 850$ kg

find: $\boxed{F = ?}$ + comment on effectiveness

* consider mass m:

① $\uparrow + \sum F_y = 0: \quad T - mg = 0 \quad \Rightarrow \quad T = (850)(9.807)$

② $T = 8.336$ kN

* consider pulley C:

③ $\uparrow + \sum F_y = 0: \quad 2(F\cos(17°)) - T = 0$

④ $F = \dfrac{(8.336)}{2\cos(17°)}$

⑤ $\boxed{F = 4.358 \text{ kN}}$.

∴ system <u>is</u> effective because applied force F is much lower than wt. of container (8.336 kN)

Problem 3.16:

FBD's:

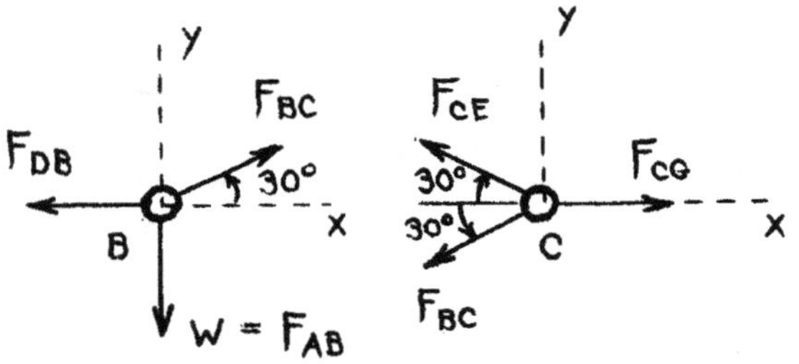

$m = 43$ kg

find: $F_{BC}, F_{DB}, F_{CE}, F_{CG}, F_{AB} = ?$

① $W = mg = (43)(9.807)$

② $W = 421.7\text{ N} = F_{AB}$

* consider ring B:

③ $\overset{+}{\rightarrow}\Sigma F_x = 0: \quad F_{BC}\cos(30°) - F_{DB} = 0$

④ $+\uparrow\Sigma F_y = 0: \quad F_{BC}\sin(30°) - 421.7 = 0 \Rightarrow F_{BC} = 843.4\text{ N}$

⑤ subing ④ into ③: $F_{DB} = (843.4)\cos(30°) \Rightarrow F_{DB} = 730.4\text{ N}$

* consider ring C:

⑥ $\overset{+}{\rightarrow}\Sigma F_x = 0: \quad F_{CG} - F_{CE}\cos(30°) - (843.4)\cos(30°) = 0$

⑦ $+\uparrow\Sigma F_y = 0: \quad F_{CE}\sin(30°) - (843.4)\sin(30°) = 0$

⑧ $F_{CE} = 843.4\text{ N}$

⑨ subing ⑧ into ⑥: $F_{CG} = 2(843.4)\cos(30°)$

⑩ $F_{CG} = 1,461\text{ N}$

Problem 3.18:

FBD's:

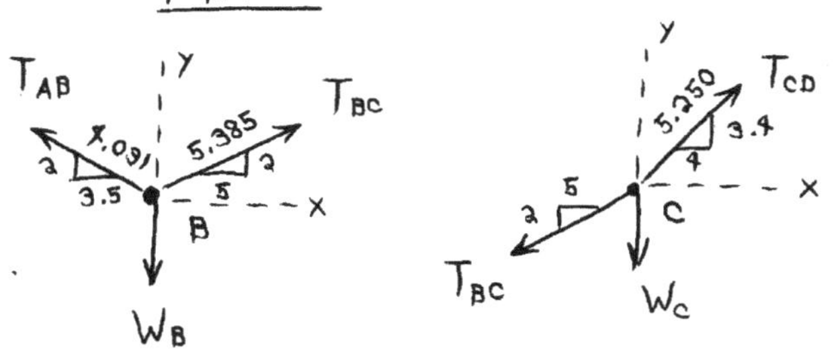

$m_B = 16$ kg

find: m_c, T_{AB}, T_{BC}, T_{CD} = ?

① $W_B = m_B g = (16)(9.807) = 156.9$ N

* consider pt. B:

② $\overset{+}{\rightarrow}\sum F_x = 0: \quad T_{BC}\left(\frac{5}{5.385}\right) - T_{AB}\left(\frac{3.5}{4.031}\right) = 0$

③ $T_{BC} = T_{AB}\,(0.9351)$

④ $\uparrow + \sum F_y = 0: \quad T_{AB}\left(\frac{2}{4.031}\right) + T_{BC}\left(\frac{2}{5.385}\right) - 156.9 = 0$

⑤ $T_{AB}(0.4962) + T_{AB}(0.9351)(0.3714) = 156.9$

⑥ $\boxed{T_{AB} = 186.0 \text{ N}}$.

⑦ $T_{BC} = (186.0)(0.9351) \Rightarrow \boxed{T_{BC} = 173.9 \text{ N}}$.

* consider pt. C:

⑧ $\overset{+}{\rightarrow}\sum F_x = 0: \quad T_{CD}\left(\frac{4}{5.250}\right) - T_{BC}\left(\frac{5}{5.385}\right) = 0$

⑨ $T_{CD} = (173.9)(1.219)$

⑩ $\boxed{T_{CD} = 212.0 \text{ N}}$.

⑪ $+\uparrow \sum F_y = 0: \quad T_{CD}\left(\frac{3.4}{5.250}\right) - T_{BC}\left(\frac{2}{5.385}\right) - W_c = 0$

⑫ $W_c = (212.0)(0.6476) - (173.9)(0.3714) = 72.70$ N

⑬ $W_c = m_c g = m_c(9.807) = 72.70$

⑭ $\boxed{m_c = 7.413 \text{ kg}}$.

Problem 3.18 - Alternate Method:

m_C ?

F_{AB}, F_{BC}, & F_{CD} ?

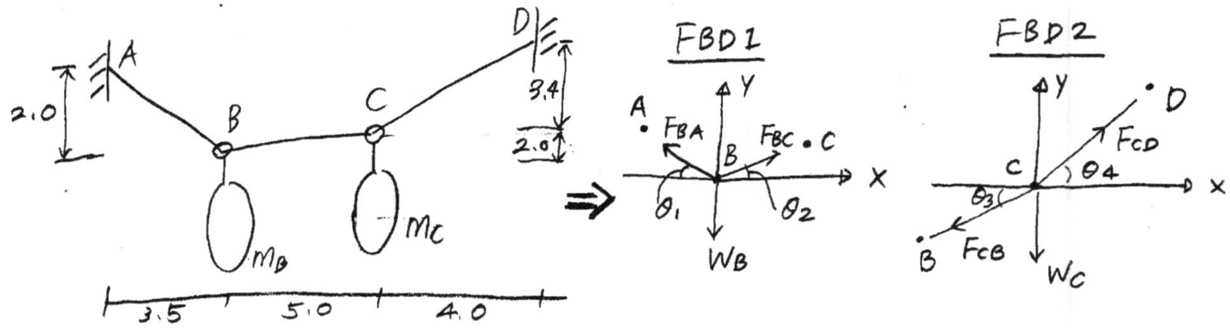

$W_B = m_B g = 16 \times 9.807 = 156.9$

$W_C = m_C g = 9.807\, m_C$

FBD 1

$\theta_1 = \tan^{-1}\left(\frac{2.0}{3.5}\right) = 29.7°$

$\theta_2 = \tan^{-1}\left(\frac{2.0}{5.0}\right) = 21.8°$

$\Sigma F_x = -F_{BA}\cos\theta_1 + F_{BC}\cos\theta_2 = 0$

$\Sigma F_y = F_{BA}\sin\theta_1 + F_{BC}\sin\theta_2 - W_B = 0$

$F_{BA} = 186.1\text{ N}$; $F_{BC} = 174.1\text{ N}$

FBD 2

$\theta_3 = \theta_2 = 21.8°$

$\theta_4 = \tan^{-1}\left(\frac{3.4}{4}\right) = 40.4°$

$\Sigma F_x = -F_{CB}\cos\theta_3 + F_{CD}\cos\theta_4 = 0$

$\Sigma F_y = F_{CD}\sin\theta_4 - F_{CB}\sin\theta_3 - W_C = 0$

$F_{CB} = F_{BC} = 174.1\text{ N}$; $F_{CD} = 212.3\text{ N}$

$W_C = 72.94\text{ N}$

$m_C = \frac{W_C}{g} = \frac{72.94}{9.807} = 7.438\text{ kg}$

$F_{BA} = 186.1\text{ N}$
$F_{BC} = 174.1\text{ N}$
$F_{CD} = 212.3\text{ N}$
$m_C = 7.438\text{ kg}$

Problem 3.19:

FBD:

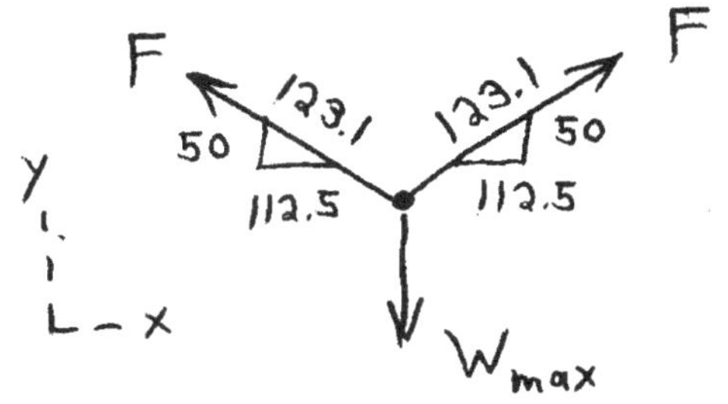

$F = k\delta$ $\quad k = 60$ lb/in

$W = 140$ lb

find: will stuntman remain dry = ?

* consider max. wt. of stuntman (W_{max}) that rope can support

① most severe loading occurs when stuntman is in center of rope & just touches water surface

$\Rightarrow L_o = 112.5$ ft ; $y = 50$ ft

② $L_d = \sqrt{(112.5)^2 + (50)^2} = 123.1$ ft

③ $\delta = L_d - L_o = ((123.1) - (112.5))(12) = 127.2$ in

④ $F = k\delta = (60)(127.2) = 7{,}632$ lb

⑤ $+\uparrow \sum F_y = 0: \quad 2F\left(\frac{50}{123.1}\right) - W_{max} = 0$

⑥ $W_{max} = 2(7{,}632)\left(\frac{50}{123.1}\right) = 6{,}200$ lb > 140 lb

∴ since wt. of stuntman is much less than W_{max}, he will remain dry.

Problem 3.19 - Alternate Method:

Problem 19

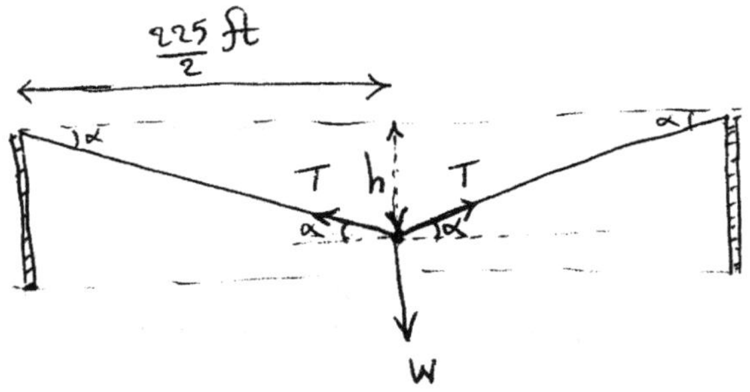

Deflection in half rope $= \delta = \frac{225}{2}\left(\frac{1}{\cos\alpha}-1\right)$

Maximum deflection occurs when the stuntman is in the middle.

$\Sigma F_y = 0 \Rightarrow W = 2T\sin\alpha \Rightarrow \boxed{T = \frac{W}{2\sin\alpha}}$ ①

Also: $\boxed{T = k\delta = k\frac{225}{2}\left(\frac{1}{\cos\alpha}-1\right)}$ ②

Substituting ① in ②: $\frac{W}{2\sin\alpha} = k\frac{225}{2}\left(\frac{1}{\cos\alpha}-1\right)$, $k = 60\frac{lb}{in} = 720\frac{lb}{ft}$, $W = 140\ lb$

$\Rightarrow \boxed{\alpha = 6.868^\circ}$

$h = \frac{225}{2}\tan\alpha \Rightarrow \boxed{h = 13.6\ ft}$

Since $h < 50\ ft$, the stuntman remains dry.

Problem 3.21:

FBD's:

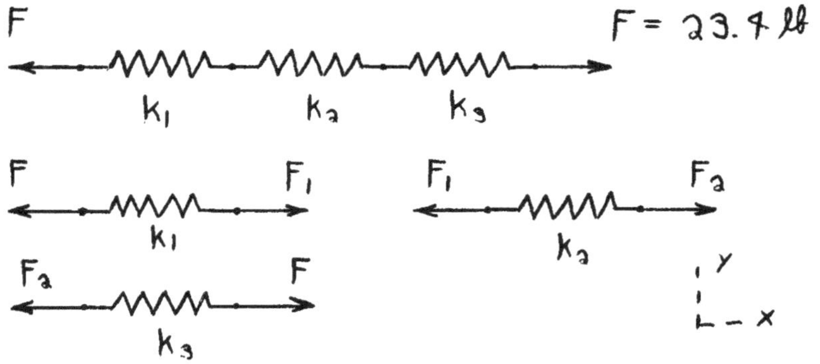

$k_1 = 20$ lb/in

$k_2 = 32$ lb/in

$k_3 = 12$ lb/in

find: $\boxed{\delta_T,\ k_{eff} = ?}$.

① from $\rightarrow \sum F_x = 0$ for each spring: $F = F_1 = F_2 = 23.4$ lb

② $F = k_1 \delta_1 \Rightarrow \delta_1 = \dfrac{F}{k_1} = \dfrac{23.4}{20} = 1.17$ in

③ $\delta_2 = \dfrac{F}{k_2} = \dfrac{23.4}{32} = 0.7313$ in

④ $\delta_3 = \dfrac{F}{k_3} = \dfrac{23.4}{12} = 1.95$ in

⑤ $\delta_T = \delta_1 + \delta_2 + \delta_3 = 1.17 + 0.7313 + 1.95 \Rightarrow \boxed{\delta_T = 3.851 \text{ in}}$.

⑥ $k_{eff} = \dfrac{F}{\delta_T} = \dfrac{23.4}{3.851} \Rightarrow \boxed{k_{eff} = 6.076 \text{ lb/in}}$.

* alternate sol'n:

① the following eqn. can be derived for springs in series:

$$\frac{1}{k_{eff}} = \sum_{i=1}^{n} \frac{1}{k_i}$$

② $\dfrac{1}{k_{eff}} = \dfrac{1}{k_1} + \dfrac{1}{k_2} + \dfrac{1}{k_3} = \dfrac{1}{20} + \dfrac{1}{32} + \dfrac{1}{12}$

③ $\boxed{k_{eff} = 6.076 \text{ lb/in}}$.

④ $\delta_T = \dfrac{F}{k_{eff}} = \dfrac{23.4}{6.076} \Rightarrow \boxed{\delta_T = 3.851 \text{ in}}$.

Problem 3.22:

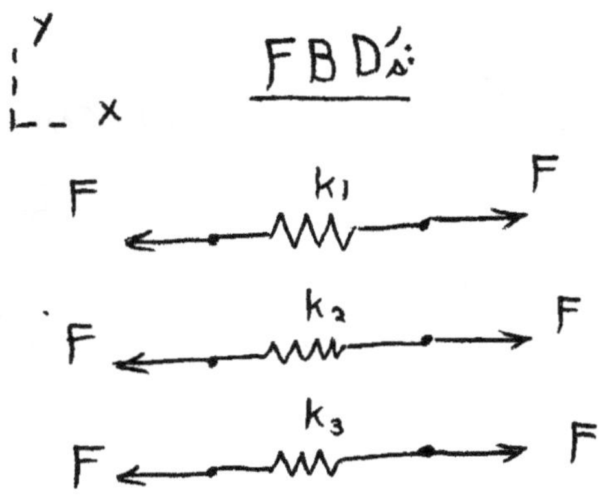

$F = 70\ N$

$k_1 = 4.4\ N/mm$

$k_2 = 7.5\ N/mm$

$k_3 = 3\ N/mm$

find: δ_T, k_e = ?

① $\overset{+}{\rightarrow}\sum F_x = 0 \Rightarrow$ each spring experiences a force

$F = 70\ N$

② $F = k\delta \Rightarrow \delta = \frac{F}{k}$

③ $\delta_T = \delta_1 + \delta_2 + \delta_3 = \frac{F}{k_1} + \frac{F}{k_2} + \frac{F}{k_3}$

④ $\delta_T = \frac{70}{4.4} + \frac{70}{7.5} + \frac{70}{3} = 15.91 + 9.333 + 23.33$

⑤ $\boxed{\delta_T = 48.57\ mm}$.

⑥ $k_e = \frac{F}{\delta_T} = \frac{70}{48.57}$

⑦ $\boxed{k_e = 1.441\ N/mm}$.

Problem 3.23:

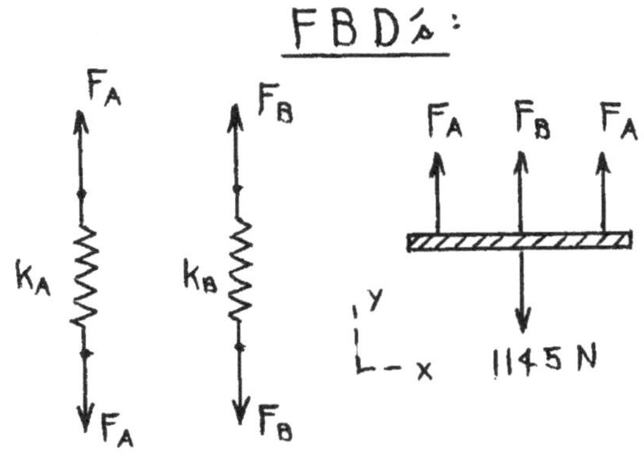

$k_A = 12\ N/mm$

$k_B = 23\ N/mm$

find: $\boxed{\delta_T,\ k_{eff} = ?}$.

① from $\uparrow + \sum F_y = 0$ for horizontal bar: $2F_A + F_B = 1145\ N$

② $F = k\delta \Rightarrow \delta = \dfrac{F}{k}$

③ from geometry: $\delta_A = \delta_B = \delta_T \Rightarrow \dfrac{F_A}{k_A} = \dfrac{F_B}{k_B}$

④ $F_B = F_A \dfrac{k_B}{k_A} = F_A \dfrac{(23)}{(12)} = (1.917)F_A$

⑤ subing ④ into ①: $2F_A + (1.917\ F_A) = 1145$

⑥ $F_A = 292.3\ N$

⑦ using ⑥ in ④: $F_B = 1.917(292.3) = 560.3\ N$

⑧ $\delta_T = \delta_A = \dfrac{F_A}{k_A} = \dfrac{292.3}{12} \Rightarrow \boxed{\delta_T = 24.36\ mm}$.

⑨ $k_{eff} = \dfrac{F_T}{\delta_T} = \dfrac{1145}{24.36} \Rightarrow \boxed{k_{eff} = 47.00\ N/mm}$.

* alternate sol'n:

① the following eqn. can be derived for springs in parallel:

$$k_{eff} = \sum_{i=1}^{n} k_i$$

② $k_{eff} = k_A + k_B + k_A = 12 + 23 + 12$

③ $\boxed{k_{eff} = 47\ N/mm}$.

④ $\delta_T = \dfrac{F_T}{k_{eff}} = \dfrac{1145}{47} \Rightarrow \boxed{\delta_T = 24.36\ mm}$.

Problem 3.28:

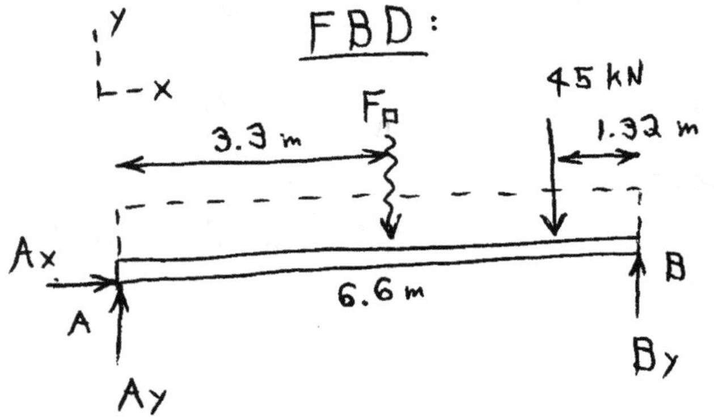

$q = 5.2 \ kN/m$

find: $A_x, A_y, B_y = ?$

① $F_\square = q \cdot L = (5.2)(6.6) = 34.32 \ kN$

② $\overset{+}{\rightarrow}\sum F_x = 0$: $A_x = 0$.

③ $\circlearrowleft + \sum M_A = 0$: $-(34.32)(3.3) - (45)(5.28) + B_y(6.6) = 0$

④ $B_y = 53.16 \ kN$.

⑤ $\uparrow + \sum F_y = 0$: $A_y + (53.16) - 34.32 - 45 = 0$

⑥ $A_y = 26.16 \ kN$.

Problem 3.31:

FBD:

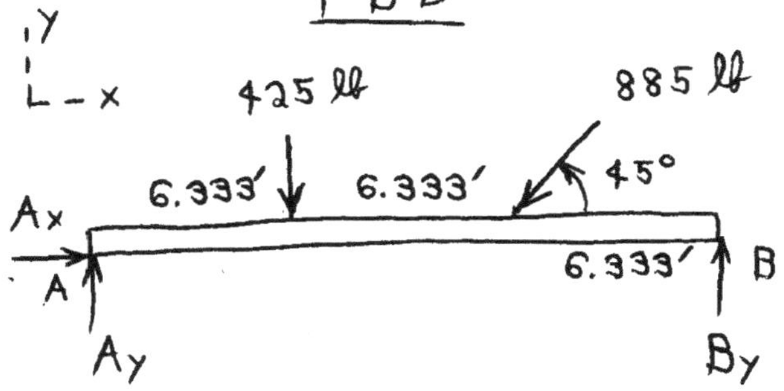

find: $A_x, A_y, B_y = ?$

① $\overset{+}{\rightarrow}\sum F_x = 0: \quad A_x - (885)\cos(45°) = 0$

② $\boxed{A_x = 625.8 \text{ lb}}$.

③ $\circlearrowleft + \sum M_A = 0:$

$$-(425)(6.333) - ((885)\sin(45°))(12.67) + B_y(19) = 0$$

④ $\boxed{B_y = 557.0 \text{ lb}}$.

⑤ $\uparrow + \sum F_y = 0: \quad A_y - 425 - (885)\sin(45°) + (557.0) = 0$

⑥ $\boxed{A_y = 491.8 \text{ lb}}$.

Problem 3.36:

FBD:

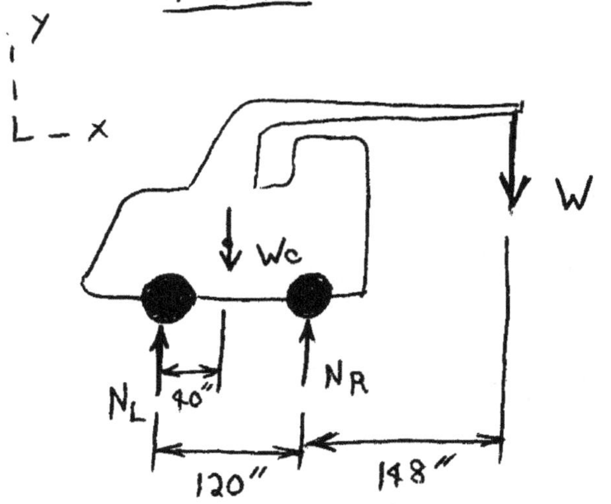

$W_c = 3.5$ ton

find: $W_{max} = ?$ before crane tips

① at tipping condition, crane pivots about right wheel

$\Rightarrow \quad N_L = 0 \;+\; W = W_{max}$

② $\circlearrowleft + \sum M_R = 0: \quad W_c(80) - W_{max}(148) = 0$

③ $W_{max} = (3.5)\dfrac{(80)}{(148)}$

④ $\boxed{W_{max} = 1.892 \text{ ton}}$

Problem 3.37:

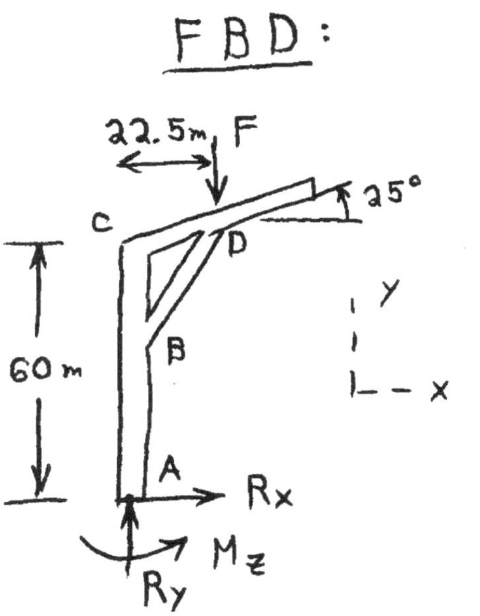

$F = 38.25$ kN

find: $\boxed{R_x,\ R_y,\ M_z = ?}$.

① $\overset{+}{\rightarrow} \sum F_x = 0:$ $\boxed{R_x = 0}$.

② $\uparrow + \sum F_y = 0:$ $R_y - 38.25 = 0$

③ $\boxed{R_y = 38.25 \text{ kN}}$.

④ $\circlearrowleft + \sum M_A = 0:$ $M_z - (38.25)(22.5) = 0$

⑤ $\boxed{M_z = 860.6 \text{ kN-m}}$.

Problem 3.39:

FBD's:

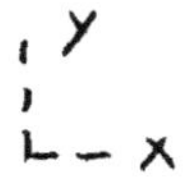

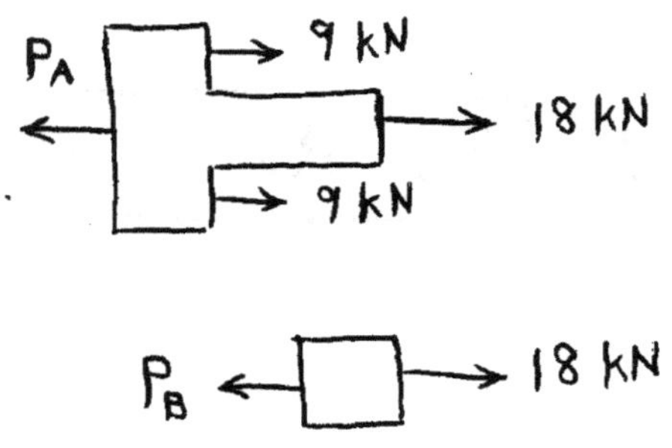

find: $P_A, P_B = ?$

* consider cut @ A-A:

① $\overset{+}{\rightarrow}\sum F_x = 0: \quad 18 + 2(9) - P_A = 0$

② $P_A = 36 \text{ kN}$.

* consider cut @ B-B:

③ $\overset{+}{\rightarrow}\sum F_x = 0: \quad P_B = 18 \text{ kN}$.

Problem 3.40:

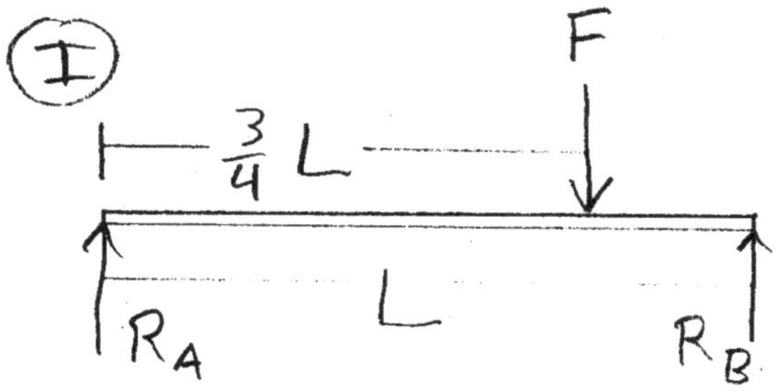

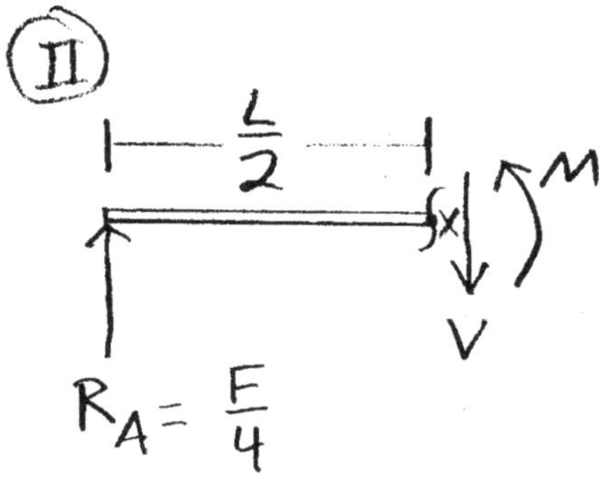

(I)

$$\Sigma M_A = L R_B - \tfrac{3}{4} L F = 0$$

$$R_B = \tfrac{3}{4} F$$

$$\Sigma F_y = R_A + R_B - F = 0$$

$$R_A = F - R_B = F/4$$

(II)

$$\Sigma F_y = R_A - V = 0$$

$$V = R_A \Rightarrow \boxed{V = F/4}$$

$$\Sigma M_x = M - (L/2)(F/4) = 0$$

$$M = (L/2)(F/4) \Rightarrow \boxed{M = FL/8}$$

Problem 3.41:

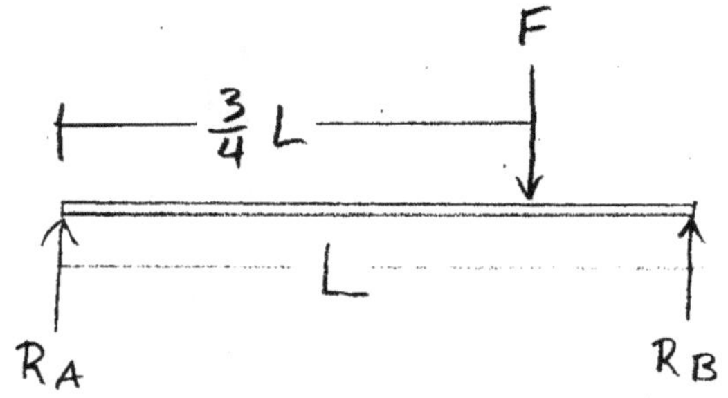

$$\Sigma M_A = L R_B - \tfrac{3}{4} L F = 0$$

$$R_B = \tfrac{3}{4} F$$

$$\Sigma F_y = R_A + R_B - F = 0$$

$$R_A = F - R_B = F/4$$

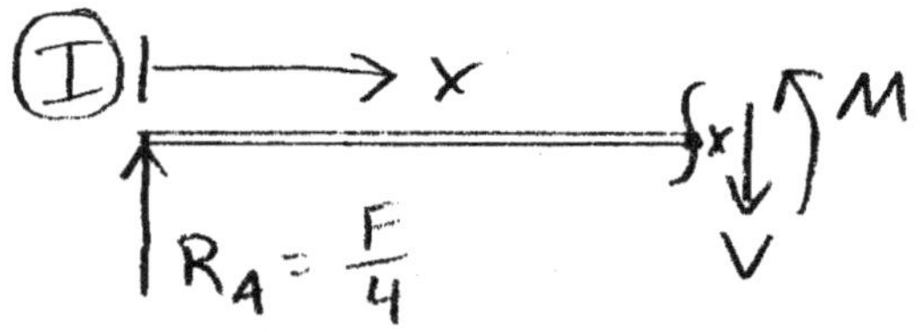

(I) $\Sigma F_y = R_A - V = 0$

$$V = R_A = F/4$$

$$\Sigma M_x = M - R_A x = 0$$

$$M = R_A x = (F/4)x$$

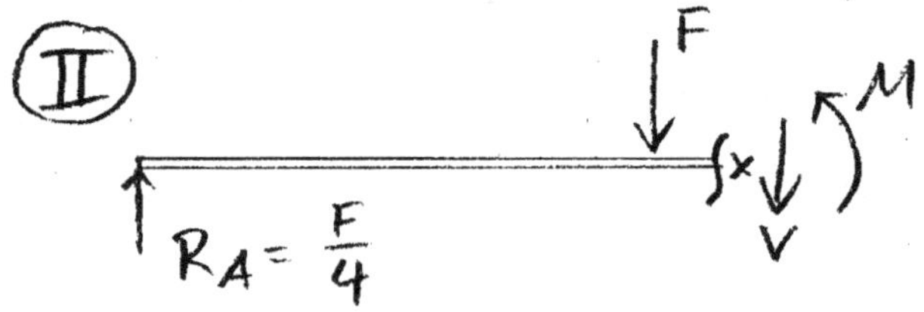

(II) $\Sigma F_y = R_A - F - V = 0$

$$V = R_A - F = -\tfrac{3}{4} F$$

$$\Sigma M_x = M - R_A x + F(x - \tfrac{3}{4}L) = 0$$

$$M = (F/4)x - F(x - \tfrac{3}{4}L)$$

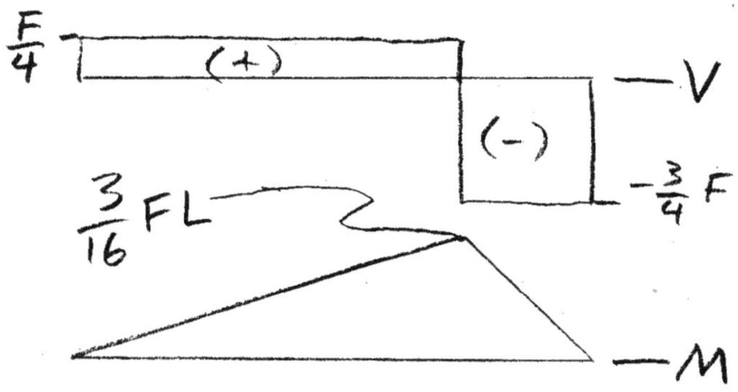

Problem 3.42:

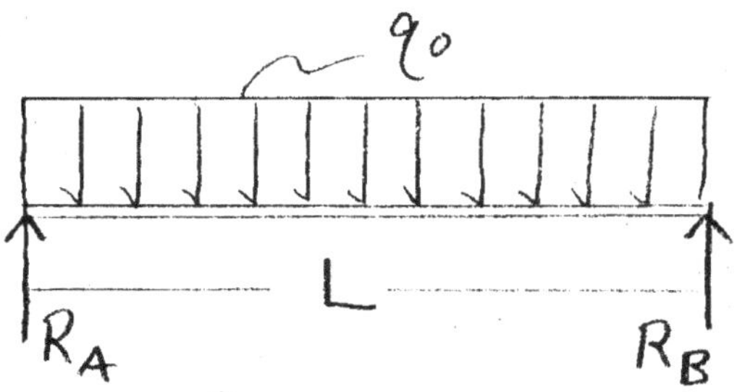

$$\Sigma M_A = R_B L - q_0(L)(L/2) = 0$$

$$R_B = q_0 L/2$$

$$\Sigma F_y = R_A + R_B - q_0 L = 0$$

$$R_A = q_0 L/2$$

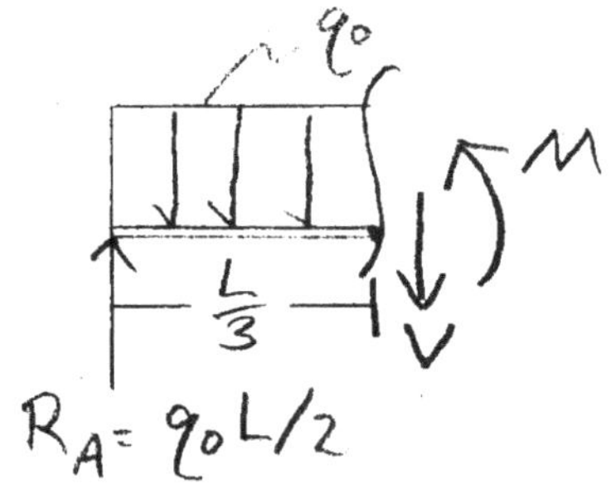

$$\Sigma F_y = R_A - q_0(L/3) - V = 0$$

$$V = q_0 L/3 - q_0 L/2 = \boxed{-q_0 L/6}$$

$$\Sigma M_x = M - R_A(L/3) + q_0(L/3)(L/6) = 0$$

$$M = R_A(L/3) - q_0 L^2/18 = q_0(L/2)(L/3) - \frac{q_0 L^2}{18}$$

$$M = q_0 L^2\left(\frac{1}{6} - \frac{1}{18}\right) = \boxed{q_0 L^2/9}$$

Problem 3.43:

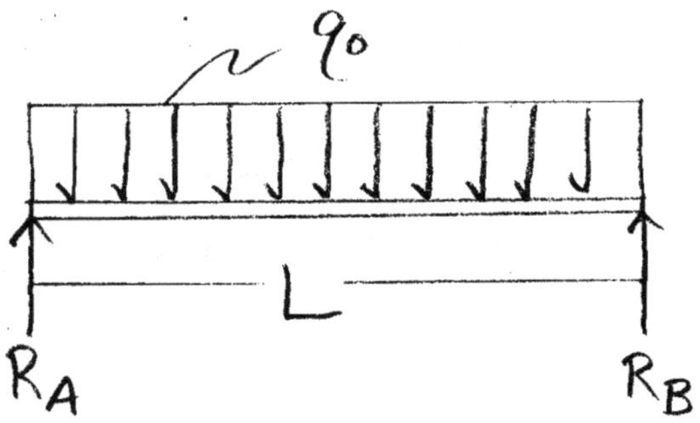

$$\Sigma M_A = R_B L - q_0 L (L/2) = 0$$

$$R_B = q_0 L/2$$

$$\Sigma F_y = R_A + R_B - q_0 L = 0$$

$$R_A = q_0 L/2$$

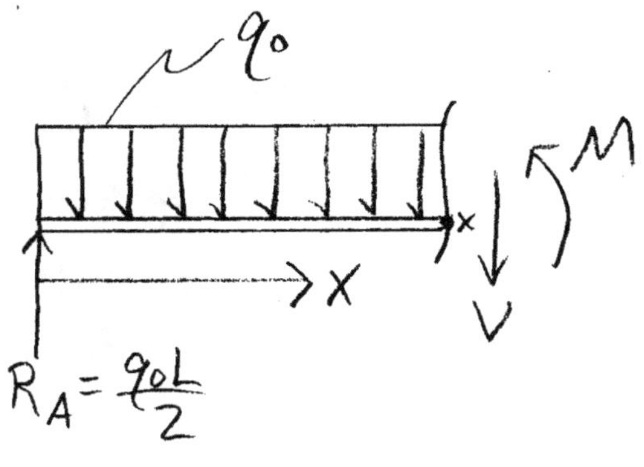

$$\Sigma F_y = R_A - q_0 X - V = 0$$

$$\boxed{V = q_0 L/2 - q_0 X}$$

$$\Sigma M_x = M - R_A X + q_0 X \left(\frac{X}{2}\right) = 0$$

$$\boxed{M = q_0 L X/2 - q_0 X^2/2}$$

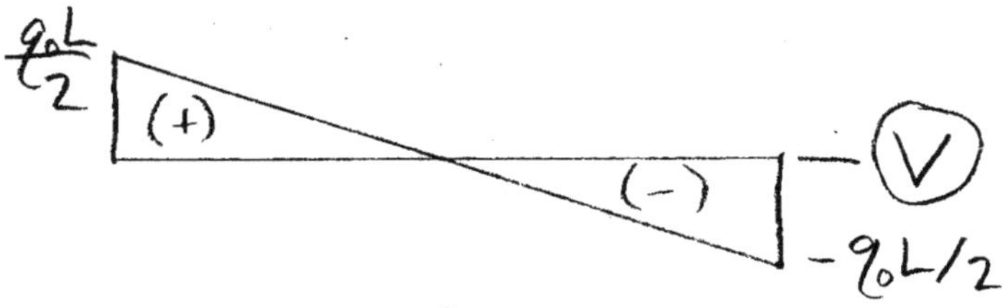

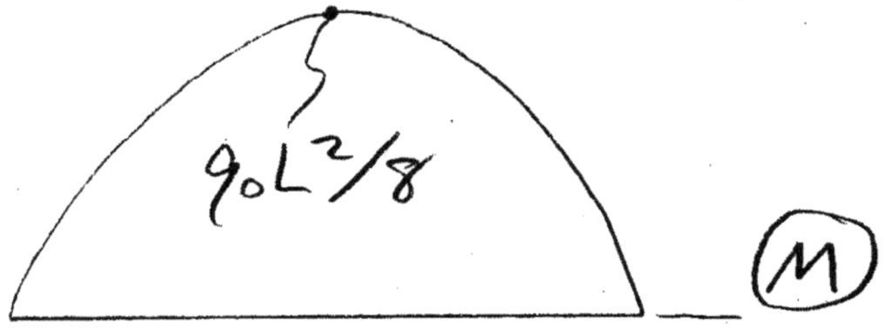

Problem 3.44:

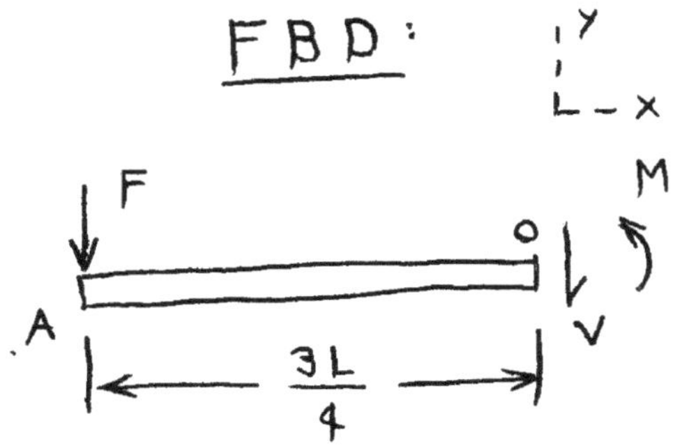

find: $\boxed{V, M = ?}$.

① $\uparrow + \sum F_y = 0$: $\boxed{V = -F}$.

② $\circlearrowleft + \sum M_o = 0$: $M + F\left(\frac{3L}{4}\right) = 0$

③ $\boxed{M = -\frac{3}{4} FL}$.

Problem 3.45:

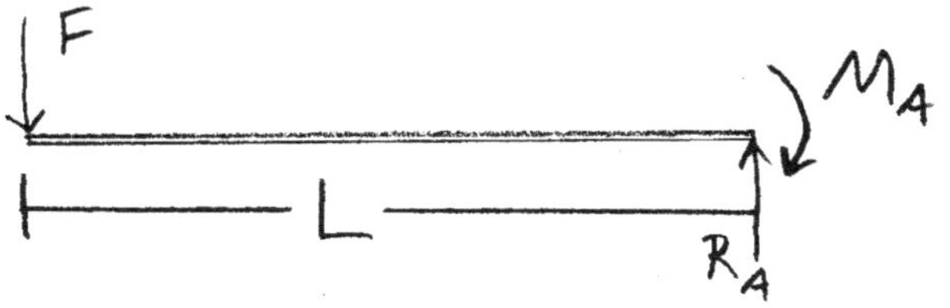

$\Sigma F_y = R_A - F = 0 \Rightarrow R_A = F$

$\Sigma M_A = M_A + FL = 0 \Rightarrow M_A = FL$

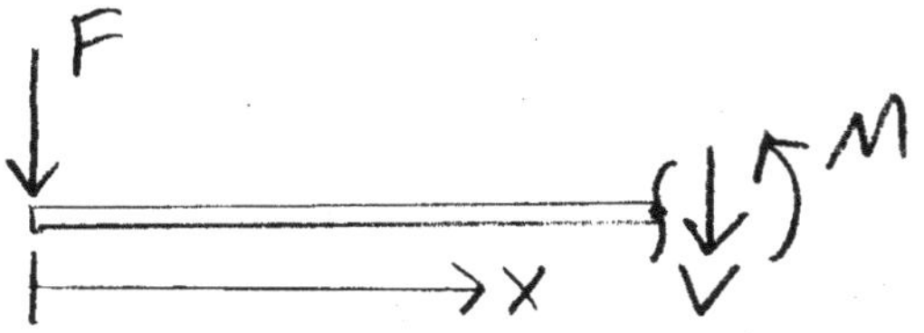

$\Sigma F_y = -F - V = 0$

$\boxed{V = -F}$

$\Sigma M_x = M + Fx = 0$

$\boxed{M = -Fx}$

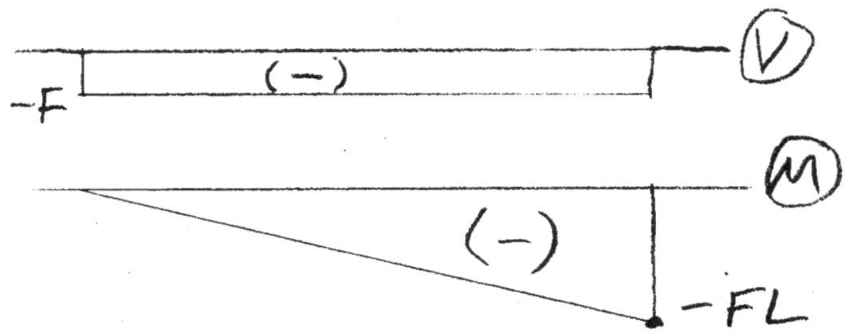

Problem 3.46:

FBD's:

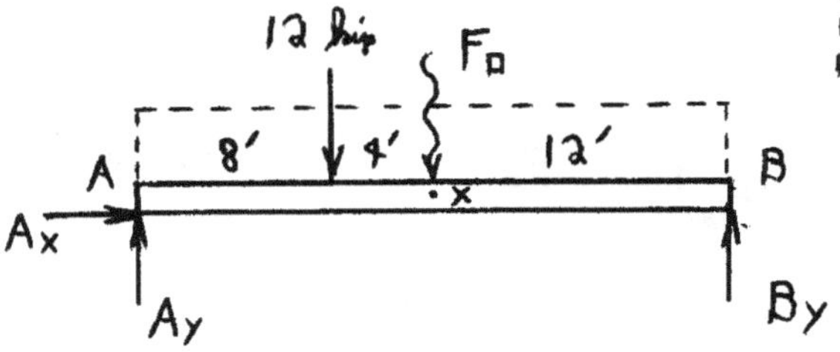

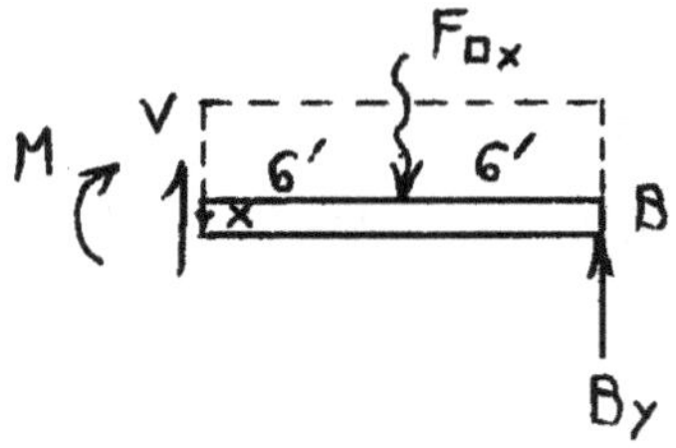

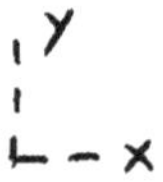

$q = 400\ ^{lb}/_{ft}$ $\quad x = 12\ ft$

find: $\boxed{V, M = ?}$ @ pt. x

* consider entire beam:

① $F_\square = qL = (400)(24)$

② $F_\square = 9.6$ kip @ 12' from each end

③ $\xrightarrow{+} \sum F_x = 0: \quad A_x = 0$

④ $\uparrow + \sum F_y = 0: \quad A_y + B_y - 12 - 9.6 = 0$

⑤ $\circlearrowleft + \sum M_A = 0: \quad -(12)(8) - (9.6)(12) + B_y(24) = 0$

⑥ $B_y = 8.8$ kip

⑦ using ⑥ in ④ + solving: $\quad A_y = 12.8$ kip

* consider segment XB:

⑧ $F_{\square x} = qL_x = (400)(12) = 4.8$ kip @ 6' from each end

⑨ $\uparrow + \sum F_y = 0: \quad V - 4.8 + (8.8) = 0$

⑩ $\boxed{V = -4 \text{ kip}}$.

⑪ $\circlearrowleft + \sum M_x = 0: \quad -M - (4.8)(6) + (8.8)(12) = 0$

⑫ $\boxed{M = 76.8 \text{ kip-ft}}$.

Problem 3.48:

FBD:

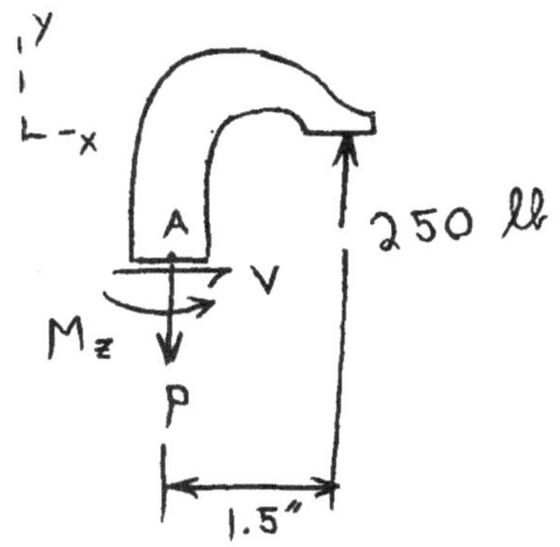

find: $\boxed{P, V, M_z = ?}$.

① $\overset{+}{\rightarrow}\sum F_x = 0$: $\boxed{V = 0}$.

② $+\uparrow \sum F_y = 0$: $\boxed{P = 250 \text{ lb}}$.

③ $\circlearrowleft + \sum M_A = 0$: $(250)(1.5) + M_z = 0$

④ $\boxed{M_z = -375 \text{ lb-in}}$.

CHAPTER 4 Axially Loaded Structural Members

4.1 A steel music wire (Gage No. 10 with a diameter of 0.024 in.) is employed to support an axial load P of 8.64 lb. If the wire was initially 22 ft in length, determine its final length and the stress and strain in the wire.

4.2 A steel music wire is tightened on a guitar by rotating the wire supporting post through an angle of 11°. If the post has a diameter d = 0.25 in. and the length of the wire is 32 in., determine the strain induced in the guitar string.

4.3 Referring to Problem 4.2, determine the additional stress induced in the guitar string as it is tightened.

4.4 A hemp rope 15 m long and 45 mm in diameter is supported from one of the roof beams in a gymnasium. Three gymnasts climb this rope together. The lead gymnast has a mass of 55 kg, the next 72-kg and the trailing gymnast 52 kg.

(a) Determine the maximum stress in this rope.
(b) Determine the stress in the rope at a location between the lead and intermediate gymnast.
(c) Determine the stress in the rope at a location between the intermediate and trailing gymnast.

4.5 A suspension bridge is to carry a roadway and traffic that may weigh up to 5,000 ton. If the safe load that can be imposed on the wire rope to be used in the construction is 175 ton, specify the number of cables to be employed. Justify your answer.

4.6 Describe the constraints on the relation $\sigma = E\varepsilon$.

4.7 Determine the design load that can be specified for Gage No. 00 steel wire with a strength of 530 MPa, if the design specification calls for the safety factor of 2.5.

4.8 Discuss the factors to consider in establishing the safety factor in a design specification.

4.9 For the cable-pulley-mass arrangement shown in the figure to the right, determine the stresses in the cable. The effective cable diameter d is 0.500 in.

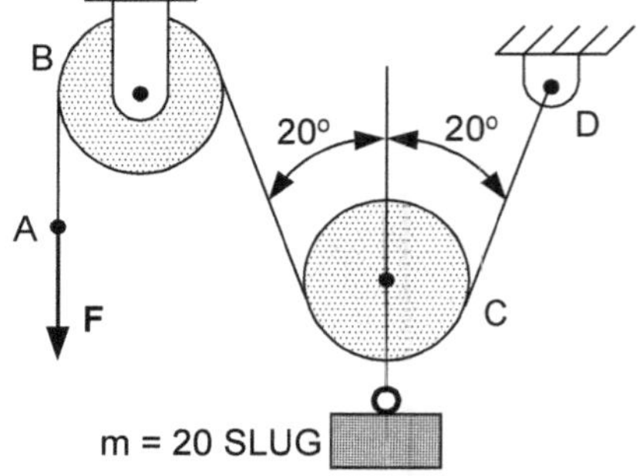

4.10 Cold drawn steel alloy wire exhibits an ultimate tensile strength of 600 MPa and a yield strength of 500 MPa. If this alloy is employed, specify the gage of the wire if it is to support a design load of 7.5 kN with a safety factor of 3.2. Specify the gage number required based on both yielding and rupturing as modes of failure.

4.11 Determine the safety factor for the wire CD in the cable-weight arrangement shown in the figure to the right. The steel cables are each 10 m long, exhibit a yield strength of 500 MPa and have an effective cross sectional area of 8 mm^2. Assume that the forces in the horizontal wires are equal in magnitude and opposite in direction (F_{CB} = – F_{CA}). Also assume $F_{CB} = (0.18)F_{CD}$.

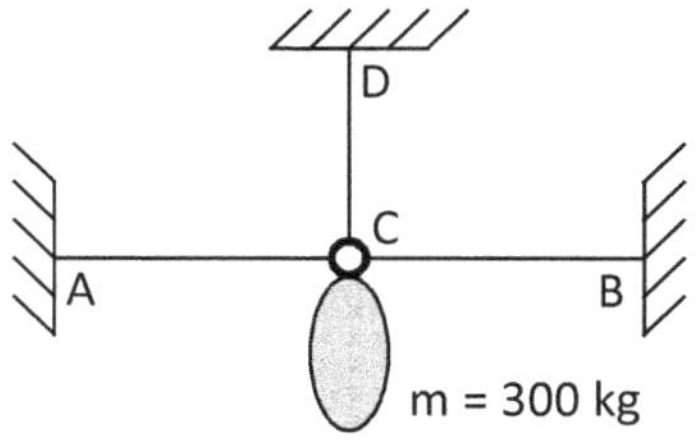

4.12 Determine the mass m, shown in the diagram to the right, that is required to yield the cable CD if the cables are each 22 ft long, exhibit a yield strength of 80ksi and have an effective cross sectional area of 0.015 $in.^2$ Assume that the forces in the horizontal wires are equal in magnitude and opposite in direction (F_{CB} = – F_{CA}). Also assume $F_{CB} = (0.25)F_{CD}$. Is the cable-weight arrangement stable after yield? Why?

4.13 The wire-mass system, shown in the figure to the left, is constructed using steel wire. Specify a suitable alloy and the gage number for the wires if the criterion for failure is based on yielding. Use the same diameter for both wires. The safety factor imposed on the design is 2.9.

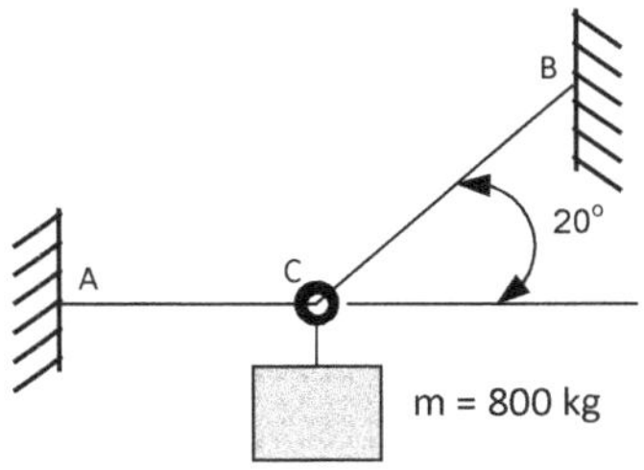

4.14 The wire-mass system, shown in the figure to the right, is constructed using aluminum wire. Specify a suitable alloy and the gage number for the wires if the criterion for failure is based on ultimate tensile strength. Use the same diameter for both wires. The safety factor imposed on the design is 1.8.

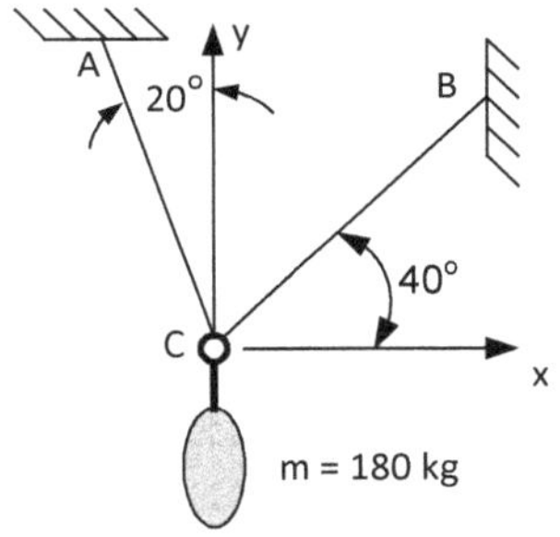

4.15 The wire-mass system ,shown in the figure to the left, is constructed using steel wire fabricated from 1010 A. A safety factor of 2.9, based on the ultimate tensile strength, is to be employed for both wires. Find the maximum weight that can be lifted (in pounds) and specify the required gage numbers for each wire.

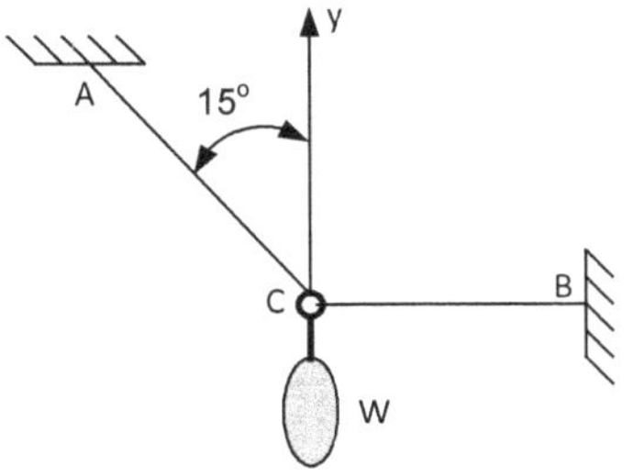

4.16 Describe in an engineering brief the differences between a wire rope and a rod.

4.17 Describe in an engineering brief the similarities between a wire rope and a rod.

4.18 A long thin steel bar with a length of 1.8 m, a width of 50 mm, and a thickness of 20 mm is subjected to an axial force of 22 kN. Determine the tensile stress and the axial deformation of the bar.

4.19 A bar fabricated from steel with a tensile strength of 54.0 ksi is subjected to an axial tensile force of 20 kip. The bar is designed with a safety factor of 3.2. Determine the design stress for the bar and the required cross sectional area.

4.20 A bar fabricated from an aluminum alloy with a tensile strength of 400 MPa is subjected to an axial tensile force of 75 kN. The bar is designed with a safety factor of 2.5. Determine the design stress for the bar and the required cross sectional area.

4.21 A bar fabricated from brass with a tensile strength of 247 MPa is subjected to an axial compressive force of 76 kN. The bar is designed with a safety factor of 2.8. Determine the design stress for the bar and the required cross sectional area. Assume the tensile and compressive stresses are equal.

4.22 A key used to lock a gear onto a shaft is subjected to a shear force of 8.0 kip. If the key is 0.25 in. wide by 0.375 in. high and 1.5 in. long, determine the shear stress acting on the key. Draw a free body diagram showing this shear stress.

4.23 If the key in Problem 4.22 is machined from a steel alloy with a tensile yield strength of 48,000 psi, determine the safety factor for the key.

4.24 A rectangular bar, shown in the figure to the right, is subjected to an axial tensile force F = 30 kN. Determine the normal stress and the shear stress on an inclined plane if the angle of the section cut is $\theta = 30°$.

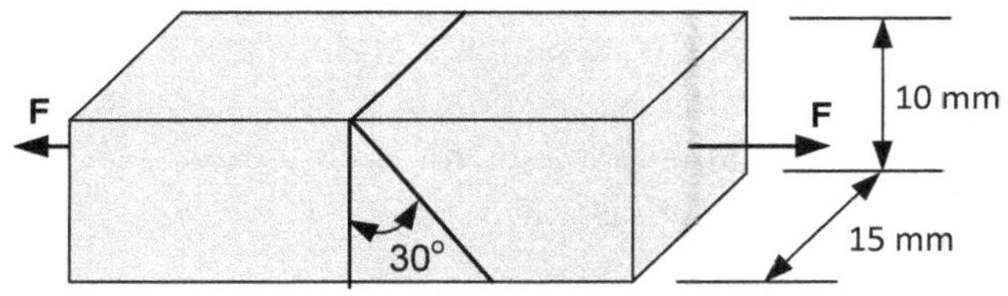

4.25 A rectangular bar, similar to the one shown in the figure to the right, is adhesively bonded along a 20° section cut. The normal stress in the bond line is limited to 2.5 ksi and the shear stress to 1.5 ksi. If the bar has a cross sectional area A = 1.75 in.^{2} and a safety factor of 3.0 is specified, determine the largest axial force that can be applied to the bar.

4.26 A rod with a circular cross section with a diameter d is fabricated from two pieces as illustrated in the figure to the right. If a force F is applied to the bar, derive the expression for the normal and shear stress acting in the plane of the adhesive joint as a function of angle ϕ.

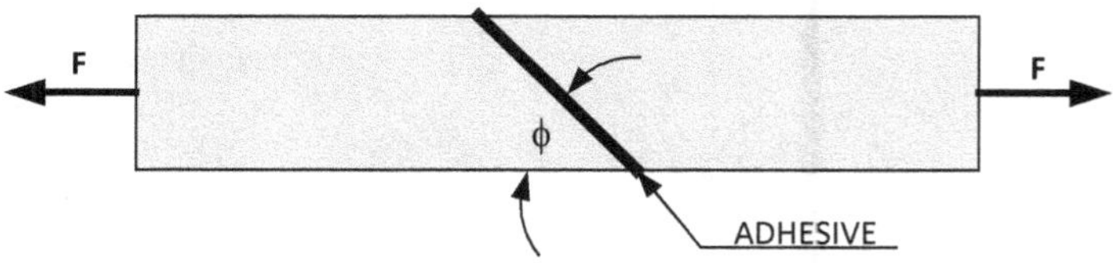

4.27 For the bar illustrated in the figure to the right, determine the shear stresses in the adhesive joint if the angle ϕ is varied from 0° to 90°. The axial force applied to the bar is 2.0 kN and its diameter is 20 mm. Also compute the normal stresses acting on the adhesive joint. Hint: Use a spreadsheet to determine the shear and tensile stresses in the adhesive joint for the range of ϕ specified.

4.28 A tensile bar, defined in the figure to the right, has a cross sectional area of 175 mm^2 and is subjected to a tensile force F. The stresses on the inclined plane A—B are σ_θ = 81 MPa and τ_θ = –27 MPa. Determine the stress σ_x, the angle θ and the force F.

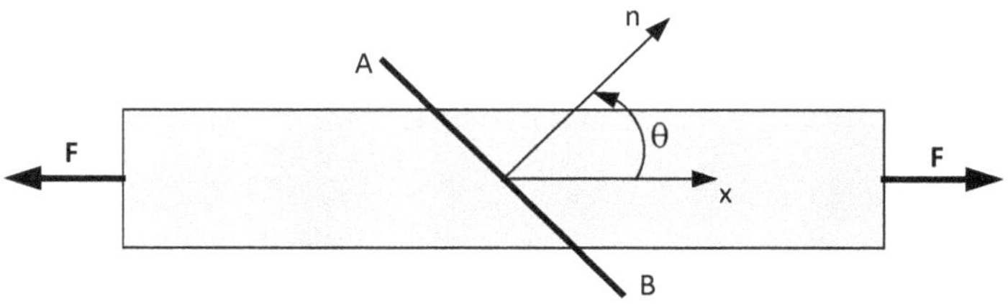

4.29 The tensile bar, defined in the figure to the right, has a cross sectional area of 175 mm^2 and is subjected to a tensile force F. The stresses on the inclined plane A—B are σ_θ = 81 MPa and τ_θ = –27 MPa. Determine the shear and normal stresses acting on an inclined plane with θ = 40°.

4.30 For the tensile bar in Problem 4.29, prepare a graph that shows the stresses σ_θ and τ_θ as a function of the angle of the inclined cut as θ varies from 0 to 90°.

4.31 Consider the tapered bar presented in the figure below. The tapered bar is subjected to an axial force of 500 kN. Prepare a graph of the axial stress σ_x as a function of x as it varies from zero to 4 m.

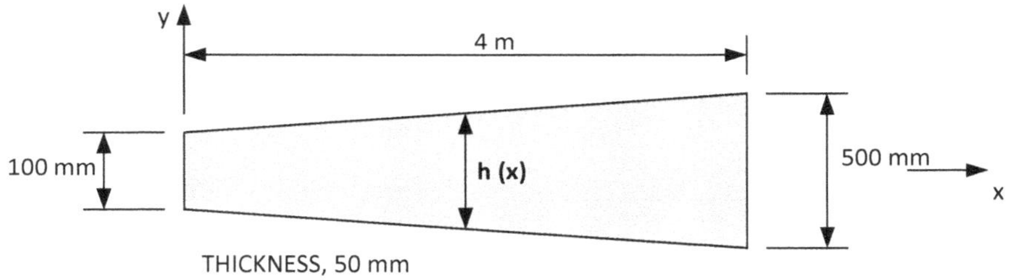

4.32 Consider the tapered bar presented in Problem 4.31. If the bar is fabricated from an aluminum alloy, determine the extension of the bar when it is subjected to an axial tensile force of 500 kN.

4.33 Consider the tapered bar presented in the figure below. If the bar is fabricated from an aluminum alloy, determine the extension of the bar when it is subjected to a tensile force of 120 kN. In this solution, do not employ Eq. (4.19). Instead, use Eq. (4.18) and perform a numerical integration on a spreadsheet.

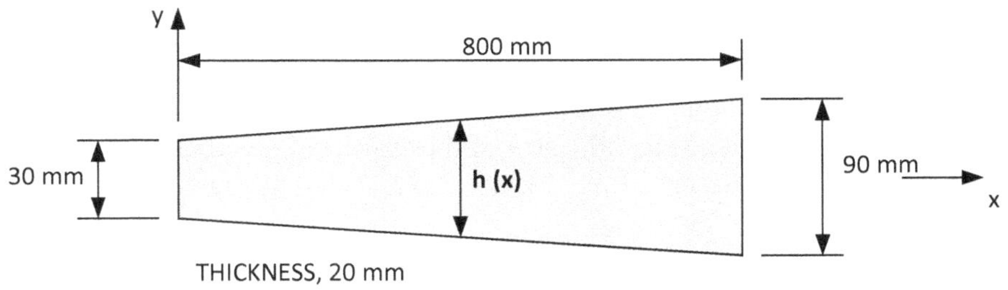

4.34 Describe the procedure employed to solve for the stresses and deflection in a stepped bar subjected to axial loading.

4.35 For the stepped bar, illustrated in the figure to the right, determine the stresses in both portions of the bar and its total deflection. The bar is fabricated from steel.

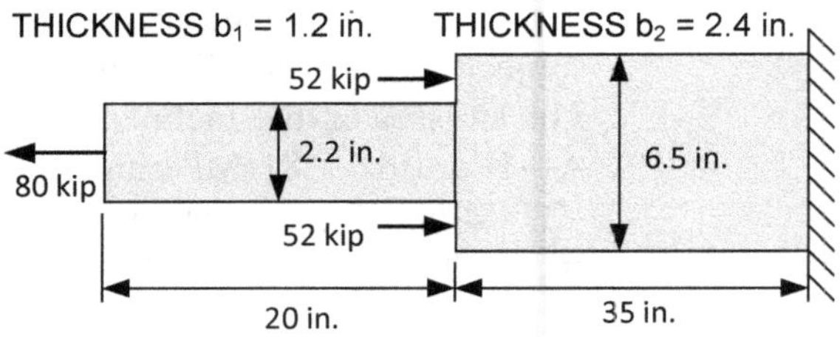

4.36 A tensile bar 40 in. long, shown in the figure below, is subjected to an axial force of 10 kip. Determine: (a) the nominal stress, (b) the stress concentration factor K, and (c) the maximum stress.

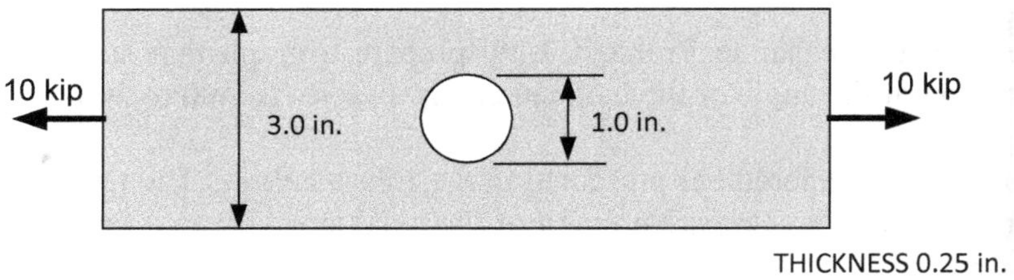

4.37 A stepped tensile bar, illustrated in the figure below, with fillets at the transition between the small and the large section is subjected to an axial tensile force of 75 kN. Determine the maximum stress at the fillets.

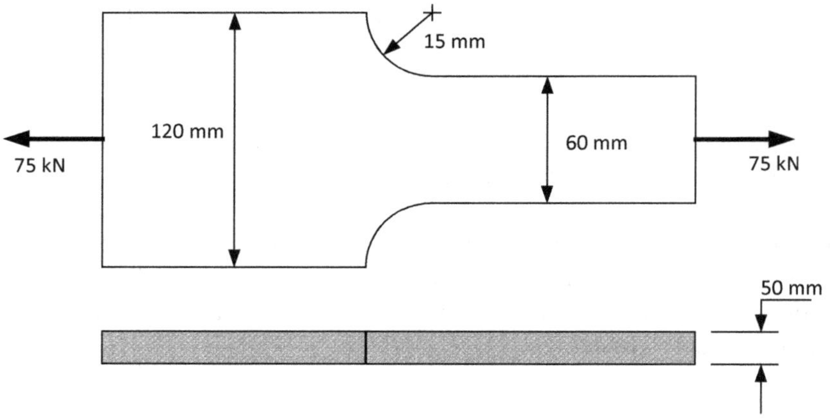

4.38 A scale model of a large structure has been fabricated from steel and tested. A strain gage on one member of the structure indicated an axial strain of ε = 1,450 μm/m. Determine the stress in the corresponding member of the prototype. The numerical parameters defining the scaling factors for the loads and the size of the structural member are w_m = 1.6 mm, w_p = 80 mm, b_m = 2.0 mm, b_p = 100 mm, L_m = 14 mm and L_p = 7.0 m. The scaling factor L = 1/10,000.

4.39 Suppose that a model of a structure is fabricated from members formed from sheet aluminum. The prototype structure is to be fabricated from steel with an open span of 200 feet. The model is geometrically scaled so that its span is four feet. The capacity of the live load on the prototype is 150 lb/ft, and the model is loaded with 2.5 lb/ft. If the model deflects a distance of 0.220 in. under full load at the center of the span, determine the deflection of the prototype under the design load.

4.40 The tensile bar, presented in the figure to the right, is fabricated with an adhesive joint inclined at an angle ϕ to the axis of the bar. Determine the optimum angle ϕ for the inclined plane if the stresses in the adhesive are not to exceed either 3.5 ksi in tension or 2.8 ksi in shear. Also determine the maximum force that can be applied to the member without exceeding the stresses in the adhesive joint.

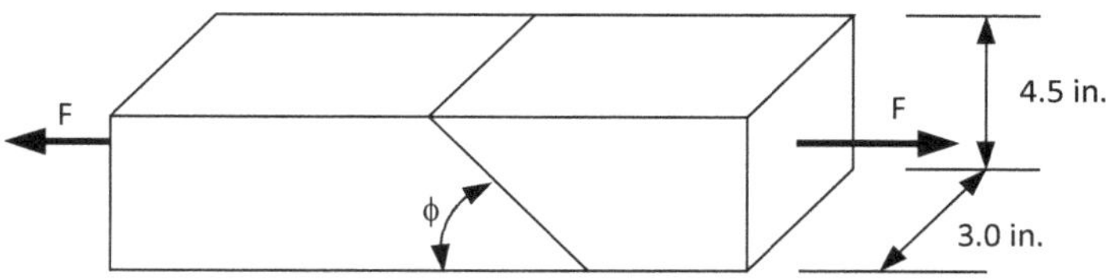

4.41 A structural member is fabricated from a solid round bar of steel with a diameter d = 50 mm. If the member is 6.0 m long, determine the maximum axial force that can be applied if the axial stress is not to exceed 175 MPa and the total elongation is not to exceed 0.14% of its length.

4.42 A steel tie rod with a diameter of 1.0 in. and length of 34.0 in. is employed to compress a brass bushing with an outside diameter of 3.0 in. and a length of 24.0 in. as shown in the figure to the right. Determine the minimum wall thickness of the bushing if the deflection of the tie rod is limited to $\delta = 0.020$ in. Note that F = 10 kip.

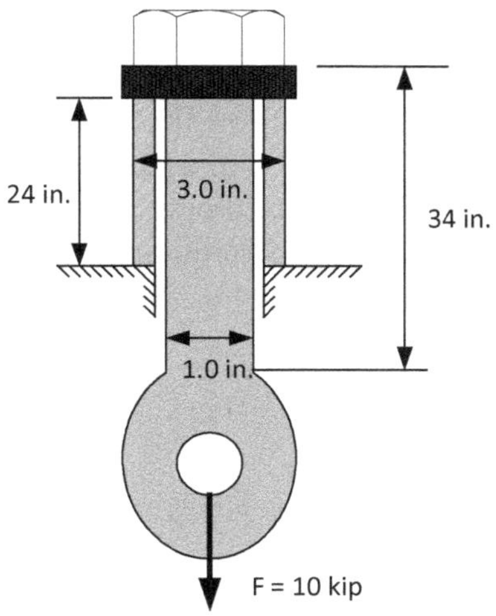

4.43 A short column with a height of 2.0 m is fabricated by adhesively bonding aluminum faceplates to a core of plastic foam as shown in the figure to the left. The foam plastic core has a square cross section with an elastic modulus of 1200 psi. Determine the stresses in the aluminum plates and the plastic foam. Also, determine the displacement of the column under the action of the applied force F = 35 kN.

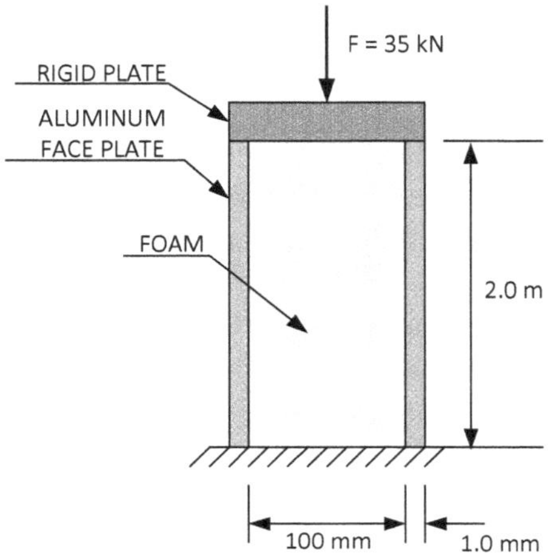

4.44 A grade 8 steel bolt, 1.0 in. in diameter (14 threads/in.) is employed to clamp a brass bushing between two rigid platens as illustrated in the figure below. The brass bushing has a 2.5 in. outside diameter and 1.5 in. inside diameter and is 12.00 in. long. After the unit is assembled with a snug fit, the nut is tightened by 1/3 of a turn. Determine the axial stresses in the bolt and the bushing. Also determine the deflection of the brass bushing.

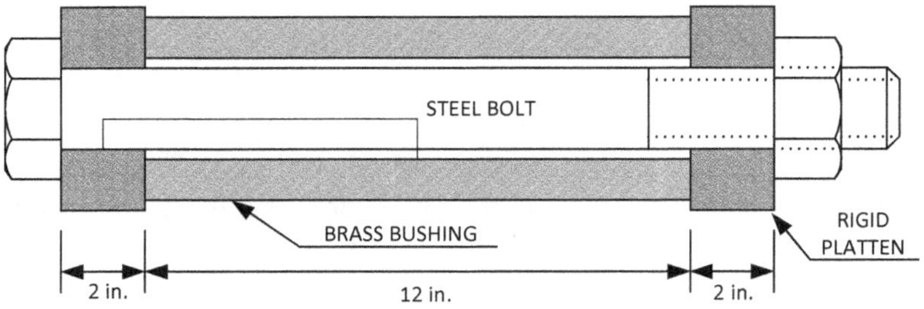

4.45 Twenty-five steel reinforcing rods with a ¼ in. diameter are placed within a high strength concrete column with a square cross section that supports an applied force of 50 kip. Determine the stresses in the steel reinforcing bars and the concrete. Also determine the amount of deflection of the column. Reference the figure to the right.

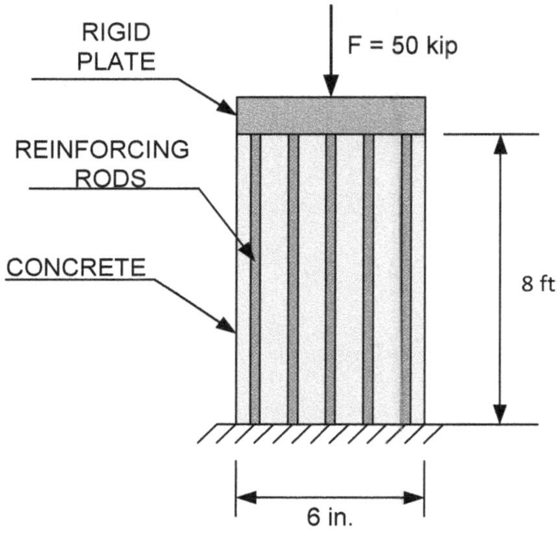

4.46 A long aluminum rod is connected to a shorter brass cylinder with a bolted flange as shown in the figure to the left. Prior to assembly a gap of δ = 0.20 mm occurred between the flange plates. Bolts were inserted and tightened bringing the flange plates together. Determine the stresses induced in both the rod and the cylinder by the assembly operation. Also determine the displacement of the face of each flange.

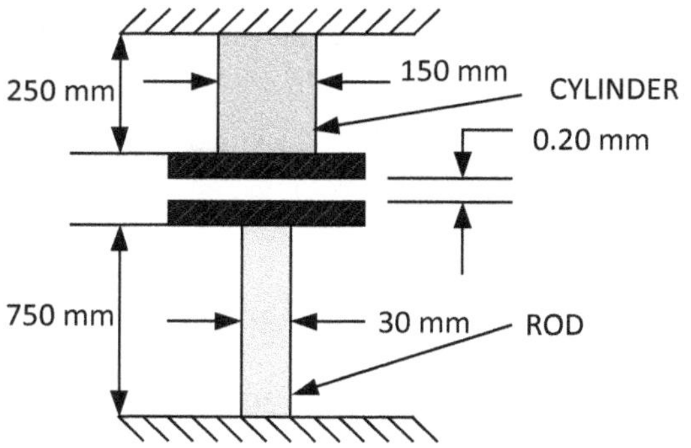

4.47 A copper and stainless steel rod are assembled between two rigid walls at an ambient temperature of 20° C as shown in the figure to the right. If the temperature is increased by an amount $\Delta T = 120$ °C, determine the thermal stresses induced in each rod. Also determine the change in length of each member.

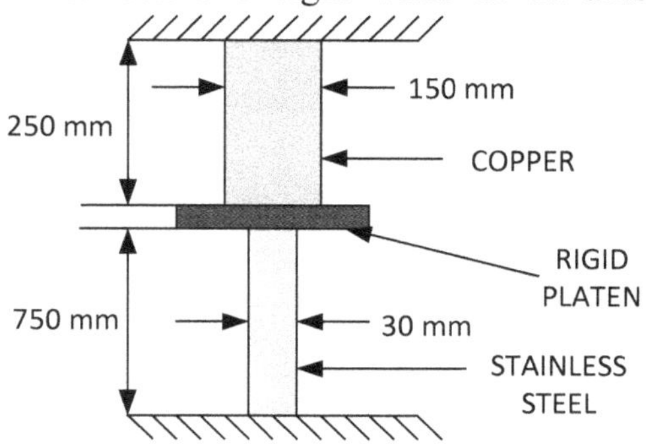

4.48 A high strength steel bolt with a diameter of 1.0 in. passes through an aluminum bushing with a cross sectional area of 3.0 in.^{2} as shown in the figure below. The unit is assembled with a snug fit at an ambient temperature 70 °F. If the temperature of the assembly increases by $\Delta T = 60$ °F, determine the thermal stresses induced in the bolt and the bushing.

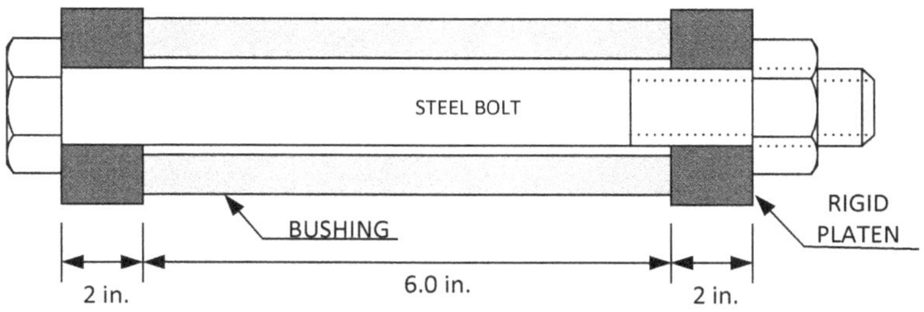

4.49 A three bar suspension system, shown in the figure to the right, is assembled at an ambient temperature of 20° C. Determine the axial stresses in each bar after a force F of 100 kN is applied to the rigid platen and the temperature is increased by $\Delta T = 80$ °C. The bars labeled A are fabricated from aluminum with a cross sectional area of 750 mm^2 and the bar labeled B is fabricated from stainless steel with a cross sectional area of 1,250 mm^2.

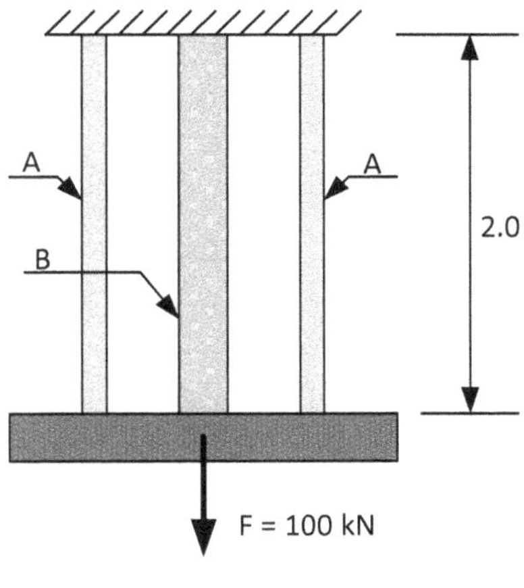

Problem 4.1:

FBD:

$L_o = 22$ ft

$d = 0.024$ in

find: $L_f, \sigma, \epsilon = ?$

* steel music wire

① $\sigma = \dfrac{P}{A} = \dfrac{P}{\frac{\pi}{4}d^2} = \dfrac{4P}{\pi d^2} = \dfrac{4(8.64)}{\pi(0.024)^2}$

② $\sigma = 1.910 \times 10^4$ psi $= 19.10$ ksi.

③ from Appendix B-1 for steel: $E = 30 \times 10^6$ psi

④ $\sigma = E\epsilon \Rightarrow \epsilon = \dfrac{\sigma}{E} = \dfrac{1.910 \times 10^4}{30 \times 10^6}$

⑤ $\epsilon = 0.0006367$.

⑥ $\epsilon = \dfrac{\delta}{L_o} = \dfrac{L_f - L_o}{L_o} \Rightarrow L_f = \epsilon L_o + L_o$

⑦ $L_f = (0.0006367)(22) + 22$

⑧ $L_f = 22.014$ ft.

Problem 4.4:

FBD's:

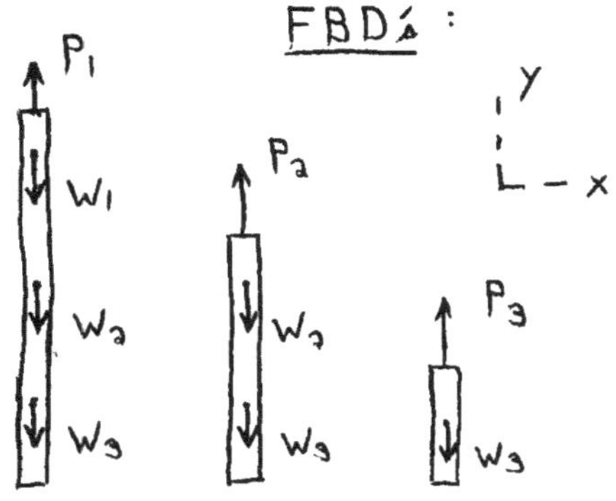

$m_1 = 55\ \text{kg}$ $\qquad L = 15\ \text{m}$

$m_2 = 72\ \text{kg}$ $\qquad d = 45\ \text{mm}$

$m_3 = 52\ \text{kg}$

find: a) $\boxed{\sigma_{max} = ?}$.

b) $\boxed{\sigma_2 = ?}$.

c) $\boxed{\sigma_3 = ?}$.

① $\uparrow + \sum F_y = 0:$

$$P_1 = W_1 + W_2 + W_3 = m_1 g + m_2 g + m_3 g = (179)(9.807)$$

$$P_2 = W_2 + W_3 = m_2 g + m_3 g = (124)(9.807)$$

$$P_3 = W_3 = m_3 g = (52)(9.807)$$

② $P_1 = 1{,}755\ \text{N}$

$P_2 = 1{,}216\ \text{N}$

$P_3 = 509.9\ \text{N}$

③ $A = \frac{\pi}{4} d^2 = \frac{\pi}{4}(45)^2 = 1{,}590\ \text{mm}^2$

④ σ_{max} occurs where P is max. $\Rightarrow$ $\sigma_{max} = \frac{P_1}{A} = \frac{1{,}755}{1{,}590}$

⑤ $\boxed{\sigma_{max} = 1.104\ \text{MPa}}$.

⑥ $\sigma_2 = \frac{P_2}{A} = \frac{1{,}216}{1{,}590}$ $\Rightarrow$ $\boxed{\sigma_2 = 0.7648\ \text{MPa} = 764.8\ \text{kPa}}$.

⑦ $\sigma_3 = \frac{P_3}{A} = \frac{509.9}{1{,}590}$ $\Rightarrow$ $\boxed{\sigma_3 = 0.3207\ \text{MPa} = 320.7\ \text{kPa}}$.

Problem 4.6:

describe : constraints on relation $\sigma = Ee$?

This equation, known as Hooke's law, is only valid for members subjected to uniaxial loading (tension or compression). It is also only valid for materials in the elastic region, where stress (σ) is linearly related to strain (e).

Problem 4.7:

Gage NO. $= 00$
Failure criterion : yield
$SF_y = 2.5$
$S_y = 530$ Mpa
Material : steel
$P_{design} = ?$

Gage NO. $= 00 \xrightarrow{\text{Appendix A}} d = 0.3310 \text{ in} \rightarrow d = 8.407 \times 10^{-3}$ m

$$A = \frac{\pi d^2}{4} = 55.52 \times 10^{-6} \text{ m}^2$$

$$SF_y = \frac{S_y}{\sigma_{design}} \rightarrow 2.5 = \frac{530 \times 10^6 \text{ (pa)}}{\sigma_{design}}$$

$$\sigma_{design} = 212 \times 10^6 \text{ (pa)}$$

$$\sigma_{design} = \frac{P_{design}}{A}$$

$$\therefore \boxed{P_{design} = 11769.3 \text{ N}}$$

Problem 4.9:

$\theta = 20^\circ$

$m = 20$ slug

$d = 0.500$ in

$\sigma = ?$

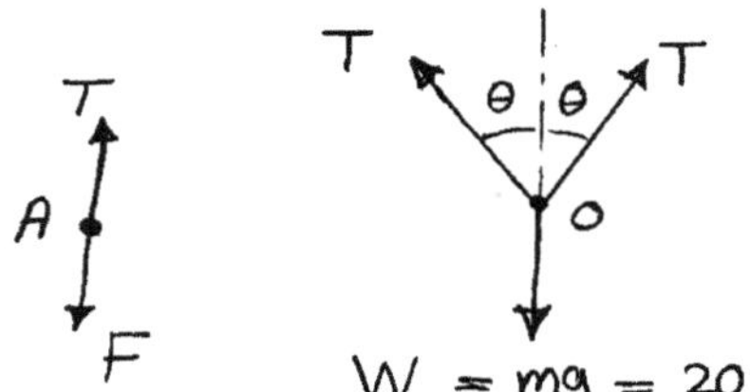

$W = mg = 20\,(\text{slug}) \times 32.17\,(\text{ft}/\text{s}^2) = 643.4\ \text{Lb}$

At point A: $\sum F_y = 0 \rightarrow T - F = 0 \rightarrow T = F$

At point O: $\sum F_y = 0 \rightarrow 2T\cos\theta - mg = 0 \rightarrow mg = 2T\cos\theta$

$T = 342.35\ \text{Lb}$

$$A = \frac{\pi d^2}{4} = 0.196\ \text{in}^2$$

$$\sigma = \frac{T}{A} = \frac{342.35\,(\text{Lb})}{0.196\,(\text{in})}$$

$\therefore$ $\boxed{\sigma = 1743.55\ \text{psi}}$

Problem 4.13:

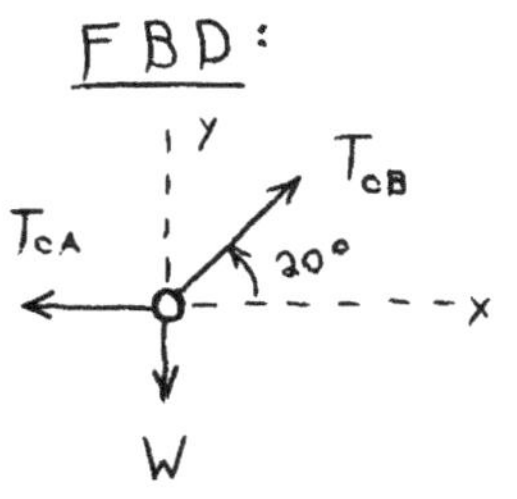

* steel wire $m = 800$ kg

$SF_y = 2.9$, for both wires

find: alloy + gage no. = ?

① $W = mg = (800)(9.807) = 7{,}846$ N

② $\uparrow + \sum F_y = 0: \quad T_{CB} \sin(20°) - 7{,}846 = 0$

③ $T_{CB} = 22{,}940$ N

④ $\overset{+}{\rightarrow} \sum F_x = 0: \quad -T_{CA} + (22{,}940)\cos(20°) = 0$

⑤ $T_{CA} = 21{,}557$ N

⑥ select material based on wire with largest tension force
⇒ T_{CB} governs design

⑦ $SF_y = \dfrac{S_y}{\sigma} = \dfrac{S_y A}{T_{CB}} = \dfrac{S_y(\pi/4\, d^2)}{T_{CB}} = \dfrac{\pi S_y d^2}{4 T_{CB}}$

⑧ $d = \sqrt{\dfrac{4 T_{CB}(SF_y)}{\pi S_y}} = \sqrt{\dfrac{4(22{,}940)(2.9)}{\pi S_y}}$

⑨ $d = \sqrt{\dfrac{84{,}704 \text{ N}}{S_y}}$

⑩ using Appendicies B-2 + A, construct table of d + gage no. for several steel alloys s.t. wire diam. ≥ d

alloy	S_y (MPa)	d (mm)	d (in)	gage no.
1010 A	200	20.58	0.8102	—
1018 A	221	19.58	0.7709	—
1020 HR	290	17.09	0.6728	—
1045 HR	414	14.30	0.5630	—
1212 HR	193	20.95	0.8248	—
4340 HR	910	9.648	0.3798	4-0s
52100 A	903	9.685	0.3813	4-0s

⑪ only 2 possible answers:

a) alloy 4340 HR, gage no. 4-0s
b) alloy 52100 A, gage no. 4-0s

Problem 4.18:

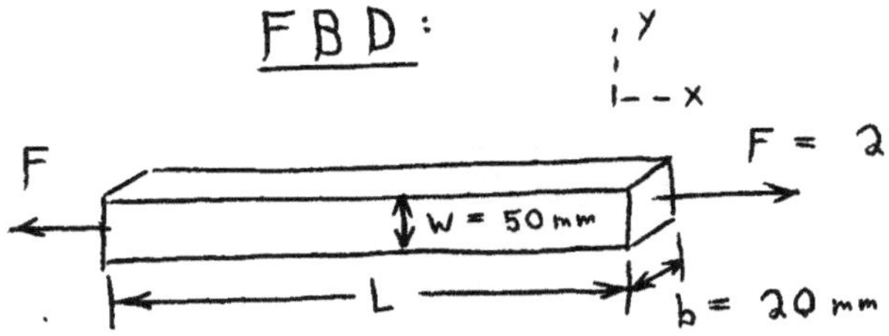

$L = 1.8\ m = 1800\ mm$

steel

find: $\sigma, \delta = ?$

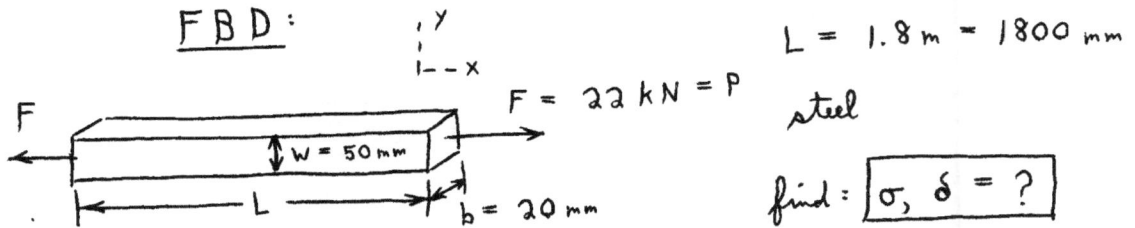

① $A = wb = (50)(20) = 1000\ mm^2$

② $\sigma = \dfrac{P}{A} = \dfrac{22{,}000\ N}{1000\ mm^2}$

③ $\boxed{\sigma = 22\ MPa}$.

④ from Appendix B-1 for steel: $E = 207\ GPa$

⑤ $\delta = \dfrac{PL}{AE} = \sigma \dfrac{L}{E} = (22\ MPa)\dfrac{(1800\ mm)}{(207{,}000\ MPa)}$

⑥ $\boxed{\delta = 0.1913\ mm}$.

Problem 4.19:

$S_u = 54.0\ ksi$

$F = 20\ kip$

$SF_u = 3.2$

$\sigma_{design} = ?$

$A = ?$

$SF_u = \dfrac{S_u}{\sigma_{design}}$

$3.2 = \dfrac{54}{\sigma_{design}} \Rightarrow \boxed{\sigma_{design} = 16.875\ ksi}$

$\sigma_{design} = \dfrac{F}{A}$

$16.875_{(ksi)} = \dfrac{20\ (kip)}{A}$

$\boxed{A = 1.185\ in^2}$

Problem 4.22:

FBD:

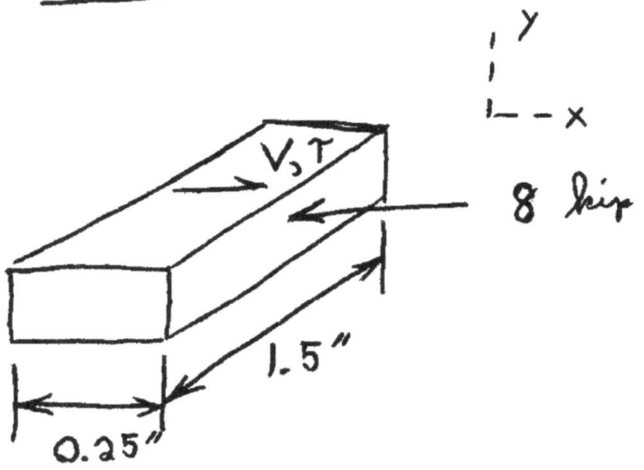

find: $\boxed{\tau = ?}$.

① $\xrightarrow{+} \sum F_x = 0$: $V = 8$ kip

② $A = (0.25)(1.5) = 0.375 \text{ in}^2$

③ $\tau = \dfrac{V}{A} = \dfrac{8}{0.375}$

④ $\boxed{\tau = 21.33 \text{ ksi}}$.

Problem 4.25:

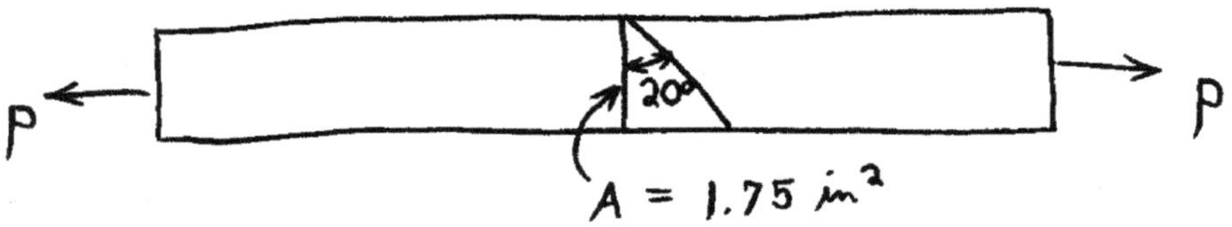

$\sigma = 2.5$ ksi

$\tau = 1.5$ ksi

F.S. = 3.0

find: $P_{max} = ?$

① $F.S. = \dfrac{\sigma_{max}}{\sigma_\theta} \Rightarrow \sigma_\theta = \dfrac{\sigma_{max}}{F.S.} = \dfrac{2.5}{3}$

② $\sigma_\theta = 0.8333$ ksi

③ $\tau_\theta = \dfrac{\tau_{max}}{F.S.} = \dfrac{1.5}{3} = 0.5$ ksi

④ $\sigma_\theta = \dfrac{P}{A}\cos^2\theta \Rightarrow P_\sigma = \dfrac{\sigma_\theta A}{\cos^2\theta}$

⑤ $P_\sigma = \dfrac{(0.8333)(1.75)}{\cos^2(20°)} = 1.651$ kip

⑥ $\tau_\theta = \dfrac{P}{A}\sin\theta\cos\theta \Rightarrow P_\tau = \dfrac{\tau_\theta A}{\sin\theta\cos\theta}$

⑦ $P_\tau = \dfrac{(0.5)(1.75)}{\sin(20°)\cos(20°)} = 2.723$ kip

⑧ choose smallest force $\Rightarrow$ $P_{max} = 1.651$ kip.

Problem 4.27:

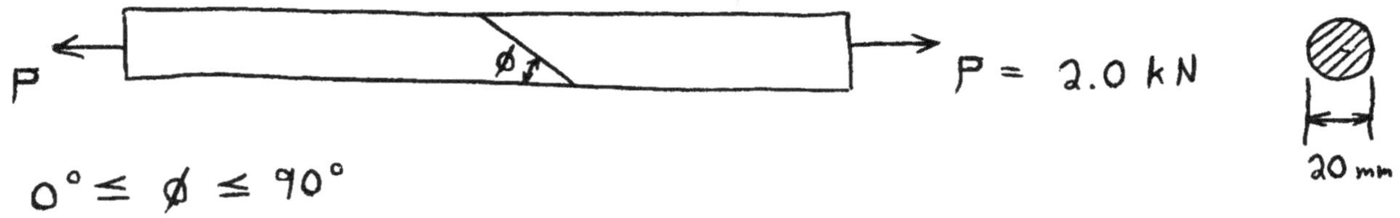

$0° \le \phi \le 90°$

find: $\boxed{\sigma_\phi + \tau_\phi \text{ vs. } \phi = ?}$

① $A = \frac{\pi}{4} d^2 = \frac{\pi}{4}(20)^2 = 314.2 \text{ mm}^2$

② recall: $\sigma_\theta = \frac{P}{A}\cos^2\theta$

$\tau_\theta = \frac{P}{A}\sin\theta\cos\theta$

③ need to write above eqns. i.t.o. ϕ

④ $\phi = 90° - \theta \Rightarrow \sin\phi = \cos\theta$

$\cos\phi = \sin\theta$

⑤ subing ④ in ②:

$\sigma_\phi = \frac{P}{A}\sin^2\phi$

$\tau_\phi = \frac{P}{A}\cos\phi\sin\phi$

⑥ $\sigma_\phi = \frac{(2000)}{(314.2)}\sin^2\phi = (6.365 \text{ MPa})\sin^2\phi$

$\tau_\phi = \frac{(2000)}{(314.2)}\cos\phi\sin\phi = (6.365 \text{ MPa})\cos\phi\sin\phi$

⑦ use Excel spreadsheet to plot $\sigma_\phi + \tau_\phi$ eqns. vs. ϕ for angles from 0° to 90° (see chart on next page)

Problem 4.27: (con't)

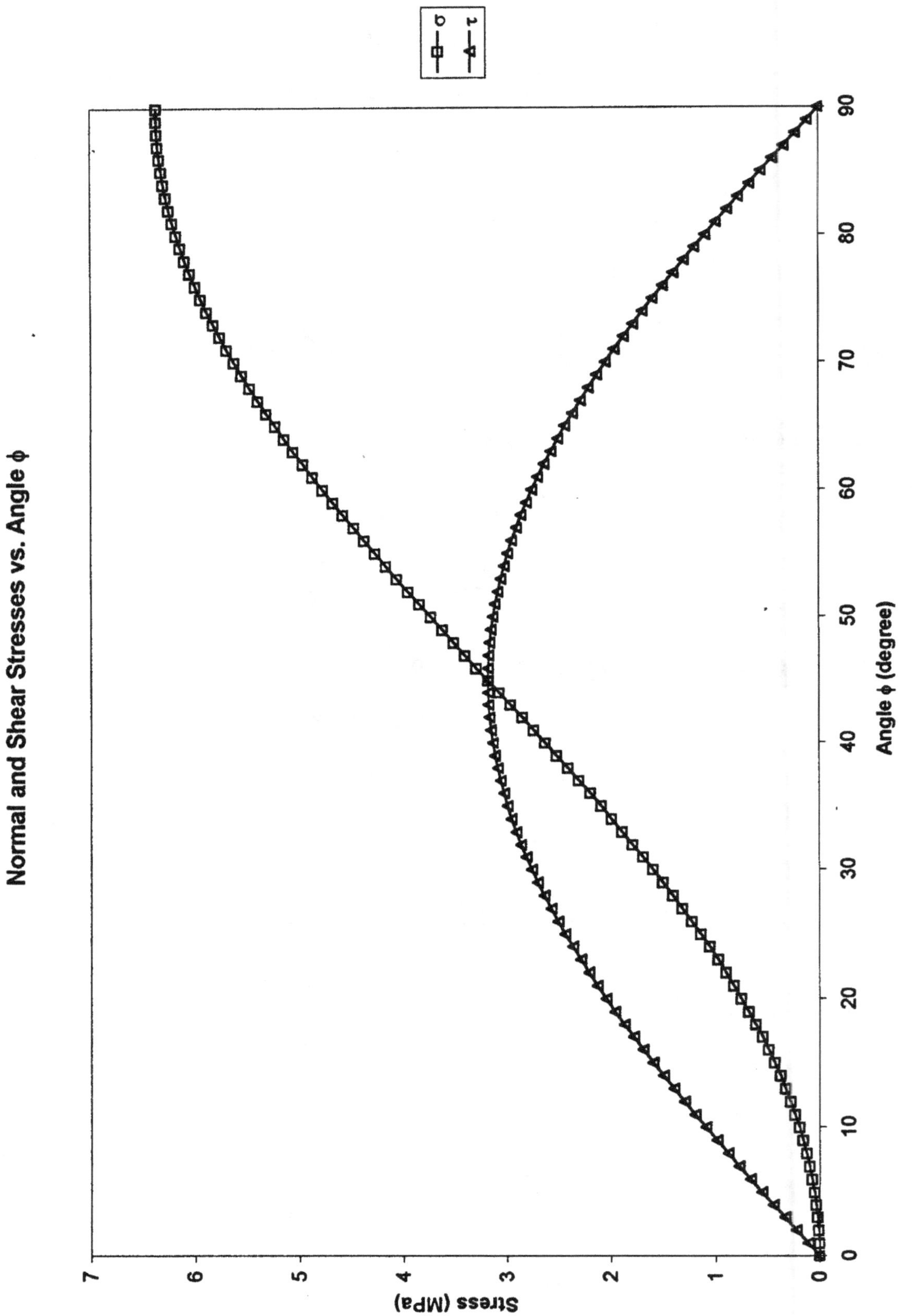

Problem 4.28:

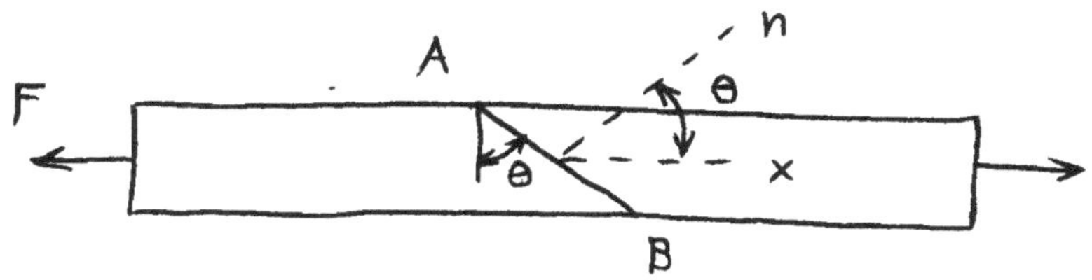

$A = 175\ mm^2$

$\sigma_\theta = 81\ MPa$

$\tau_\theta = -27\ MPa$

find: $\boxed{\sigma_x,\ \theta,\ F = ?}$

① $\sigma_\theta = 81\ MPa = \dfrac{F}{A}\cos^2\theta = \sigma_x \cos^2\theta$

② $\tau_\theta = -27\ MPa = \dfrac{F}{A}\sin\theta\cos\theta = \sigma_x \sin\theta \cos\theta$

③ $\dfrac{\tau_\theta}{\sigma_\theta} = \dfrac{-27}{81} = \dfrac{\sigma_x \sin\theta\cos\theta}{\sigma_x \cos^2\theta}$

④ $-0.3333 = \dfrac{\sin\theta}{\cos\theta} = \tan\theta$

⑤ $\boxed{\theta = 161.57° \ \text{or}\ -18.43°}$

⑥ using ⑤ in ①:

$\sigma_x = \dfrac{\sigma_\theta}{\cos^2\theta} = \dfrac{81}{\cos^2(161.57°)}$

⑦ $\boxed{\sigma_x = 90.0\ MPa}$

⑧ $F = \sigma_x A = (90\ MPa)(175\ mm^2)$

⑨ $\boxed{F = 15,750\ N = 15.75\ kN}$

Problem 4.32:

tapered bar:

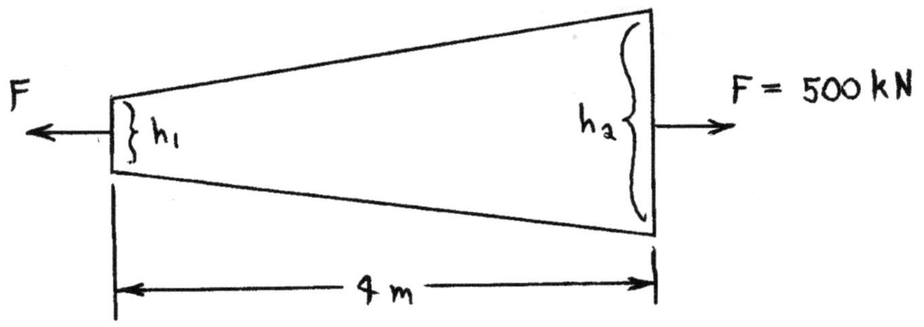

$h_1 = 100$ mm

$h_2 = 500$ mm

$b = 50$ mm

* aluminum

find: $\boxed{\delta = ?}$.

① from Appendix B for aluminum: $E = 72\,GPa = 72\,\frac{kN}{mm^2}$

② $\delta = \frac{FL}{Eb}\left(\frac{1}{h_2 - h_1}\right)\ln\left(\frac{h_2}{h_1}\right) = \frac{(500)(4000)}{(72)(50)}\left(\frac{1}{500-100}\right)\ln\left(\frac{500}{100}\right)$

③ $\boxed{\delta = 2.235\ mm}$.

Problem 4.35:

stepped bar:

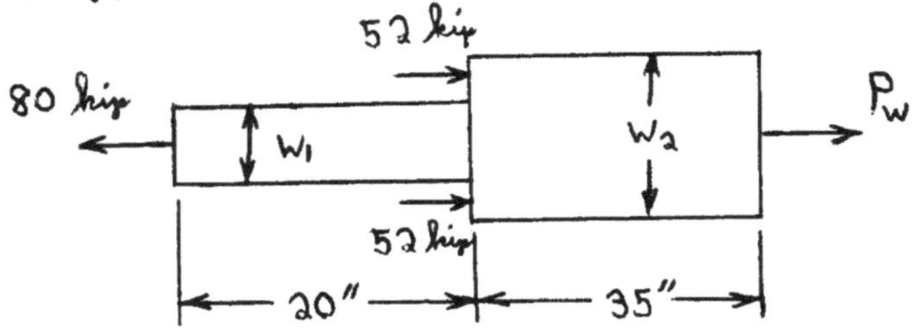

$w_1 = 2.2''$ $b_1 = 1.2''$

$w_2 = 6.5''$ $b_2 = 2.4''$

* steel

find: $\sigma_1, \sigma_2, \delta_T = ?$

① $\overset{+}{\rightarrow}\sum F_x = 0: \; -80 + 2(52) + P_w = 0 \;\Rightarrow\; P_w = -24$ kip

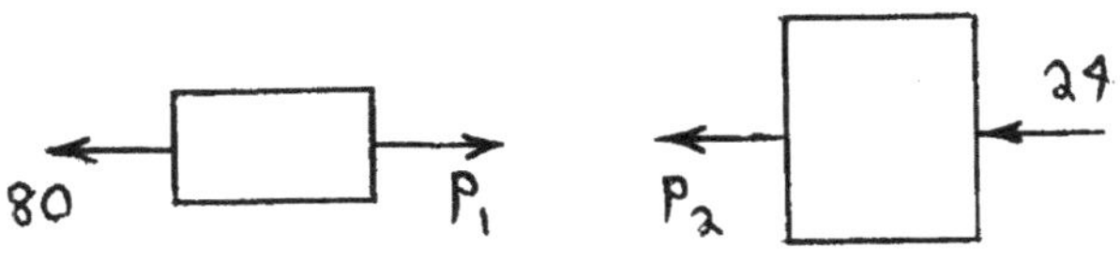

② $\overset{+}{\rightarrow}\sum F_x = 0$:

$P_1 = 80$ kip

$P_2 = -24$ kip

③ $A_1 = (2.2)(1.2) = 2.64 \text{ in}^2$

$A_2 = (6.5)(2.4) = 15.6 \text{ in}^2$

④ $\sigma_1 = \dfrac{P_1}{A_1} = \dfrac{80}{2.64} \Rightarrow \boxed{\sigma_1 = 30.30 \text{ ksi}}$.

$\sigma_2 = \dfrac{P_2}{A_2} = \dfrac{-24}{15.6} \Rightarrow \boxed{\sigma_2 = -1.538 \text{ ksi}}$.

⑤ from Appendix B for steel: $E = 30 \times 10^6$ psi $= 30{,}000$ ksi

⑥ $\delta = \dfrac{PL}{AE} = \sigma \dfrac{L}{E}$

⑦ $\delta_T = \delta_1 + \delta_2 = \sigma_1 \dfrac{L_1}{E} + \sigma_2 \dfrac{L_2}{E} = (30.30)\dfrac{(20)}{(30{,}000)} + (-1.538)\dfrac{(35)}{(30{,}000)}$

⑧ $\delta_T = 0.0202 - 0.001794$

⑨ $\boxed{\delta_T = 0.01841 \text{ in}}$.

Problem 4.36:

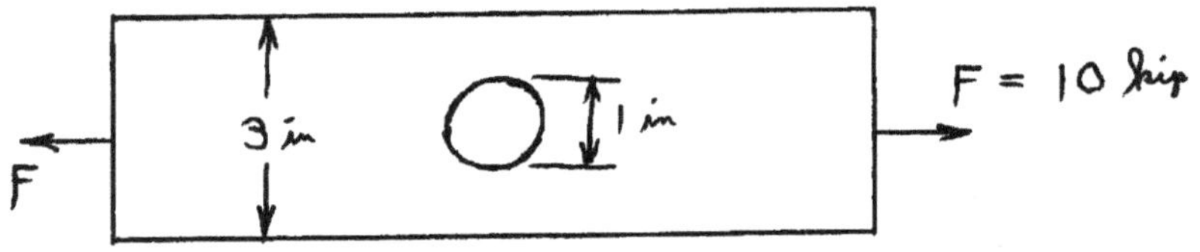

$b = 0.25$ in

find: σ_{nom}, K, σ_{max} = ?

① $\sigma_{nom} = \dfrac{F}{(w-d)b} = \dfrac{10}{(3-1)(0.25)}$

② $\boxed{\sigma_{nom} = 20 \text{ ksi}}$.

③ $\dfrac{d}{w} = \dfrac{1}{3} = 0.3333$

④ use fig. 4.20 $\Rightarrow$ $K = 2.33$

⑤ $\sigma_{max} = K\,\sigma_{nom} = (2.33)(20)$

⑥ $\boxed{\sigma_{max} = 46.6 \text{ ksi}}$.

Problem 4.37:

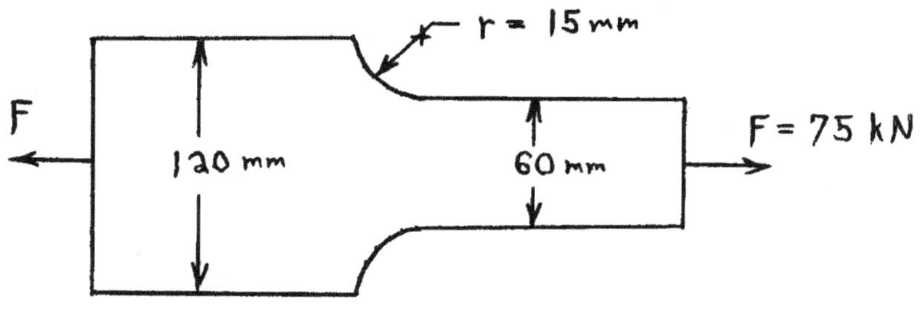

$b = 50$ mm

find: σ_{max} = ?

① $\sigma_{nom} = \dfrac{F}{W_1 \cdot b} = \dfrac{75{,}000}{(60)(50)} = 25$ MPa

② $\dfrac{r}{W_1} = \dfrac{15}{60} = 0.25$; $\dfrac{W_2}{W_1} = \dfrac{120}{60} = 2.0$

③ use fig. 4.22 $\Rightarrow$ $K = 1.75$

④ $\sigma_{max} = K\,\sigma_{nom} = (1.75)(25)$

⑤ $\boxed{\sigma_{max} = 43.75 \text{ MPa}}$.

Problem 4.41:

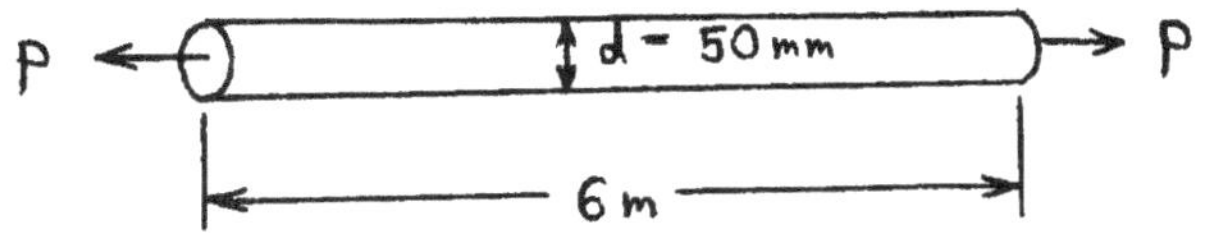

$\sigma \leq 175$ MPa

$\delta \leq 0.14\%$ of L

* steel

find: $\boxed{P_{max} = ?}$.

① $\delta_{max} = 0.0014\,(6000) = 8.4$ mm

② from Appendix B: $E = 207$ GPa

③ $A = \frac{\pi}{4} d^2 = \frac{\pi}{4}(50)^2 = 1963\ \text{mm}^2$

* consider stress limit:

④ $\sigma = \frac{P}{A} \Rightarrow P = \sigma A = (175)(1963) = 343.5$ kN

* consider elongation limit:

⑤ $\delta = \frac{PL}{AE} \Rightarrow P = \frac{\delta AE}{L} = \frac{(8.4)(1963)(207)}{6000} = 568.9$ kN

⑥ choose smallest P $\Rightarrow$ $\boxed{P_{max} = 343.5\ \text{kN}}$.

Problem 4.42:

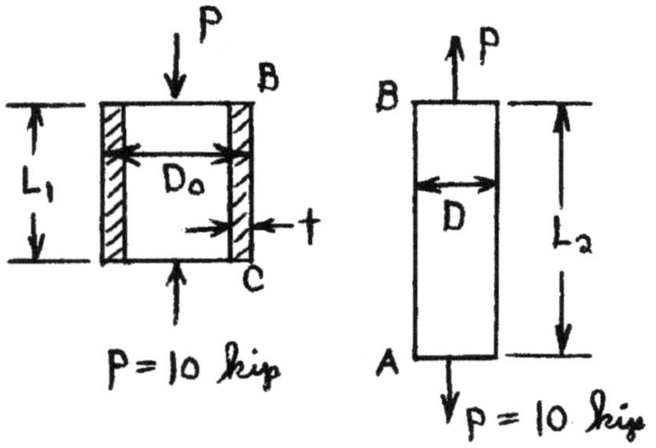

* steel rod ; brass bushing

$D_o = 3$ in $D = 1$ in

$L_1 = 24$ in $L_2 = 34$ in

$\delta_A \leq 0.020$ in

find : $t_{min} = ?$

① from Appendix B: $E_s = 30{,}000$ ksi ; $E_b = 16{,}000$ ksi

② $A_2 = \frac{\pi}{4} D^2 = \frac{\pi}{4}(1)^2 = 0.7854$ in^2

③ $\delta_{AB} = \frac{PL_2}{A_2 E_s} = \delta_A - \delta_B \Rightarrow \delta_B = \delta_A - \frac{PL_2}{A_2 E_s}$

④ $\delta_B = 0.02 - \frac{(10)(34)}{(0.7854)(30{,}000)} = 0.005570$ in

⑤ $\delta_{BC} = \frac{PL_1}{A_1 E_b} = \delta_B - \delta_C^{\,0} = \delta_B$

⑥ $A_1 = \frac{PL_1}{\delta_B E_b} = \frac{(10)(24)}{(0.005570)(16{,}000)} = \frac{\pi}{4}((3)^2 - D_i^2)$

⑦ $D_i = 2.360$ in

⑧ $t_{min} = \frac{1}{2}(D_o - D_i) = \frac{1}{2}(3 - 2.36)$

⑨ $t_{min} = 0.320$ in

Problem 4.43:

FBD's:

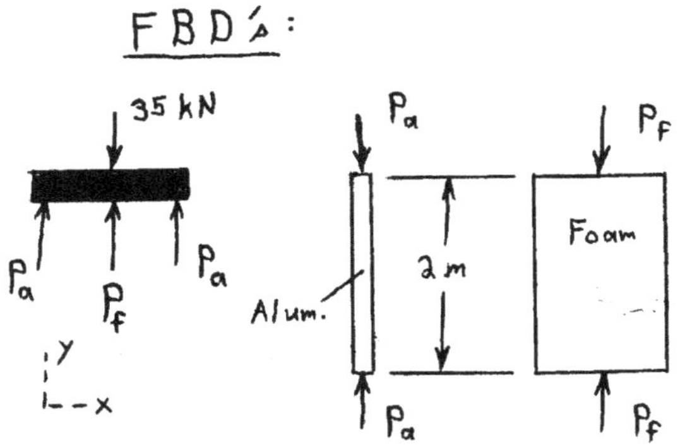

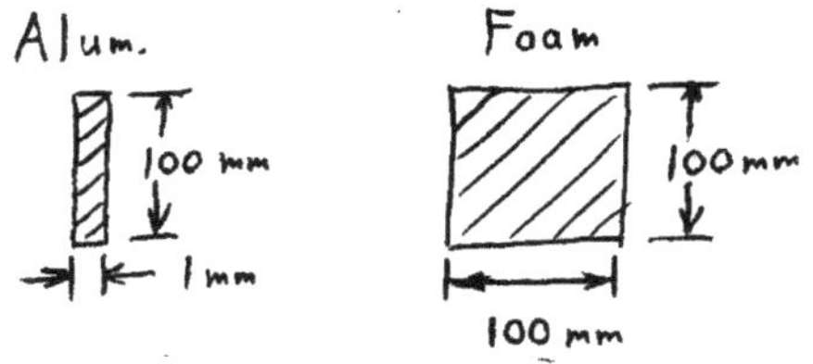

$E_f = 1200$ psi $\qquad L_a = L_f = 2\text{m}$

find: $\boxed{\sigma_a,\ \sigma_f,\ \delta = ?}$.

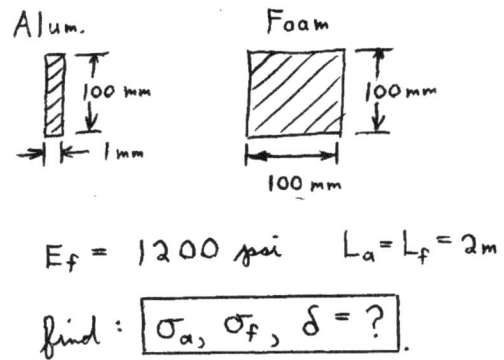

① $\uparrow + \sum F_y = 0: \quad 2P_a + P_f = 35\text{ kN} = 35{,}000\text{ N}$

② from deformation considerations:

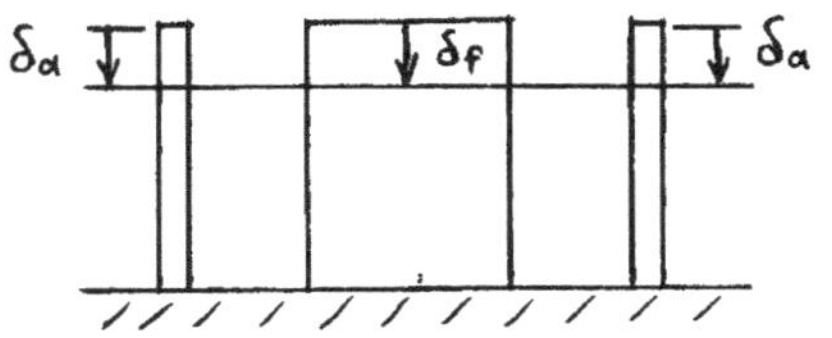

$\delta_a = \delta_f = \delta$

③ $\dfrac{P_a L}{A_a E_a} = \dfrac{P_f L}{A_f E_f} \quad \Rightarrow \quad P_a = P_f \dfrac{A_a E_a}{A_f E_f}$

④ from Appendix B: $E_a = 72\text{ GPa} = 72{,}000\text{ MPa}$

⑤ $E_f = (1200\text{ psi})\left(\dfrac{\text{ksi}}{1000\text{ psi}}\right)\left(\dfrac{6.895\text{ MPa}}{\text{ksi}}\right) = 8.274\text{ MPa}$

⑥ $P_a = P_f \dfrac{(1)(100)(72{,}000)}{(100)^2(8.274)} \quad \Rightarrow \quad P_a = 87.02\,P_f$

⑦ using ⑥ in ①: $\quad 2(87.02\,P_f) + P_f = 35{,}000$

⑧ $P_f = 200.0\text{ N}$

⑨ $P_a = 87.02\,(200) = 17{,}400\text{ N}$

⑩ $\sigma_a = \dfrac{P_a}{A_a} = \dfrac{17{,}400}{(1)(100)} \quad \Rightarrow \quad \boxed{\sigma_a = 174\text{ MPa (c)}}$.

$\sigma_f = \dfrac{P_f}{A_f} = \dfrac{200}{(100)^2} \quad \Rightarrow \quad \boxed{\sigma_f = 0.020\text{ MPa} = 20\text{ kPa (c)}}$.

⑪ $\delta = \dfrac{P_a L}{A_a E_a} = \sigma_a \dfrac{L}{E_a} = (174)\dfrac{(2000)}{(72{,}000)} \quad \Rightarrow \quad \boxed{\delta = 4.833\text{ mm} \downarrow}$.

Problem 4.44:

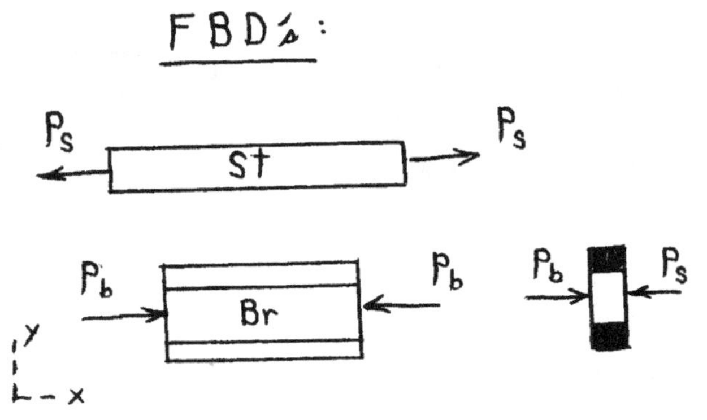

$d_s = 1$ in $\quad L_s = 16$ in

$d_{bo} = 2.5$ in $\quad L_b = 12$ in

$d_{bi} = 1.5$ in

* bolt : 14 thread / in

* nut tightened 1/3 turn

find : $\boxed{\sigma_s,\ \sigma_b,\ \delta_b = ?}$.

① $\overset{+}{\rightarrow}\sum F_x = 0: \quad P_b = P_s = P$

② as nut is tightened, bolt elongates + bushing contracts:

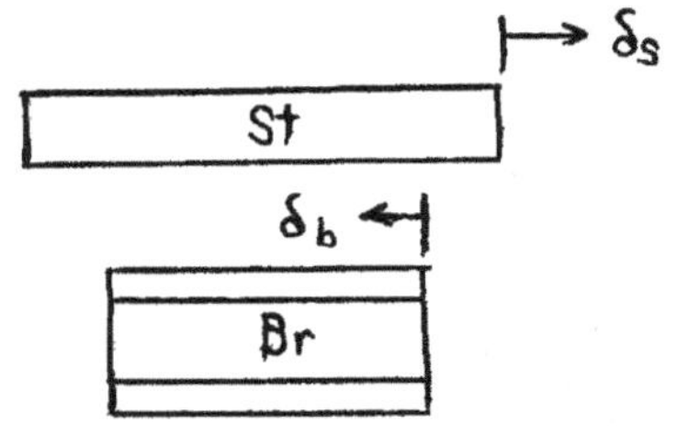

③ $\delta_{tot} = \frac{1}{3}\left(\frac{1}{14}\right) = 0.02381$ in

④ $\delta_{tot} = \delta_s + \delta_b$

⑤ $\delta_{tot} = \dfrac{P L_s}{A_s E_s} + \dfrac{P L_b}{A_b E_b}$

⑥ from Appendix B: $E_s = 30 \times 10^6$ psi ; $E_b = 16 \times 10^6$ psi

⑦ $A_s = \frac{\pi}{4}(1)^2 = 0.7854$ in^2

$A_b = \frac{\pi}{4}\left((2.5)^2 - (1.5)^2\right) = 3.142$ in^2

⑧ subing into ⑤: $0.02381 = \dfrac{P(16)}{(0.7854)(30\times10^6)} + \dfrac{P(12)}{(3.142)(16\times10^6)}$

⑨ $P = 25{,}940$ lb $= 25.94$ kip

⑩ $\sigma_s = \dfrac{P}{A_s} = \dfrac{25.94}{0.7854} \Rightarrow \boxed{\sigma_s = 33.03 \text{ ksi (T)}}$.

$\sigma_b = \dfrac{P}{A_b} = \dfrac{25.94}{3.142} \Rightarrow \boxed{\sigma_b = 8.256 \text{ ksi (C)}}$.

⑪ $\delta_b = \dfrac{P L_b}{A_b E_b} = \sigma_b \dfrac{L_b}{E_b} = (8.256)\dfrac{(12)}{(16\times10^3)}$

⑫ $\boxed{\delta_b = 0.006192 \text{ in (shorter)}}$.

Problem 4.45:

FBD's:

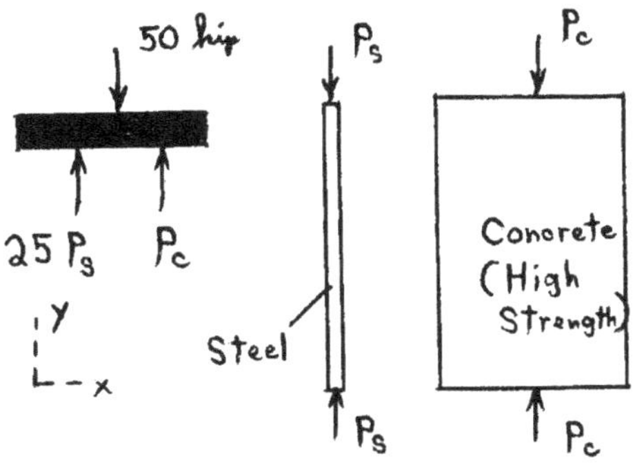

$L_s = L_c = 8$ ft

$w_c = t_c = 6$ in

$d_s = 0.25$ in

$N = 25$ rods

find: $\boxed{\sigma_s,\ \sigma_c,\ \delta = ?}$

① $\uparrow + \Sigma F_y = 0: \quad 25P_s + P_c = 50$ kip

② consider deformation:

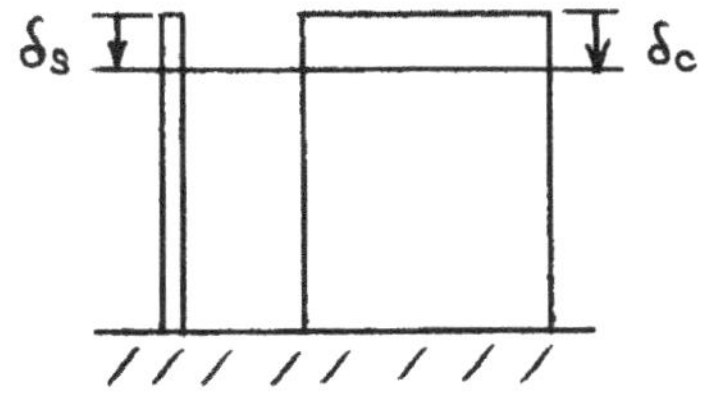

③ $\delta_s = \delta_c = \delta$

④ $$\frac{P_s L}{A_s E_s} = \frac{P_c L}{A_c E_c}$$

⑤ $$P_s = P_c \frac{A_s E_s}{A_c E_c}$$

⑥ from Appendix B: $E_s = 30 \times 10^6$ psi ; $E_c = 4.5 \times 10^6$ psi

⑦ $A_s = \frac{\pi}{4}(0.25)^2 = 0.04909$ in^2

$A_c = w_c t_c - 25A_s = (6)^2 - 25(0.04909) = 34.77$ in^2

⑧ subbing into ⑤: $$P_s = P_c \frac{(0.04909)(30 \times 10^6)}{(34.77)(4.5 \times 10^6)}$$

⑨ $P_s = 0.009412\, P_c$

⑩ using ⑨ in ①: $25(0.009412\, P_c) + P_c = 50$ kip

⑪ $P_c = 40.48$ kip

⑫ $P_s = 0.009412(40.48) = 0.3810$ kip

⑬ $$\sigma_s = \frac{P_s}{A_s} = \frac{0.3810}{0.04909} \Rightarrow \boxed{\sigma_s = 7.761 \text{ ksi (C)}}$$

$$\sigma_c = \frac{P_c}{A_c} = \frac{40.48}{34.77} \Rightarrow \boxed{\sigma_c = 1.164 \text{ ksi (C)}}$$

⑭ $$\delta = \sigma_c \frac{L_c}{E_c} = (1.164)\frac{(8)(12)}{(4.5 \times 10^3)} \Rightarrow \boxed{\delta = 0.02483 \text{ in} \downarrow}$$

Problem 4.46:

FBD's:

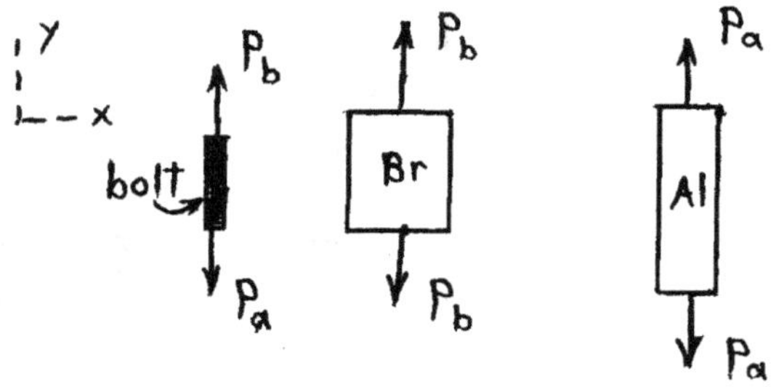

$d_b = 150$ mm $L_b = 250$ mm

$d_a = 30$ mm $L_a = 750$ mm

$\delta_{gap} = 0.2$ mm

find: $\sigma_a, \sigma_b, \delta_a, \delta_b = ?$

① $+\uparrow \sum F_y = 0: \quad P_b = P_a = P$

② consider deformation:

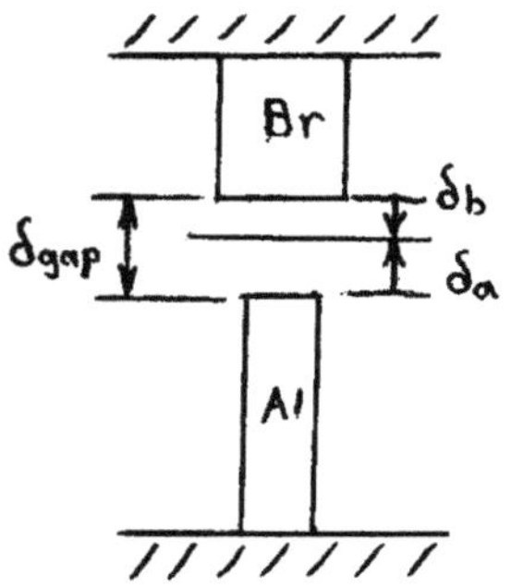

③ $\delta_{gap} = \delta_a + \delta_b$

④ $\delta_{gap} = \dfrac{PL_a}{A_a E_a} + \dfrac{PL_b}{A_b E_b}$

⑤ $A_a = \dfrac{\pi}{4}(30)^2 = 706.9 \text{ mm}^2$

$A_b = \dfrac{\pi}{4}(150)^2 = 17{,}670 \text{ mm}^2$

⑥ from Appendix B:

$E_a = 72$ GPa; $E_b = 110$ GPa

⑦ subing into ④: $0.2 = \dfrac{P(750)}{(706.9)(72{,}000)} + \dfrac{P(250)}{(17{,}670)(110{,}000)}$

⑧ $P = 13{,}460$ N

⑨ $\sigma_a = \dfrac{P}{A_a} = \dfrac{13{,}460}{706.9} \Rightarrow$ $\boxed{\sigma_a = 19.04 \text{ MPa (T)}}$

$\sigma_b = \dfrac{P}{A_b} = \dfrac{13{,}460}{17{,}670} \Rightarrow$ $\boxed{\sigma_b = 0.7617 \text{ MPa} = 761.7 \text{ kPa (T)}}$

⑩ $\delta_a = \dfrac{PL_a}{A_a E_a} = \dfrac{(13{,}460)(750)}{(706.9)(72{,}000)} \Rightarrow$ $\boxed{\delta_a = 0.1983 \text{ mm} \uparrow}$

$\delta_b = \dfrac{PL_b}{A_b E_b} = \dfrac{(13{,}460)(250)}{(17{,}670)(110{,}000)} \Rightarrow$ $\boxed{\delta_b = 0.001731 \text{ mm} \downarrow}$

Problem 4.47:

FBD's:

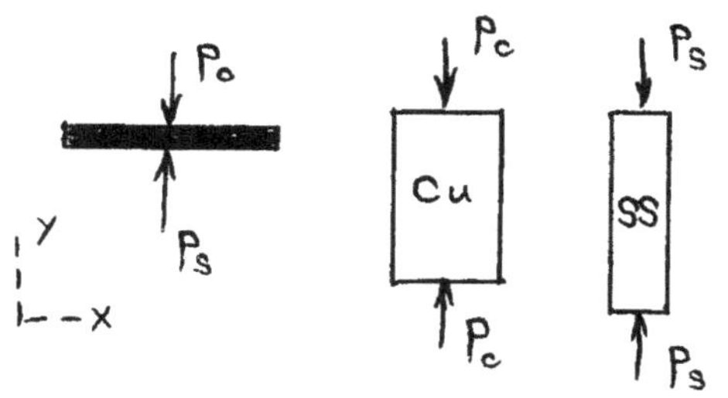

$d_c = 150$ mm $L_c = 250$ mm

$d_s = 30$ mm $L_s = 750$ mm

$T_i = 20°C$ $\Delta T = 120°C$

find: $\boxed{\sigma_c,\ \sigma_s,\ \delta_c,\ \delta_s = ?}$

① $\uparrow + \Sigma F_y = 0: \quad P_c = P_s = P$

② consider deformation:

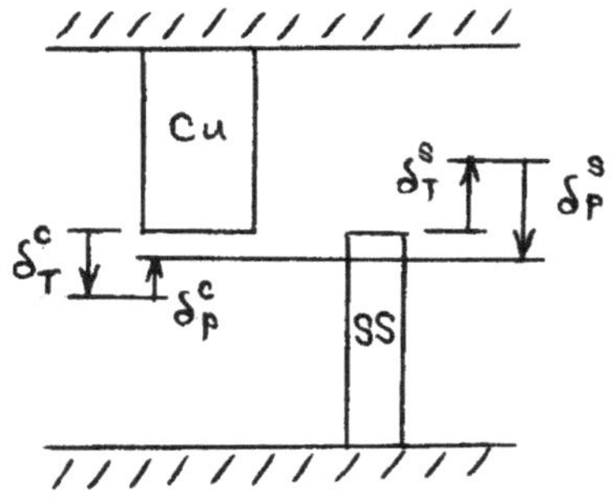

③ $\delta_T^c - \delta_P^c + \delta_T^s - \delta_P^s = 0$

④ $\delta_T = \alpha(\Delta T)L$

$\delta_P = \dfrac{PL}{AE}$

⑤ from Appendix B + Table 4.1:

$E_c = 121$ GPa ; $E_s = 190$ GPa

$\alpha_c = 16.9 \times 10^{-6}/°C$

$\alpha_s = 17.3 \times 10^{-6}/°C$

⑥ $\delta_T^c = \alpha_c(\Delta T)L_c = (16.9 \times 10^{-6})(120)(250) = 0.507$ mm

$\delta_T^s = \alpha_s(\Delta T)L_s = (17.3 \times 10^{-6})(120)(750) = 1.557$ mm

⑦ $A_c = \frac{\pi}{4}(150)^2 = 17{,}670 \text{ mm}^2$; $A_s = \frac{\pi}{4}(30)^2 = 706.9 \text{ mm}^2$

⑧ subing into ③:

$$0.507 - \frac{P(250)}{(17{,}670)(121{,}000)} + 1.557 - \frac{P(750)}{(706.9)(190{,}000)} = 0$$

⑨ $P = 362{,}000 \text{ N} = 362.0 \text{ kN}$

⑩ $\sigma_c = \dfrac{P}{A_c} = \dfrac{362{,}000}{17{,}670} \Rightarrow \boxed{\sigma_c = 20.49 \text{ MPa (C)}}$

$\sigma_s = \dfrac{P}{A_s} = \dfrac{362{,}000}{706.9} \Rightarrow \boxed{\sigma_s = 512.1 \text{ MPa (C)}}$

⑪ $\delta_c = \delta_T^c - \delta_P^c = 0.507 - \dfrac{(362{,}000)(250)}{(17{,}670)(121{,}000)} = 0.507 - 0.04233$

⑫ $\boxed{\delta_c = 0.4647 \text{ mm (longer)}}$

⑬ $\delta_c + \delta_s = 0 \Rightarrow \boxed{\delta_s = 0.4647 \text{ mm (shorter)}}$

Problem 4.48:

FBD's:

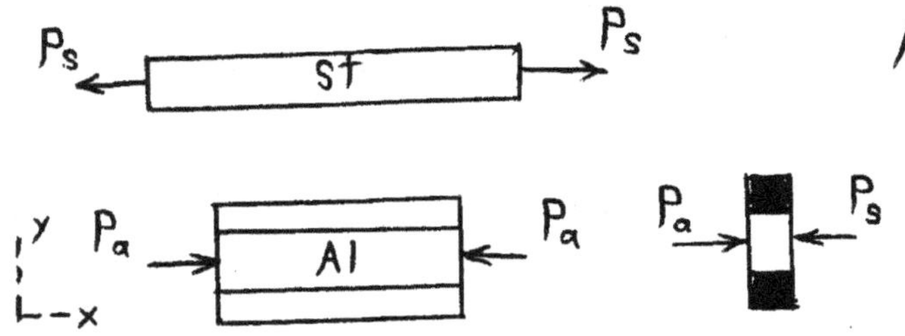

$d_s = 1$ in $L_s = 10$ in

$A_a = 3$ in^2 $L_a = 6$ in

$T_i = 70°F$ $\Delta T = 60°F$

find : $\boxed{\sigma_s, \sigma_a = ?}$.

① $\overset{+}{\rightarrow} \Sigma F_x = 0: \quad P_a = P_s = P$

② consider deformation :

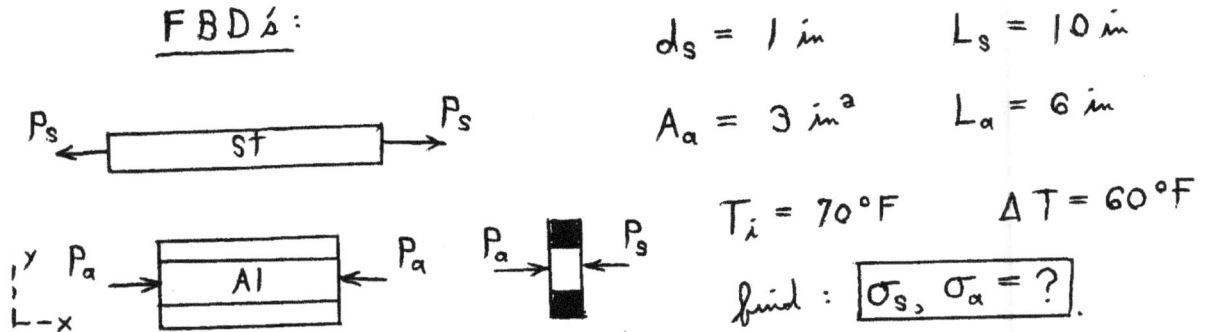

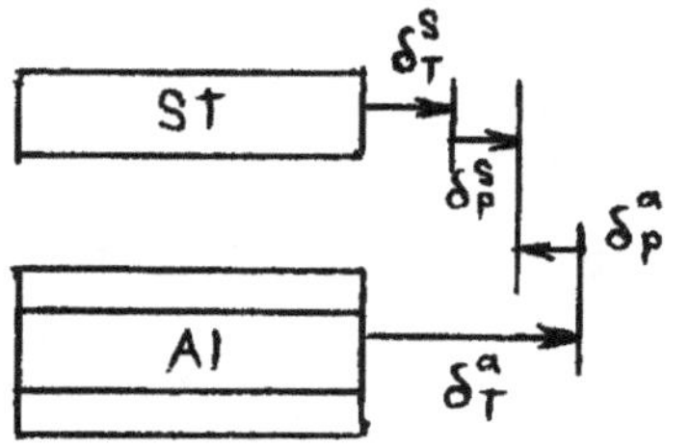

③ $\delta_T^s + \delta_P^s = \delta_T^a - \delta_P^a$

④ $\delta_T = \alpha (\Delta T) L \; ; \quad \delta_P = \dfrac{PL}{AE}$

⑤ from Appendix B + Table 4.1:

$E_s = 30 \times 10^6$ psi ; $E_a = 10.4 \times 10^6$ psi

$\alpha_s = 6.3 \times 10^{-6}/°F$; $\alpha_a = 12.9 \times 10^{-6}/°F$

⑥ $\delta_T^s = \alpha_s (\Delta T) L_s = (6.3 \times 10^{-6})(60)(10) = 0.00378$ in

$\delta_T^a = \alpha_a (\Delta T) L_a = (12.9 \times 10^{-6})(60)(6) = 0.004644$ in

⑦ $A_s = \dfrac{\pi}{4}(1)^2 = 0.7854$ in^2

⑧ subing into ③:

$$0.00378 + \frac{P(10)}{(0.7854)(30 \times 10^6)} = 0.004644 - \frac{P(6)}{(3)(10.4 \times 10^6)}$$

⑨ $P = 1401$ lb

⑩ $\sigma_s = \dfrac{P}{A_s} = \dfrac{1401}{0.7854} \Rightarrow \boxed{\sigma_s = 1784 \text{ psi (T)}}$.

$\sigma_a = \dfrac{P}{A_a} = \dfrac{1401}{3} \Rightarrow \boxed{\sigma_a = 467 \text{ psi (C)}}$.

Problem 4.49:

FBD's:

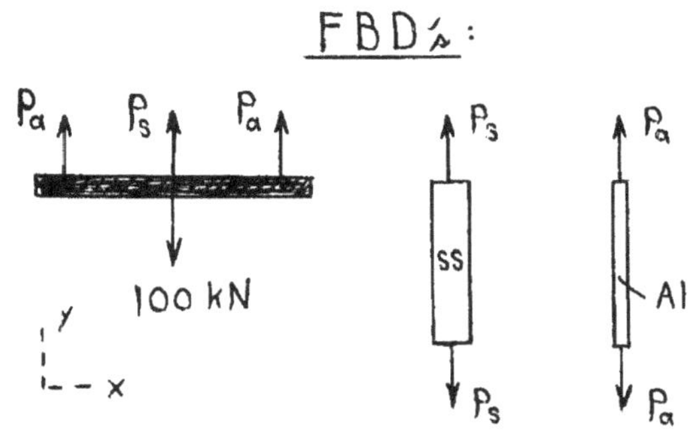

a = aluminum   s = stainless steel

$A_a = 750\ mm^2$ $T_i = 20°C$

$A_s = 1250\ mm^2$ $\Delta T = 80°C$

$L_a = L_s = 2\ m$ find: $\sigma_a, \sigma_s = ?$

① from symmetry, the force in each Al bar is the same (P_a)

② $+\uparrow \Sigma F_y = 0: \quad 2P_a + P_s = 100\ kN = 100{,}000\ N$

③ consider deformation:

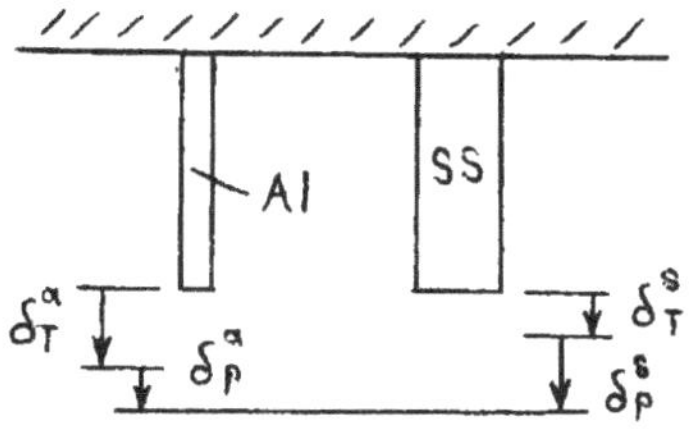

③ $\delta_T^a + \delta_P^a = \delta_T^s + \delta_P^s$

④ $\delta_T = \alpha(\Delta T)L \ ; \quad \delta_P = \dfrac{PL}{AE}$

⑤ from Appendix B:

$E_a = 72\ GPa \ ; \quad \alpha_a = 23.2 \times 10^{-6}/°C$

$E_s = 190\ GPa ; \quad \alpha_s = 17.3 \times 10^{-6}/°C$

⑥ $\delta_T^a = \alpha_a(\Delta T)L_a = (23.2 \times 10^{-6})(80)(2000) = 3.712\ mm$

$\delta_T^s = \alpha_s(\Delta T)L_s = (17.3 \times 10^{-6})(80)(2000) = 2.768\ mm$

⑦ subing into ③:

$$3.712 + \frac{P_a(2000)}{(750)(72{,}000)} = 2.768 + \frac{P_s(2000)}{(1250)(190{,}000)}$$

⑧ $P_s = 4.398\ P_a + 112{,}100\ N$

⑨ subing ⑧ in ②: $2P_a + (4.398\ P_a + 112{,}100) = 100{,}000$

⑩ $P_a = -1891\ N$

⑪ $P_s = 4.398(-1891) + 112{,}100 = 103{,}800\ N$

⑫ $\sigma_a = \dfrac{P_a}{A_a} = \dfrac{1891}{750} \Rightarrow$ $\boxed{\sigma_a = 2.521\ MPa\ (C)}$

$\sigma_s = \dfrac{P_s}{A_s} = \dfrac{103{,}800}{1250} \Rightarrow$ $\boxed{\sigma_s = 83.04\ MPa\ (T)}$

CHAPTER 5 PROBLEMS

5.1 Sketch a stress-strain diagram for a relatively brittle material and identify the ultimate tensile strength on the diagram.

5.2 Sketch a stress-strain diagram for a relatively ductile material and identify the yield strength and ultimate tensile strength on the diagram.

5.3 Sketch a stress-strain diagram for a ductile material that does not exhibit a well defined yield point. Identify on the diagram the yield strength and the ultimate tensile strength.

5.4 Write a test plan for conducting a standard tensile test.

5.5 Determine the yield strength and the ultimate tensile strength for the material represented by the stress-strain diagram shown in the figure to the right.

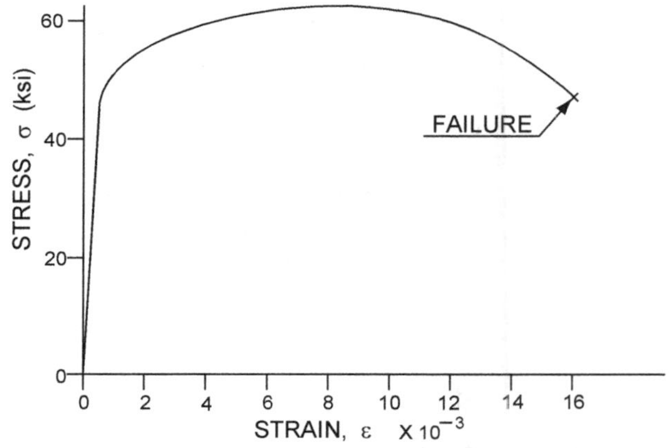

5.6 Determine the yield strength and the ultimate tensile strength for the material represented by the stress-strain diagram shown in the figure to the left.

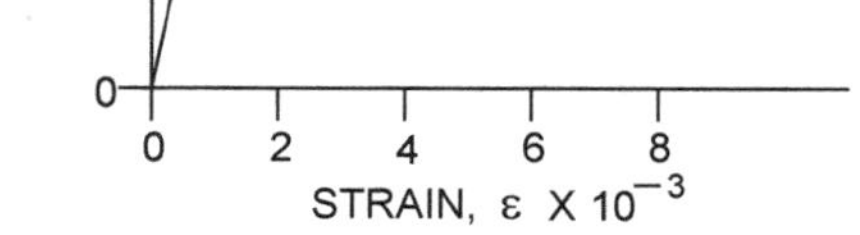

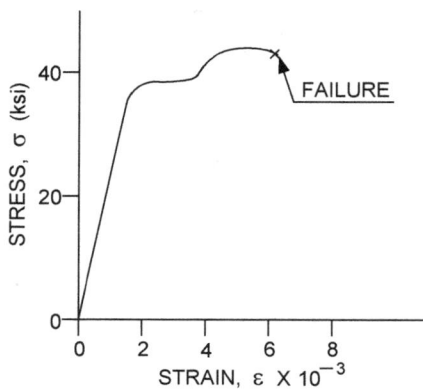

5.7 Determine the percent elongation and the percent reduction in area if measurements of $L_f = 66.4$ mm and $d_f = 10.52$ mm were made during a standard tensile test. The initial length and diameter of the standard tensile specimen is $L_o = 50$ mm and $d_o = 12.7$ mm.

Material Properties

5.8 In conducting a tensile test of an aluminum alloy, you adjust the signals from the load cell and the extensometer to zero and then load the specimen. You dwell at several loads in the elastic region of the stress-strain response, and record readings of the load F and the stretch δ. The values of F and δ are plotted on an F-δ diagram to obtain the slope $\Delta F/\Delta\delta = (16{,}000 \text{ N})/(100 \times 10^{-3} \text{ mm})$. Determine the elastic modulus of the aluminum alloy if the initial length and diameter of the tensile specimen is $L_o = 50$ mm and $d_o = 12.7$ mm.

5.9 If a steel alloy exhibits a Poisson's ratio of $\nu = 0.30$, determine the diameter of a tensile specimen when it is subjected to an elastic stress of $\sigma = 52.6$ ksi. The dimensions L_o and d_o for the specimen are 2.00 in. and 0.500 in., respectively.

5.10 A long (3 m) sheet of an aluminum alloy is stretched in a manufacturing process until it is 3.8 m in length. If the sheet was initially 0.6 m wide and 2 mm thick, determine its new width and thickness. In the plastic regime, Poisson's ratio is 0.5 for all metallic materials because volume is conserved in the plastic deformation process.

5.11 A long (3.8 m) sheet of an aluminum alloy is stretched in a manufacturing process until it is 4.6 m in length. If a central hole initially 75 mm in diameter was drilled in the sheet prior to stretching, determine its new dimensions. In the plastic regime, Poisson's ratio is 0.5 for all metallic materials because volume is conserved in the plastic deformation process.

5.12 Prove that volume of an object is conserved in a uniaxial stress state if $\nu = ½$.

5.13 If a steel alloy is linearly elastic until the applied axial stress equals the yield strength, determine the strain at yield for a structural steel with a yield strength of 38 ksi and a modulus of elasticity of 30×10^6 psi.

5.14 A long, thin, aluminum-alloy bar is subjected to an axial stress of 227 MPa. Determine the strain in the axial and transverse directions of the bar.

5.15 A spherical pressure vessel, fabricated from a steel alloy, is subjected to a biaxial stress field with $\sigma_x = \sigma_y = 20.6$ ksi. Determine the strains ε_x and ε_y.

5.16 An electrical strain gage is placed with an arbitrary orientation on a spherical pressure vessel fabricated from a titanium alloy. The gage provides a strain measurement $\varepsilon = 1{,}140 \times 10^{-6}$. Determine the stresses σ_x and σ_y in terms of MPa.

5.17 Prove that the ratio of hoop strain to axial strain for a cylindrical pressure vessel is given by:

$$\varepsilon_h/\varepsilon_a = (2 - \nu)/(1 - 2\nu).$$

5.18 An orthogonal pair of electrical strain gages is placed on a cylindrical pressure vessel fabricated from an aluminum alloy. The strain gages provide measurements of the axial strain $\varepsilon_a = 264 \times 10^{-6}$ and hoop strain $\varepsilon_h = 1{,}232 \times 10^{-6}$. Determine the axial stress σ_a and the hoop stress σ_h in terms of MPa.

5.19 A standard tensile specimen ruptured with an applied force F = 11.3 kip. Your measurement of the diameter at the neck region after failure gives $d_f = 0.366$ in. Determine the engineering stress and the true stress at failure.

5.20 Determine the engineering strain and the true strain for a tensile specimen with L_o = 2.00 in. if the final length, L_f is given in the table shown below:

Final Length L_f (in.)	Engineering Strain	True Strain
2.1		
2.2		
2.3		
2.4		
2.5		
2.6		
2.7		
2.8		
2.9		

5.21 Determine the engineering strain and the true strain for a tensile specimen with L_o = 50 mm if the final length L_f is given in the table below:

Final Length L_f (mm)	Engineering Strain	True Strain
53		
56		
59		
62		
65		
68		
71		
74		
77		

5.22 Complete the conversion table that relates engineering strain and true strain.

Engineering Strain	True Strain
0.001	
0.002	
0.005	
0.010	
0.020	
0.050	
0.100	
0.200	
0.500	
1.000	

5.23 The maximum stress imposed at the root of a tooth on a spur gear is σ_{max} = 78.9 ksi and the minimum stress σ_{min} = 0. Neglecting the influence of the cyclic mean stress on the fatigue life, determine the anticipated life of the gear, if the $S_f - N$ curve presented to the right characterizes its fatigue properties.

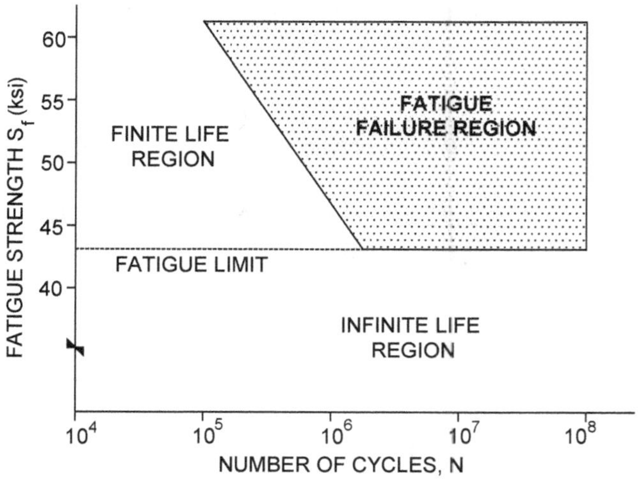

Material Properties

5.24 A rule for approximating the fatigue strength is $S_f = S_u/2$. Using this rule, determine the maximum stress that can be imposed on a structural member with a dynamic load that is twice the static load as illustrated in the figure below. The structural member is fabricated from steel with $S_u = 340$ MPa. Neglect the influence of the cyclic mean stresses on the fatigue behavior.

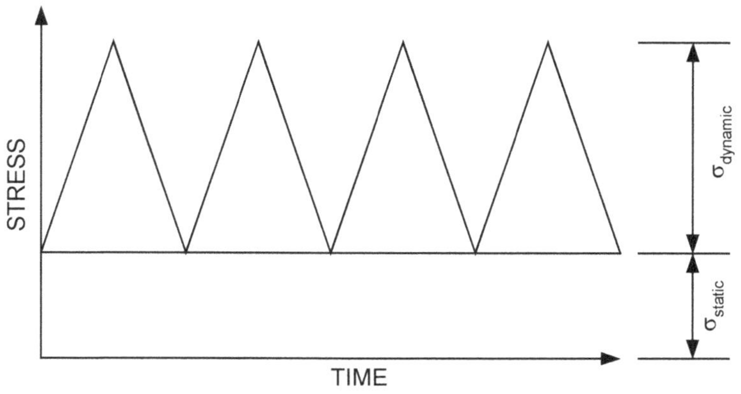

5.25 Reconsider Problem 5.23 taking into account the effect of the cyclic mean stress.

5.26 Reconsider Problem 5.24 taking into account the effect of the cyclic mean stress.

5.27 Repeat Problem 5.24 with the dynamic load equal to the static load.

5.28 Repeat Problem 5.24 with the dynamic load equal to one half of the static load.

5.29 Beginning with Eq. (5.9) derive Eq. (5.10).

5.30 Convert the stress-strain curve shown in the figure to the right to a true stress-true strain curve.

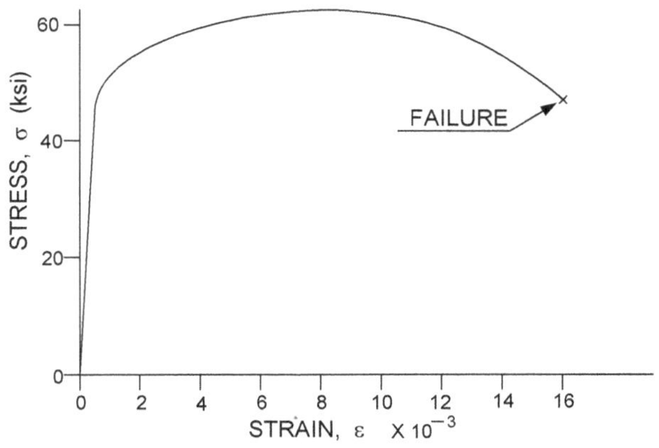

5.31 Convert the stress-strain curve shown in the figure to the left to a true stress-true strain curve.

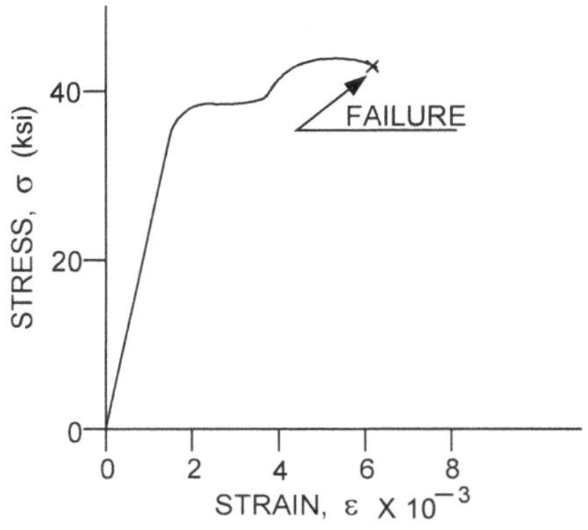

5.32 At the proportional limit in a tensile test, a 2 in. gage length had elongated by 0.003 in. and the diameter of the standard specimen was smaller by 0.00026 in. The load measured by a load cell fitted to the tensile machine at the proportional limit was 5,000 lb. Determine the modulus of elasticity, Poisson's ratio and the proportional limit for this material.

5.33 A rod with a diameter of 1.00 in. and length of 8.0 ft undergoes an extension of 0.220 in. when subjected to an axial force of 54.0 kip. The diameter of the rod decreases by 0.0007 in. at this load. Determine the modulus of elasticity, Poisson's ratio and the shear modulus for the rod's material.

5.34 A standard tensile specimen was fitted with an extensometer with a 2.00-inch gage length and tested until failure. The force and extension measured during the test is presented in the table below. The diameter of the specimen at the fracture neck was 0.413 in. Analyze the data using a spreadsheet and determine the following quantities.

a. Modulus of elasticity
b. Yield strength (0.2% offset)
c. Ultimate strength
d. Fracture stress
e. Strain at fracture
f. Percent elongation
g. Percent reduction in area.

Load (kip)	Extension (in.)	Load (kip)	Extension (in.)
0	0	10.8	0.251
2.0	0.0021	11.2	0.282
4.0	0.0039	11.6	0.303
6.0	0.0058	11.6	0.318
7.0	0.0072	11.5	0.327
7.8	0.0084	11.2	0.351
8.3	0.0280	10.8	0.367
8.7	0.0530	10.1	0.394
9.2	0.0860	9.7	0.402
9.8	0.1450	9.6	0.420
10.4	0.2220		

5.35 A standard tensile specimen 12.7 mm in diameter was fitted with an extensometer with a 50 mm gage length and tested until failure. The force and extension measured during the test is presented in the table below. The diameter of the specimen at the fracture neck was 9.0 mm. Analyze the data using a spreadsheet and determine the following quantities.

a. Modulus of elasticity
b. Yield strength (0.2% offset)
c. Ultimate strength
d. Fracture stress
e. Strain at fracture
f. Percent elongation
g. Percent reduction in area.

Load (kN)	Extension (mm.)	Load (kN)	Extension (mm)
0	0	35.0	3.54
5.0	0.010	37.0	4.52
10.0	0.019	38.0	5.53
15.0	0.029	38.0	6.18
20.0	0.042	36.0	6.66
25.0	0.048	33.0	7.35
26.0	0.50	30.0	7.86
27.0	0.95	27.0	8.42
29.0	1.50	23.0	8.78
31.5	2.21	19.0	9.32

Problem 5.5:

given: $\sigma - \epsilon$ diagram (P 5.5) find: $S_y, S_u = ?$

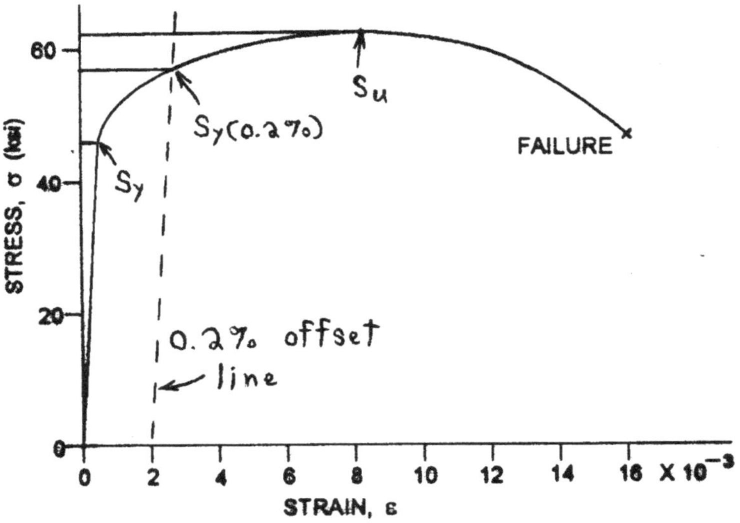

① see figure for location of S_y, $S_{y(0.2\%)}$, + S_u

② from figure:

$S_y = 46.1$ ksi

$S_{y(0.2\%)} = 56.7$ ksi

$S_u = 62.2$ ksi

Problem 5.6:

given: $\sigma - \epsilon$ diagram (P 5.6) find: $S_y, S_u = ?$

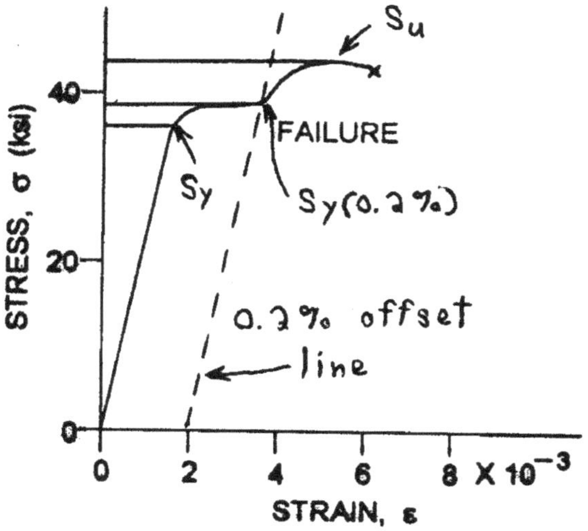

① see figure for location of S_y, $S_{y(0.2\%)}$, + S_u

② from figure:

$S_y = 36$ ksi

$S_{y(0.2\%)} = 38.5$ ksi

$S_u = 44$ ksi

Problem 5.7:

std. tensile specimen

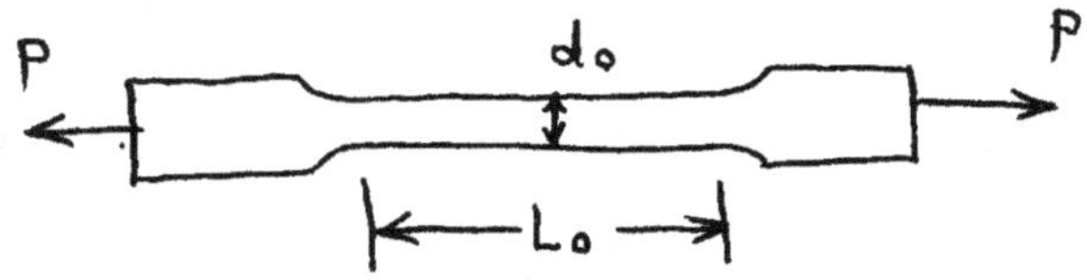

$L_o = 50$ mm $\quad d_o = 12.7$ mm

$L_f = 66.4$ mm $\quad d_f = 10.52$ mm

find: $\boxed{\%e,\ \%A = ?}$.

① $\%e = \dfrac{L_f - L_o}{L_o}(100\%) = \dfrac{66.4 - 50}{50}(100\%)$

② $\boxed{\%e = 32.8\ \%}$.

③ $\%A = \dfrac{A_o - A_f}{A_o}(100\%) = \dfrac{d_o^2 - d_f^2}{d_o^2}(100\%)$

④ $\%A = \dfrac{(12.7)^2 - (10.52)^2}{(12.7)^2}(100\%)$

⑤ $\boxed{\%A = 31.38\ \%}$.

Problem 5.9:

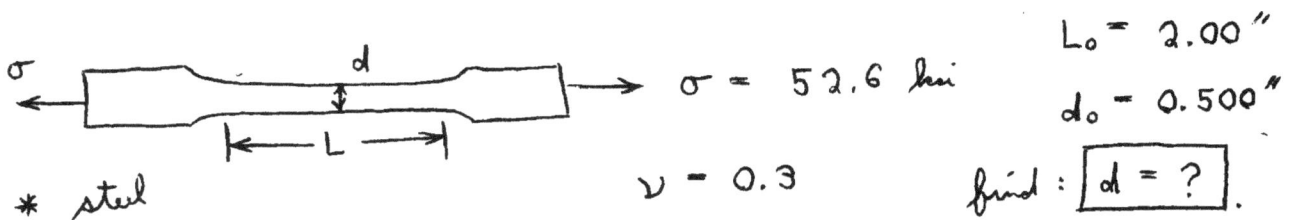

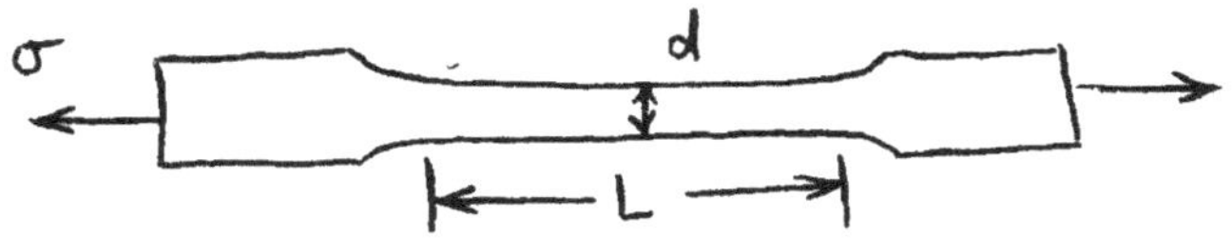

$\sigma = 52.6$ ksi

$L_o = 2.00''$

$d_o = 0.500''$

* steel

$\nu = 0.3$

find: $\boxed{d = ?}$.

① from Appendix B-1 for steel: $E = 30 \times 10^6$ psi

② $$\epsilon = \frac{\sigma}{E} = \frac{52,600}{30 \times 10^6} = 0.0017533 = \epsilon_a$$

③ $$\epsilon_t = -\nu \epsilon_a = -(0.3)(0.0017533) = -0.000526$$

④ $$\epsilon_t = \frac{\delta_d}{d_o} = \frac{d - d_o}{d_o} = \frac{d}{d_o} - 1 \Rightarrow d = d_o(\epsilon_t + 1)$$

⑤ $$d = (0.500)(-0.000526 + 1)$$

⑥ $$\boxed{d = 0.49974 \text{ in}}.$$

Problem 5.12:

given : uniaxial deformation ; $\nu = 0.5$

prove : volume is conserved

① consider rectangular block subjected to uniaxial force :

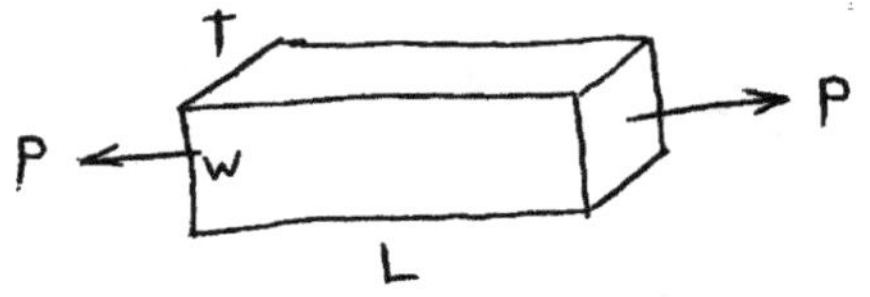

② original dimensions : L_o, w_o, t_o

dimensions under load P : L, w, t

③ original volume : $V_o = L_o w_o t_o$

④ volume under load P : $V = L w t$

⑤ recall : $\epsilon_a = \dfrac{L - L_o}{L_o} \Rightarrow L = L_o(1 + \epsilon_a)$

$\epsilon_t = \dfrac{w - w_o}{w_o} \Rightarrow w = w_o(1 + \epsilon_t)$

$\epsilon_t = \dfrac{t - t_o}{t_o} \Rightarrow t = t_o(1 + \epsilon_t)$

⑥ subing ⑤ in ④ : $V = L_o(1 + \epsilon_a) w_o(1 + \epsilon_t) t_o(1 + \epsilon_t)$

⑦ $V = V_o(1 + \epsilon_a)(1 + \epsilon_t)^2$

⑧ recall : $\epsilon_t = -\nu\epsilon_a$

⑨ $V = V_o(1 + \epsilon_a)(1 - \nu\epsilon_a)^2$

⑩ $V = V_o[1 + (1 - 2\nu)\epsilon_a + (\nu^2 - 2\nu)\epsilon_a^2 + \nu^2\epsilon_a^3]$

⑪ subing $\nu = 0.5$ into ⑩ + ignoring effects of higher-order strain terms ($\epsilon_a^2 + \epsilon_a^3$) :

$V = V_o[1 + 0]$

$\therefore V = V_o \Rightarrow$ volume is conserved

Problem 5.12 - Alternate Method:

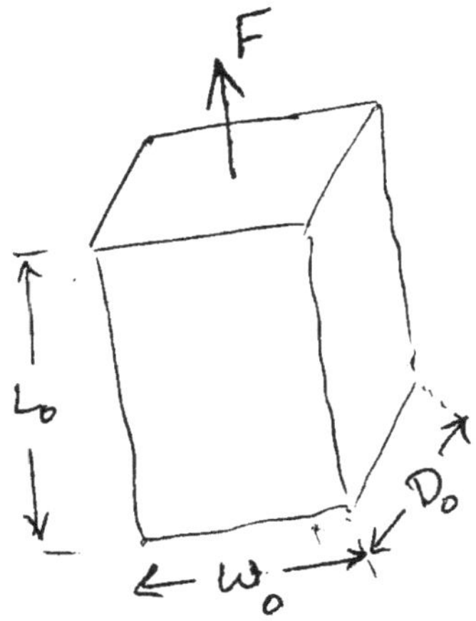

$$V_o = w_o L_o D_o$$

Axial: $\varepsilon_a = \dfrac{\Delta L}{L_o}$

Transverse: $\varepsilon_t = \dfrac{\Delta w}{w_o} = \dfrac{\Delta D}{D_o} = -\nu \varepsilon_a$

$$\therefore \quad \boxed{\frac{\Delta w}{w_o} = \frac{\Delta D}{D_o} = -\nu \frac{\Delta L}{L_o} = \varepsilon_t} \quad (1)$$

$$\frac{\Delta V}{V_o} = \frac{(w_o + \Delta w)(D_o + \Delta D)(L_o + \Delta L) - w_o L_o D_o}{w_o L_o D_o}$$

$$\Rightarrow \boxed{\frac{\Delta V}{V_o} = \left(1 + \frac{\Delta w}{w_o}\right)\left(1 + \frac{\Delta D}{D_o}\right)\left(1 + \frac{\Delta L}{L_o}\right) - 1} \quad (2)$$

$$(1), (2) \Rightarrow \frac{\Delta V}{V_o} = (1 + \varepsilon_t)(1 + \varepsilon_t)\left(1 - \frac{1}{\nu}\varepsilon_t\right) - 1$$

$$= \cancel{1} + \varepsilon_t + \varepsilon_t - \frac{1}{\nu}\varepsilon_t + \varepsilon_t^2 - \frac{2}{\nu}\varepsilon_t^2 - \frac{1}{\nu}\varepsilon_t^3 - \cancel{1}$$

Since $\varepsilon_t \ll 1 \Rightarrow \dfrac{\Delta V}{V_o} \simeq \left(2 - \dfrac{1}{\nu}\right)\varepsilon_t$

$\therefore$ For $\nu = \dfrac{1}{2} \Rightarrow \dfrac{\Delta V}{V_o} \simeq \left(2 - \dfrac{1}{(1/2)}\right)\varepsilon_t = 0 \Rightarrow \boxed{\Delta V = 0}$ ▨

Problem 5.13:

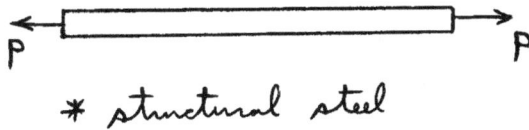

* structural steel

$\sigma = S_y = 38\ ksi = \sigma_y$

$E = 30 \times 10^6\ psi = 30{,}000\ ksi$

find : $\boxed{e_y = ?}$.

① $e_y = \dfrac{\sigma_y}{E} = \dfrac{38}{30{,}000}$

② $\boxed{e_y = 0.001267}$.

Problem 5.14:

$\sigma = 227\ MPa$

find : $\boxed{e_a,\ e_t = ?}$.

* aluminum alloy

① from Appendix B-1 for aluminum :

$E = 72\ GPa\ ; \quad \nu = 0.32$

② $e_a = \dfrac{\sigma}{E} = \dfrac{227}{72{,}000} \Rightarrow \boxed{e_a = 0.003153}$.

③ $e_t = -\nu e_a = -(0.32)(0.003153)$

④ $\boxed{e_t = -0.001009}$.

Problem 5.15:

spherical pressure vessel: * steel

$\sigma_x = \sigma_y = 20.6$ ksi find: $\boxed{\epsilon_x, \epsilon_y = ?}$

① from Appendix B: $E = 30,000$ ksi ; $\nu = 0.30$

② $\epsilon_x = \frac{1}{E}(\sigma_x - \nu\sigma_y) = \frac{1}{30,000}((20.6) - (0.3)(20.6))$

③ $\boxed{\epsilon_x = 480.7 \times 10^{-6} = 480.7\,\mu\epsilon}$

④ $\epsilon_y = \frac{1}{E}(\sigma_y - \nu\sigma_x) = \frac{1}{30,000}((20.6) - (0.3)(20.6))$

⑤ $\boxed{\epsilon_y = 480.7 \times 10^{-6} = 480.7\,\mu\epsilon}$

Problem 5.18:

strain gages on cylindrical pressure vessel:

$\epsilon_a = 264 \times 10^{-6}$ * aluminum alloy

$\epsilon_h = 1232 \times 10^{-6}$ find: $\boxed{\sigma_a, \sigma_h = ?}$ (MPa)

① from Appendix B: $E = 72$ GPa; $\nu = 0.32$

② $\sigma_a = \frac{E}{1-\nu^2}(\epsilon_a + \nu\epsilon_h) = \frac{72\times10^3}{1-(0.32)^2}(264 + (0.32)(1232))(10^{-6})$

③ $\boxed{\sigma_a = 52.8 \text{ MPa}}$

④ $\sigma_h = \frac{E}{1-\nu^2}(\epsilon_h + \nu\epsilon_a) = \frac{72\times10^3}{1-(0.32)^2}(1232 + (0.32)(264))(10^{-6})$

⑤ $\boxed{\sigma_h = 105.6 \text{ MPa}}$

Problem 5.19:

std. tensile specimen:

$L_o = 2.00$ in

$d_o = 0.500$ in

$P_f = 11.3$ kip $\quad d_f = 0.366$ in

find: $\boxed{\sigma, \sigma_T = ?}$ @ failure

① at failure:

$$\sigma = \frac{P_f}{A_o} = \frac{4P_f}{\pi d_o^2} = \frac{4(11.3)}{\pi(0.5)^2} \Rightarrow \boxed{\sigma = 57.55 \text{ ksi}}.$$

② at failure:

$$\sigma_T = \frac{P_f}{A_f} = \frac{4P_f}{\pi d_f^2} = \frac{4(11.3)}{\pi(0.366)^2} \Rightarrow \boxed{\sigma_T = 107.4 \text{ ksi}}.$$

Problem 5.22:

given: value of ϵ ranging from 0.001 to 1.000

find: $\boxed{\epsilon_T = ?}$.

① convert ϵ to ϵ_T using the eqn.: $\epsilon_T = \ln(1+\epsilon)$

② completing table:

ϵ	ϵ_T
0.001	0.00100
0.002	0.00200
0.005	0.00499
0.010	0.00995
0.020	0.0198

ϵ	ϵ_T
0.050	0.0488
0.100	0.0953
0.200	0.182
0.500	0.405
1.000	0.693

Problem 5.23:

spur gear tooth:

$\sigma_{max} = 78.9$ ksi

$\sigma_{min} = 0$

* neglect effect of σ_m

find: $\boxed{N = ?}$. (until failure)

① $\sigma_a = \frac{\sigma_{max} - \sigma_{min}}{2} = \frac{78.9 - 0}{2} = 39.45$ ksi

② from $S_f - N$ curve:

$S_e = 43$ ksi

③ since $\sigma_a = 39.45$ ksi < 43 ksi $= S_e$, then gear will

never fail in fatigue

$\therefore$ $\boxed{N = \infty \text{ (infinite life)}}$.

Problem 5.24:

steel: $S_u = 340\ MPa$ * neglect effect of σ_m

$\sigma_{dynamic} = 2\sigma_{static}$ find: $\boxed{\sigma_{max} = ?}$.

① consider loading details:

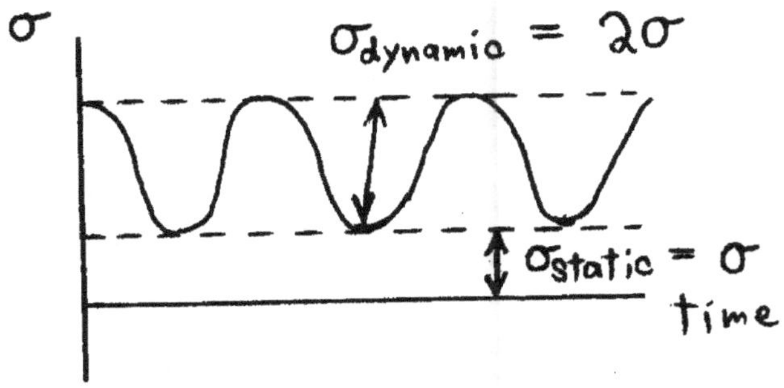

② from figure:

$\sigma_{min} = \sigma$

$\sigma_{max} = \sigma + 2\sigma = 3\sigma$

$\sigma_m = \frac{\sigma_{max} + \sigma_{min}}{2} = \frac{3\sigma + \sigma}{2} = 2\sigma$

$\sigma_a = \frac{\sigma_{max} - \sigma_{min}}{2} = \frac{3\sigma - \sigma}{2} = \sigma$

③ $S_e = \frac{S_u}{2} = \frac{340}{2} = 170\ MPa$

④ $\sigma_a = S_e = 170\ MPa$

⑤ $\sigma = 170\ MPa$

⑥ $\sigma_{max} = 3(170) = 510\ MPa > 340\ MPa = S_u$

$\therefore \boxed{\sigma_{max} = 340\ MPa}$.

Problem 5.26:

steel: $S_u = 340 \text{ MPa}$ find: $\boxed{\sigma_{max} = ?}$.

$\sigma_{dynamic} = 2\sigma_{static}$

① consider loading details:

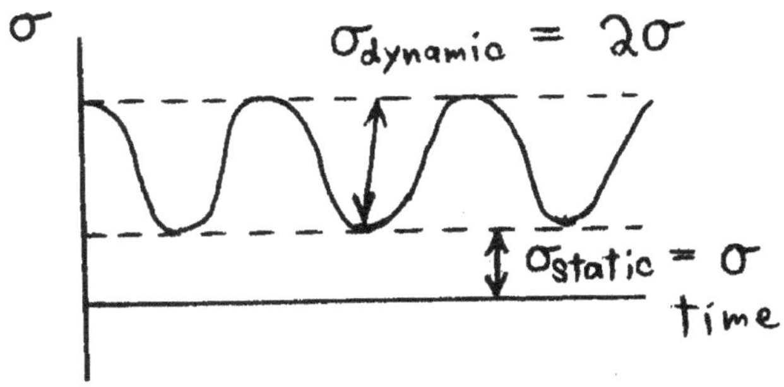

② from figure:

$\sigma_{min} = \sigma$

$\sigma_{max} = \sigma + 2\sigma = 3\sigma$

$\sigma_m = \dfrac{\sigma_{max} + \sigma_{min}}{2} = \dfrac{3\sigma + \sigma}{2} = 2\sigma$

$\sigma_a = \dfrac{\sigma_{max} - \sigma_{min}}{2} = \dfrac{3\sigma - \sigma}{2} = \sigma$

③ $S_e = \dfrac{S_u}{2} = \dfrac{340}{2} = 170 \text{ MPa}$

④ $\sigma_a = S_e\left(1 - \dfrac{\sigma_m}{S_u}\right) \Rightarrow \sigma = (170)\left(1 - \dfrac{2\sigma}{340}\right)$

⑤ $\sigma = 170 - \dfrac{340\sigma}{340} = 170 - \sigma$

⑥ $\sigma = 85 \text{ MPa}$

⑦ $\sigma_{max} = 3(85) = 255 \text{ MPa} < 340 \text{ MPa} = S_u$

$\therefore \boxed{\sigma_{max} = 255 \text{ MPa}}$.

Problem 5.33:

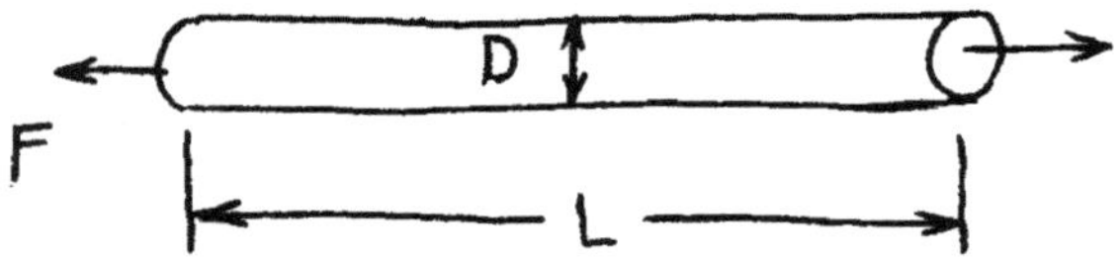

$L = 8\ ft = 96\ in$

$D = 1\ in$

$\delta = 0.22\ in$

$\Delta = -0.0007\ in$

find: $E, \nu, G = ?$

① $A = \frac{\pi}{4} D^2 = \frac{\pi}{4}(1)^2 = 0.7854\ in^2$

② $\sigma = \frac{P}{A} = \frac{54}{0.7854} = 68.75\ ksi$

③ $e_a = \frac{\delta}{L} = \frac{0.22}{96} = 0.002292$

④ $E = \frac{\sigma}{\epsilon_a} = \frac{68.75}{0.002292} \Rightarrow E = 30{,}000\ ksi$.

⑤ $e_t = \frac{\Delta}{D} = \frac{-0.0007}{1} = -0.0007$

⑥ $\nu = -\frac{e_t}{e_a} = -\frac{(-0.0007)}{0.002292} \Rightarrow \nu = 0.3054$.

⑦ $G = \frac{E}{2(1+\nu)} = \frac{30{,}000}{2(1+0.3054)}$

⑧ $G = 11{,}490\ ksi$.

Problem 5.34:

std. tensile specimen:

$d_o = 0.50$ in

$L_o = 2.0$ in

$d_f = 0.413$ in

* see P − δ table in text

find:

a) $E = ?$

b) $\sigma_{y(0.2\%)} = ?$

c) $\sigma_u = ?$

d) $\sigma_f = ?$

e) $\epsilon_f = ?$

f) $\%e = ?$

g) $\%A = ?$

① $A_o = \frac{\pi}{4} d_o^2 = \frac{\pi}{4}(0.5)^2 = 0.19635 \text{ in}^2$

② recall: $\sigma = \frac{P}{A_o}$; $e = \frac{\delta}{L_o}$

③ use eqns. in ② to convert P − δ data to σ − ε data:

Load (kip)	Extension (in)	ε	σ (ksi)
0.0	0.0000	0.00000	0.00
2.0	0.0021	0.00105	10.19
4.0	0.0039	0.00195	20.37
6.0	0.0058	0.00290	30.56
7.0	0.0072	0.00360	35.65
7.8	0.0084	0.00420	39.72
8.3	0.0280	0.01400	42.27
8.7	0.0530	0.02650	44.31
9.2	0.0860	0.04300	46.86
9.8	0.1450	0.07250	49.91
10.4	0.2220	0.11100	52.97

Load (kip)	Extension (in)	ε	σ (ksi)
10.8	0.2510	0.12550	55.00
11.2	0.2820	0.14100	57.04
11.6	0.3030	0.15150	59.08
11.6	0.3180	0.15900	59.08
11.5	0.3270	0.16350	58.57
11.2	0.3510	0.17550	57.04
10.8	0.3670	0.18350	55.00
10.1	0.3940	0.19700	51.44
9.7	0.4020	0.20100	49.40
9.6	0.4200	0.21000	48.89

④ use data to construct σ − ε diagram in MS Excel

⑤ insert trend line thru first 6 data pts. (elastic region) to get slope ⇒ slope = $E = 9{,}888$ ksi $= 9.888 \times 10^6$ psi

⑥ read pts. directly from σ − ε diagram:

$\sigma_u = 59.08$ ksi ; $\sigma_f = 48.89$ ksi ; $\epsilon_f = 0.2100$

⑦ construct 0.2% offset line on σ − ε diagram to determine $\sigma_{y(0.2\%)}$:

$\sigma_{y(0.2\%)} = 40.21$ ksi

⑧ $\%e = \frac{\delta_f}{L_o}(100\%) = \frac{0.420}{2.0}(100\%) \Rightarrow \%e = 21.00\%$

⑨ $\%A = \frac{A_o - A_f}{A_o}(100\%) = \left(1 - \frac{d_f^2}{d_o^2}\right)(100\%)$

⑩ $\%A = \left(1 - \frac{(0.413)^2}{(0.5)^2}\right)(100\%) \Rightarrow \%A = 31.77\%$

Problem 5.34: (con't)

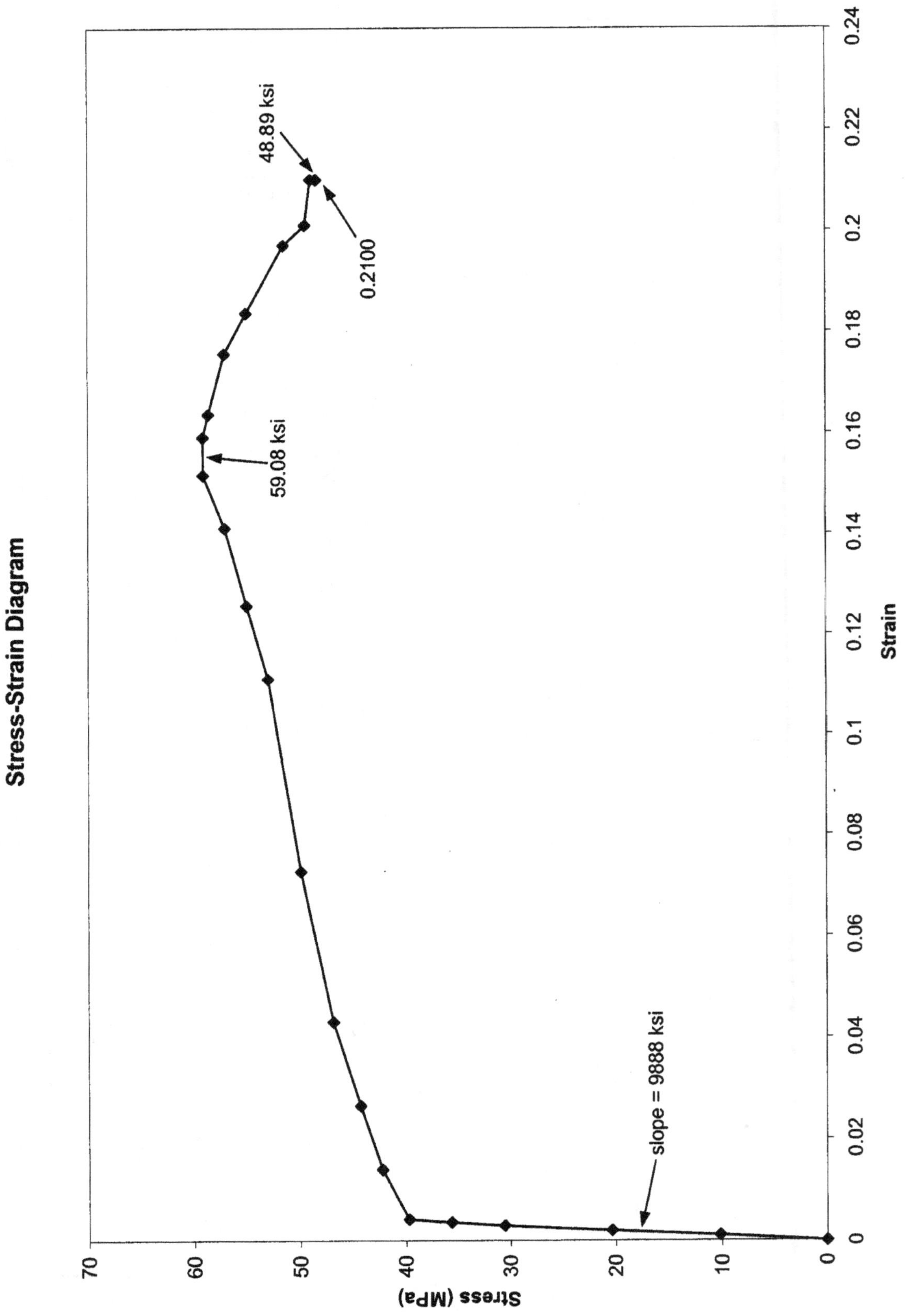

Problem 5.34: (con't)

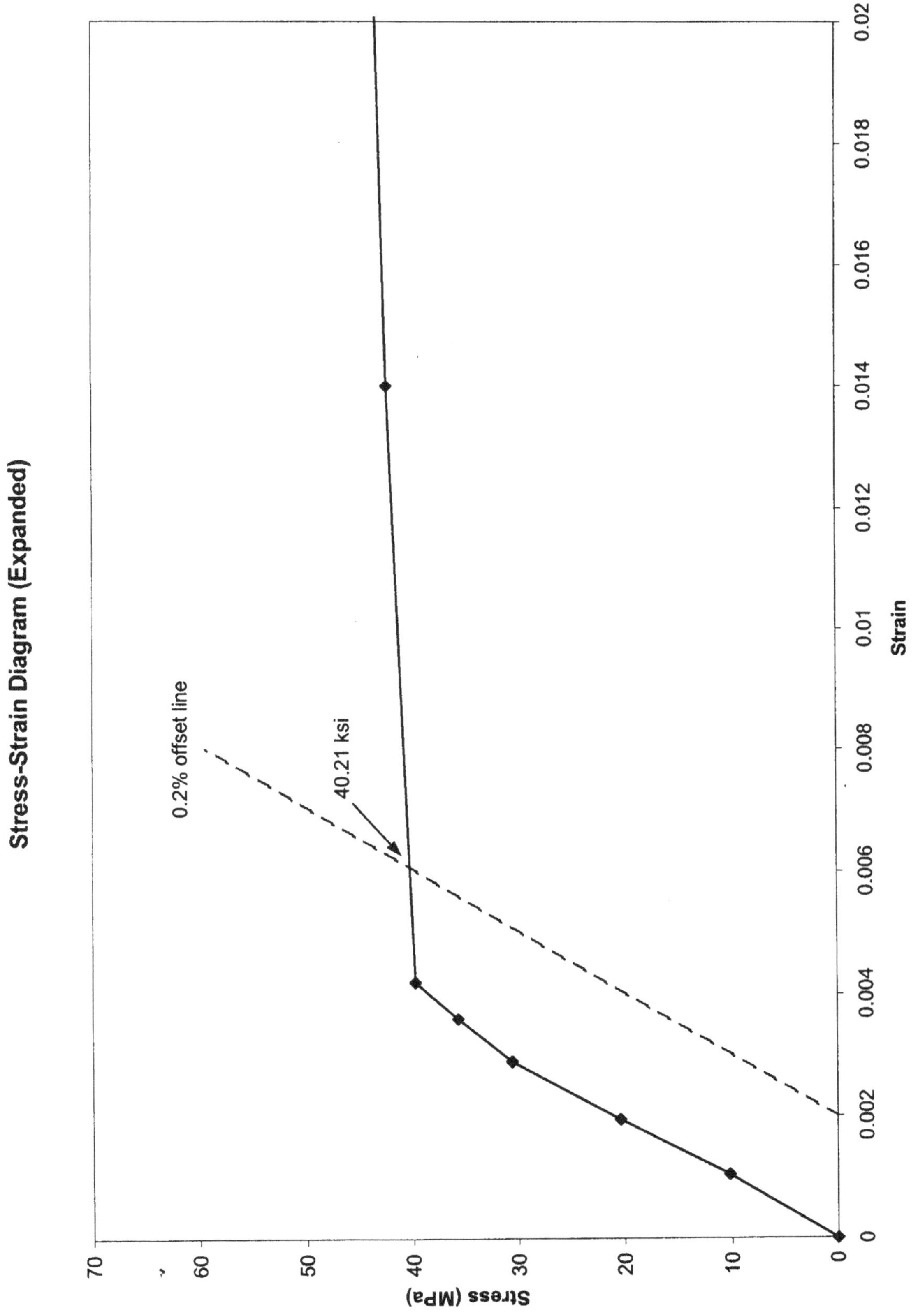

Problem 5.35:

std. tensile specimen:

$d_o = 12.7$ mm

$L_o = 50$ mm

$d_f = 9.0$ mm

* see P–δ table in text

find: a) $E = ?$ b) $\sigma_{y(0.2\%)} = ?$ c) $\sigma_u = ?$ d) $\sigma_f = ?$ e) $\epsilon_f = ?$ f) $\%e = ?$ g) $\%A = ?$

① $A_o = \frac{\pi}{4} d_o^2 = \frac{\pi}{4}(12.7)^2$

② $A_o = 126.68 \text{ mm}^2$

③ recall: $\sigma = \frac{P}{A_o}$; $\epsilon = \frac{\delta}{L_o}$

④ use eqns. in ③ to convert P–δ data to σ–ε data:

Load (kN)	Extension (mm)	ε	σ (MPa)
0.0	0.000	0.00000	0.00
5.0	0.010	0.00020	39.47
10.0	0.019	0.00038	78.94
15.0	0.029	0.00058	118.41
20.0	0.042	0.00084	157.88
25.0	0.048	0.00096	197.35
26.0	0.500	0.01000	205.24
27.0	0.950	0.01900	213.14
29.0	1.500	0.03000	228.92
31.5	2.210	0.04420	248.66
35.0	3.540	0.07080	276.29

Load (kN)	Extension (mm)	ε	σ (MPa)
37.0	4.520	0.09040	292.07
38.0	5.530	0.11060	299.97
38.0	6.180	0.12360	299.97
36.0	6.660	0.13320	284.18
33.0	7.350	0.14700	260.50
30.0	7.860	0.15720	236.82
27.0	8.420	0.16840	213.14
23.0	8.780	0.17560	181.56
19.0	9.320	0.18640	149.98

⑤ use data to construct σ–ε diagrams in Excel

⑥ insert trend line thru first 6 data pts. (elastic region) to get slope ⇒ slope = $E = 199{,}600$ MPa $= 199.6$ GPa.

⑦ read pts. directly from σ–ε diagram:

$\sigma_u = 300.0$ MPa; $\sigma_f = 150.0$ MPa; $\epsilon_f = 0.1864$.

⑧ construct 0.2% offset line on σ–ε diagram to determine σ_y:

$\sigma_{y(0.2\%)} = 199.6$ MPa.

⑨ $\%e = \frac{\delta_f}{L_o}(100\%) = \frac{9.32}{50}(100\%) \Rightarrow \%e = 18.64\%$.

⑩ $A_f = \frac{\pi}{4}(9)^2 = 63.62 \text{ mm}^2$

⑪ $\%A = \frac{A_o - A_f}{A_o}(100\%) = \frac{126.68 - 63.62}{126.68}(100) \Rightarrow \%A = 49.78\%$.

Problem 5.35: (con't)

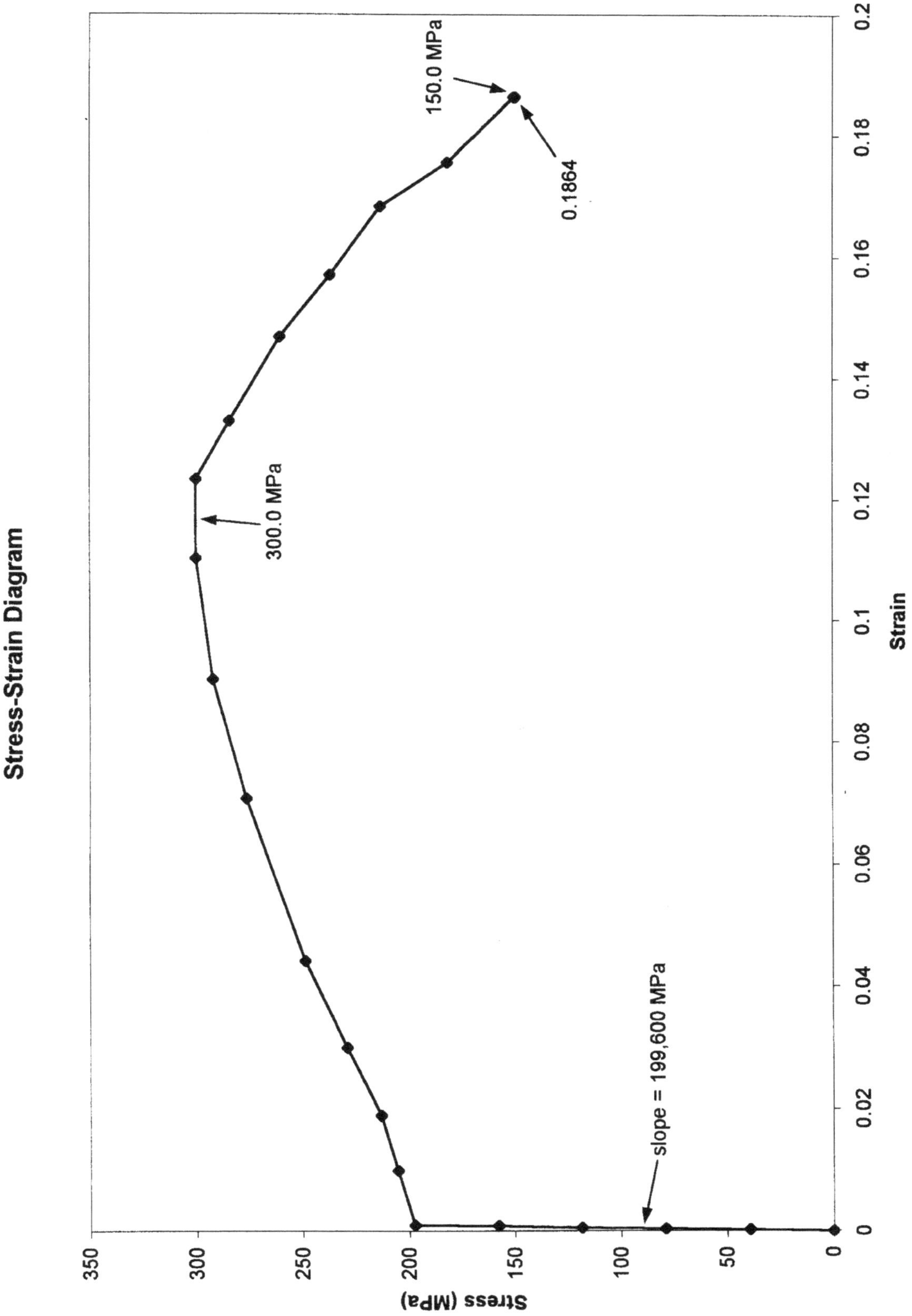

Problem 5.35: (con't)

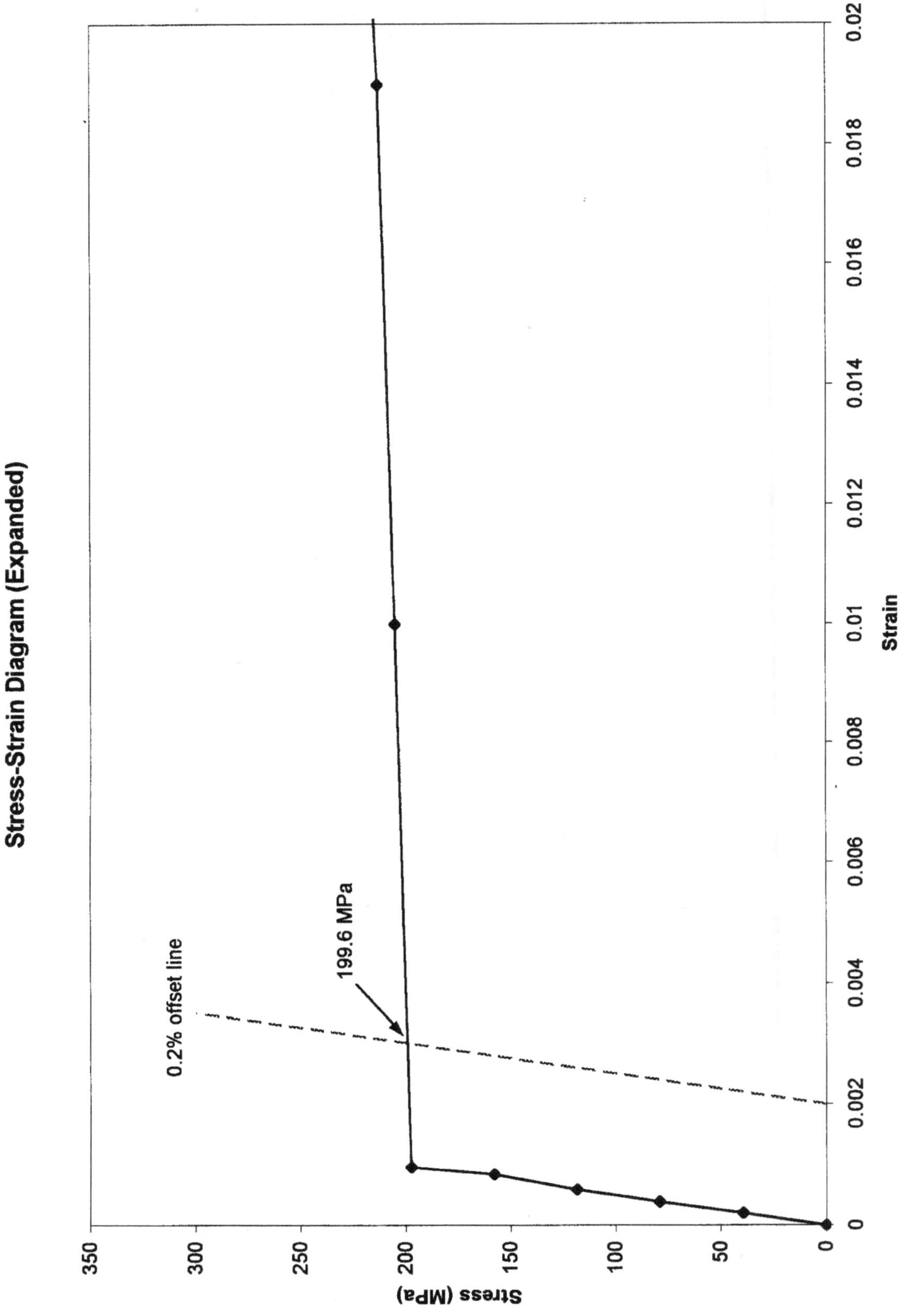

CHAPTER 6 PROBLEMS

6.1 Why is a truss used in constructing a bridge or a roof covering for a large structure?

6.2 What is the inherent geometric form that is used repeatedly in all truss designs? Why?

6.3 If a truss is fabricated from 37 members, determine the number of joints that will be required in constructing the truss.

6.4 Why is it necessary to employ an odd number of members in constructing a truss?

6.5 Prepare a sketch of a joint using a gusset plate for connecting four uniaxial bars in a truss structure. Identify all the components included in the joint that you have designed.

6.6 Why is it important to restrict the application of external loads on a truss to only the joints?

6.7 Consider the four-unit Howe truss shown in the figure to the right. If the external forces are as specified in this figure, use the method of joints to determine the forces in members AC, BC, CD and CE.

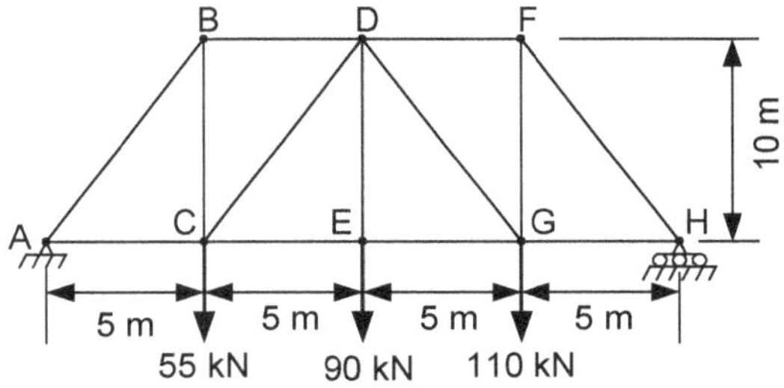

6.8 Consider the four-unit Howe truss shown in the figure to the left. If the external forces are as specified in this figure, use the method of joints to determine the forces in members AC, BC, CD and CE.

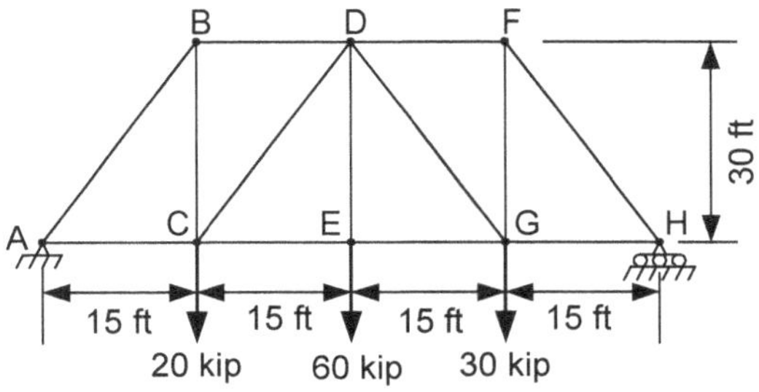

6.9 Consider the four-unit Howe truss shown in the figure to the right. If the external forces are as specified in this figure, use the method of joints to determine the forces in members GE, GD, GF and GH.

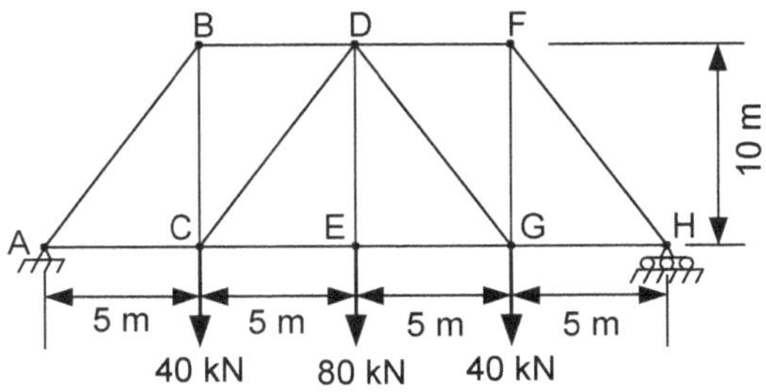

6.10 Consider the four-unit Howe truss shown in the figure to the right. If the external forces are as specified in this figure, use the method of joints to determine the forces in members AC, BC, CD and CE.

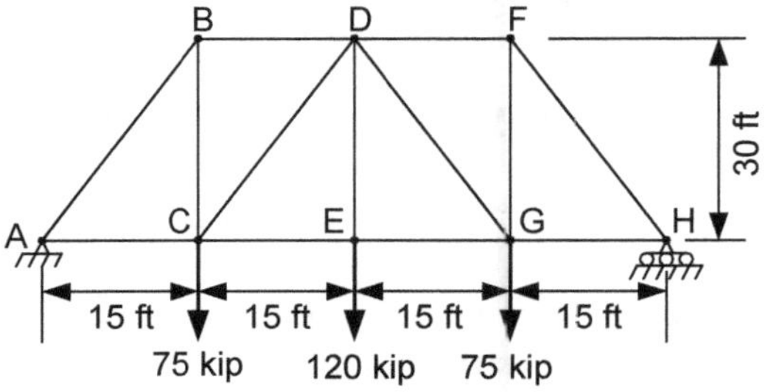

6.11 If the safety factor SF for each of the members in the truss of Problem 6.7 is specified as 3.0, determine the minimum required cross sectional area for members AC, BC, CD and CE. Note the yield strength S_y of the hot rolled structural steel used in fabricating the truss is 200 MPa.

6.12 If the safety factor SF for each of the members in the truss of Problem 6.8 is specified as 3.2, determine the minimum required cross sectional area for members AB, AC, BD and BC. Note the yield strength S_y of the hot rolled structural steel used in fabricating the truss is 30 ksi.

6.13 A scissors truss, illustrated in the figure to the right, is loaded with three forces at the joints B, C and D. If the truss is fabricated from 1018 A steel, determine the size of members AB, AF, BC and BF. The safety factor is specified as 3.0.

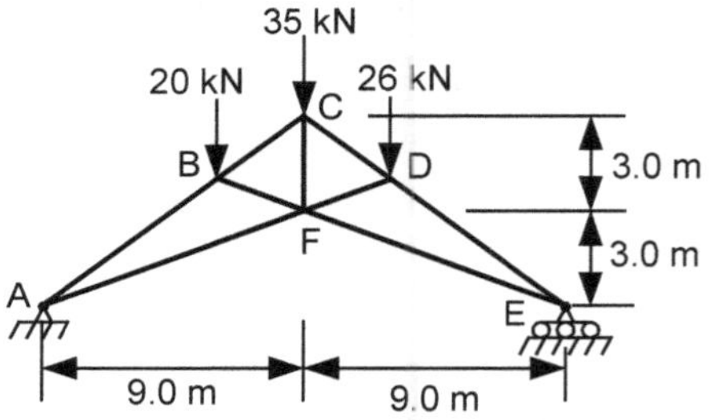

6.14 A scissors truss, illustrated in the figure to the left, is loaded with three forces at the joints B, C and D. If the truss is fabricated from 1020 HR steel, determine the size of members AB, AF, BC and BF. The safety factor is specified as 2.9.

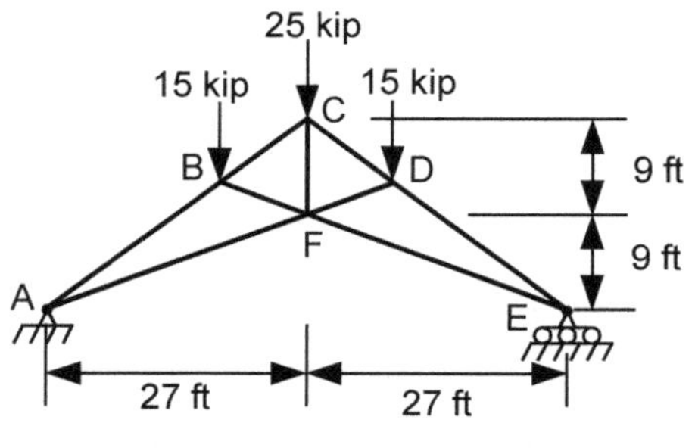

6.15 For the Howe truss, shown in the figure to the right, use the method of sections to determine the forces in the members BD, CD and CE.

6.16 For the Howe truss, shown in the figure to the right, use the method of sections to determine the forces in the members DF, DG and EG.

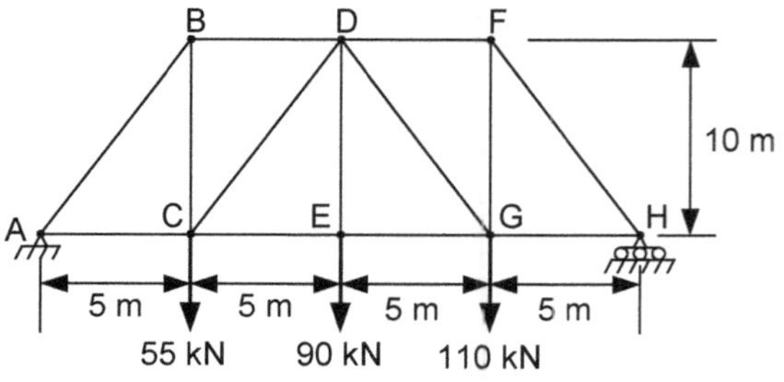

Trusses

6.17 For the truss, defined in the figure to the right, use the method of sections to determine the forces in members DF, DB and DE. The external forces applied at joints C, E, G, I, and K are each equal to 15 kip.

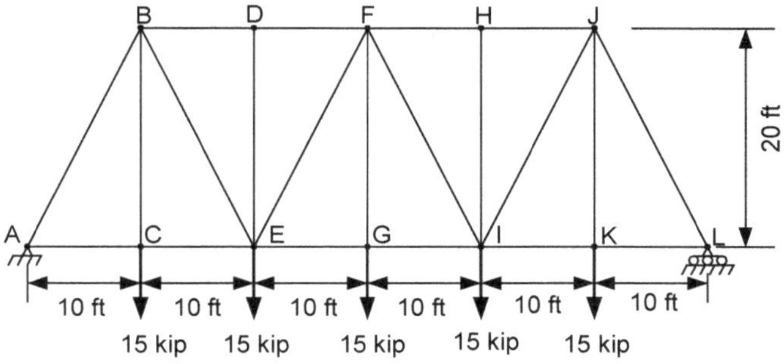

6.18 If the cross sectional areas of all of the members in the truss defined in Problem 6.17 are equal to A = 2.3 in.^{2}, determine the safety factor for structural members DF, DB and DE. The yield strength of the steel used in the truss members is 35 ksi.

6.19 By inspection identify the zero-force members in the truss defined in Problem 6.17.

6.20 For the truss, defined in the figure to the right, use the method of joints to determine the forces in members DC, DE and DF.

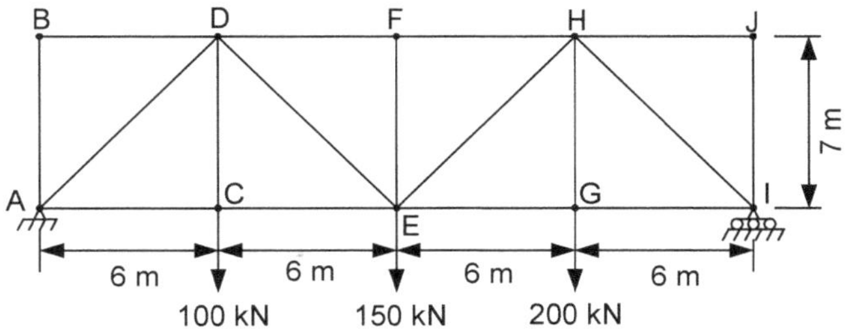

6.21 If the cross sectional areas of all of the members in the truss defined in Problem 6.20 are equal to 3,000 mm^2, determine the safety factor for the members DC, DE and DF. Note the yield strength of the steel used in the truss is 310 MPa.

6.22 Use the method of sections with the truss, shown in the figure for Problem 6.20, to determine the forces in members FH, EH, and EG.

6.23 Let s/h be a variable in the truss structure defined in the figure to the left. Determine the forces in the members AC and AD as a function of the s/h ratio. We suggest you use a spreadsheet to perform the calculations. Consider the ratio s/h over the range from 0.5 to 2.0 varying in steps of 0.1. Prepare a graph of the results for the forces in members AC and AD as a function of s/h.

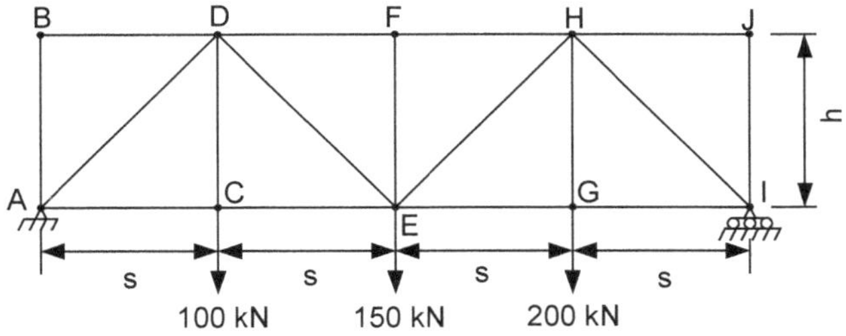

6.24 Determine the forces in members CE, DE and DF of the bowstring truss shown in the figure to the right.

The span s = 12 ft.

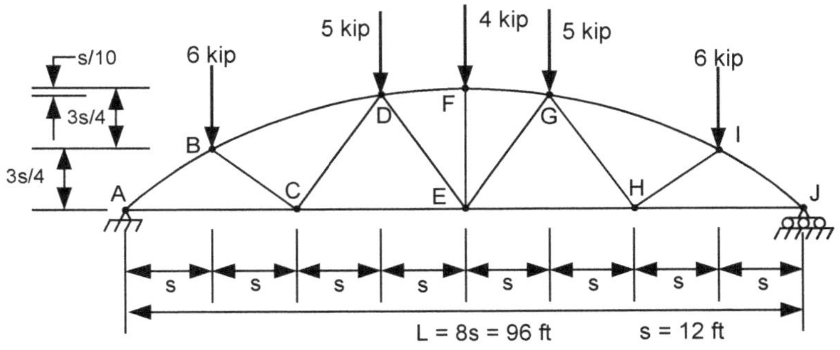

6.25 Determine the forces in members AC, CD, CE and DF of the inclined truss shown in the figure to the right.

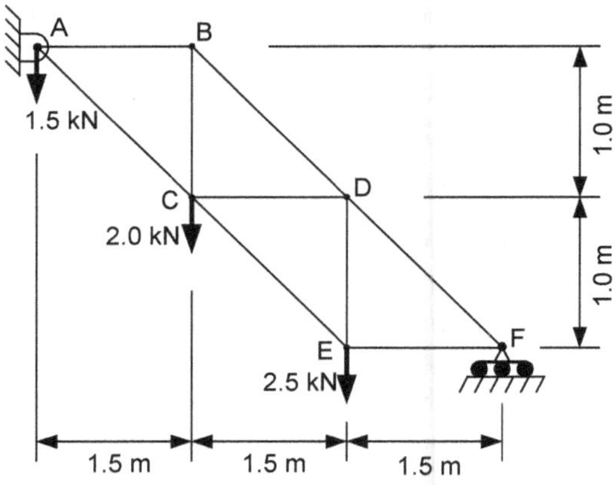

6.26 Determine the forces in members CD, CJ, and KJ of the truss shown in the figure below.

6.27 Determine the forces in members DE, EJ, and JI of the truss shown in the figure to the right.

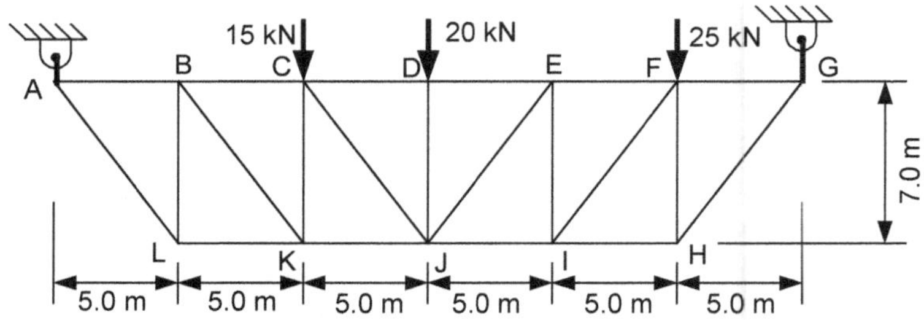

6.28 Determine the forces in members EF, IF, and IH, of the truss shown in the figure below.

6.29 Determine the forces in members BC, BK, and LK, of the truss shown in the figure to the right.

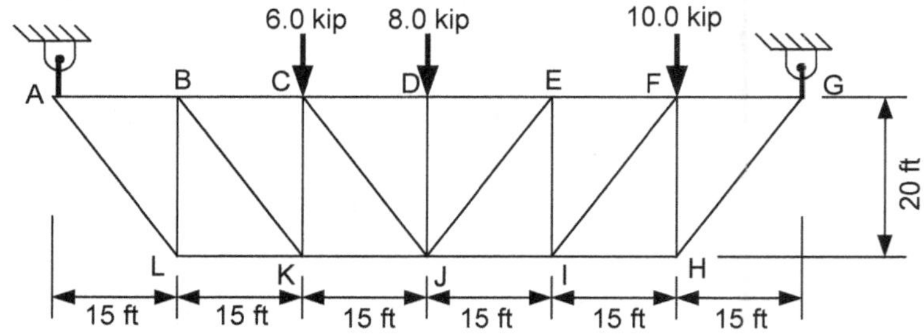

6.30 For the cantilever truss, shown in the figure to the right, determine the stress in member BC if its cross sectional area is 2.5 in^2. Also determine the safety factor for this truss member if it is fabricated from 1020 HR steel.

6.31 For the cantilever truss, shown in the figure to the right, determine the stress in member AB if its cross sectional area is 3.5 in^2. Also determine the safety factor for this truss member if it is fabricated from 1020 HR steel.

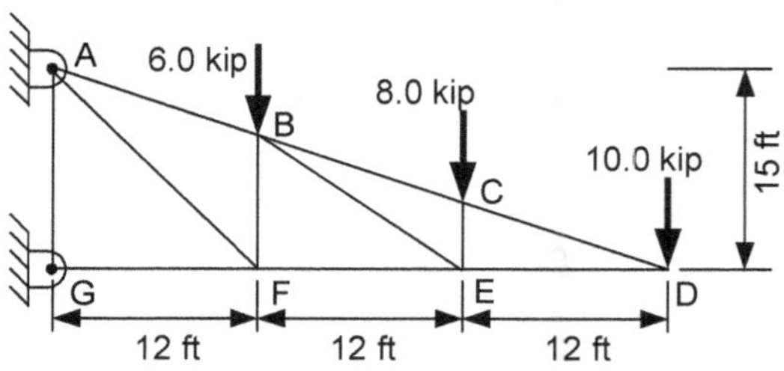

6.32 For the cantilever truss, shown in the figure to the right, determine the force in member GF. Specify the minimum cross sectional area if the member is to exhibit a safety factor of 3.0 relative to the yield strength of its material. This truss member is fabricated from 1020 HR steel.

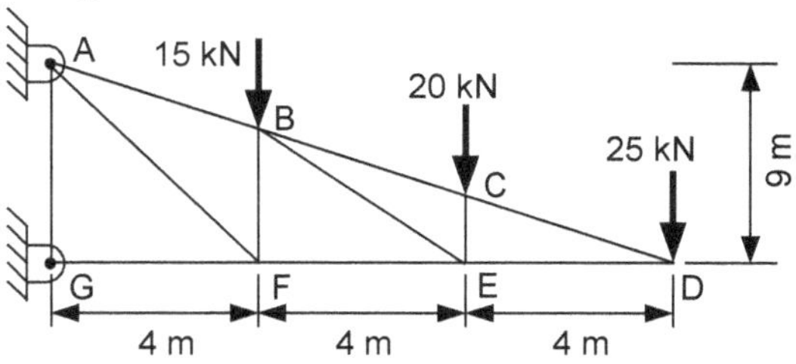

6.33 For the cantilever truss, shown in the figure to the right, determine the stress in member CD if its cross sectional area is 900 mm^2. Also determine the safety factor for this truss member if it is fabricated from 1020 HR steel.

6.34 For the cantilever truss shown in Problem 6.33, determine the force in member ED. Specify the minimum cross sectional area if this member is to exhibit a safety factor of 3.0 relative to the yield strength of its material. This truss member is fabricated from 1212 HR steel.

6.35 For the truss, presented in the figure below, determine the cross sectional area required for members AB, BD, DE, EF and CF if they are fabricated from 1020 HR steel and a safety factor of 3.2 is specified.

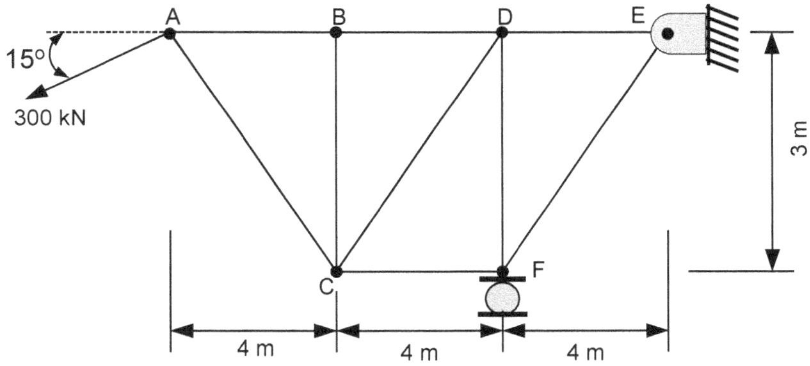

Problem 6.8:

FBD:

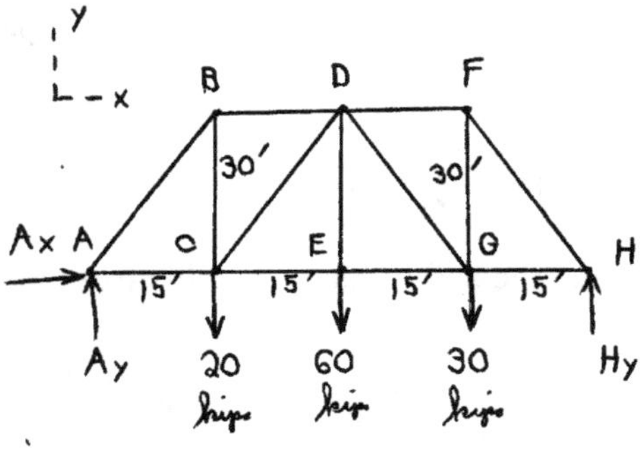

find: P_{AC}, P_{BC}, P_{CD}, + P_{CE} = ?

* consider entire truss:

① $\overset{+}{\rightarrow}\sum F_x = 0: \quad A_x = 0$

② $\circlearrowleft + \sum M_A = 0:$

$-(20)(15) - (60)(30) - (30)(45) + H_y(60) = 0$

③ $H_y = 57.5$ kips

④ $\uparrow + \sum F_y = 0: \quad A_y + (57.5) - 110 = 0 \Rightarrow A_y = 52.5$ kips

* consider joint A:

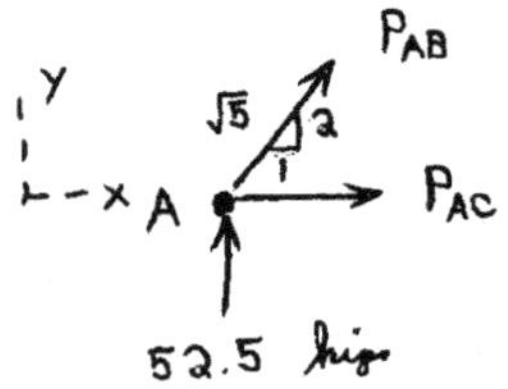

⑤ $\uparrow + \sum F_y = 0: \quad \frac{2}{\sqrt{5}} P_{AB} + 52.5 = 0$

⑥ $P_{AB} = -58.70$ kips $= 58.70$ kips (C)

⑦ $\overset{+}{\rightarrow}\sum F_x = 0: \quad \frac{1}{\sqrt{5}} P_{AB} + P_{AC} = 0$

⑧ $P_{AC} = -\frac{1}{\sqrt{5}}(-58.70) \Rightarrow$ $P_{AC} = 26.25$ kips (T)

* consider joint B:

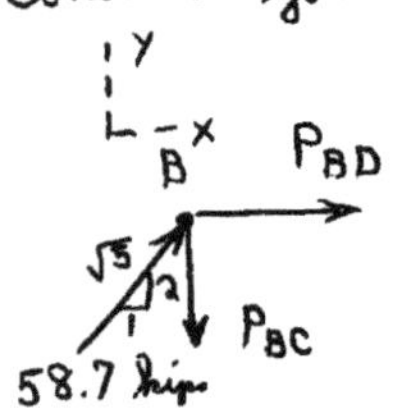

⑨ $\overset{+}{\rightarrow}\sum F_x = 0: \quad \frac{1}{\sqrt{5}}(58.7) + P_{BD} = 0$

⑩ $P_{BD} = -26.25$ kips $= 26.25$ kips (C)

⑪ $+\uparrow \sum F_y = 0: \quad \frac{2}{\sqrt{5}}(58.7) - P_{BC} = 0$

⑫ $P_{BC} = 52.5$ kips (T)

* consider joint C:

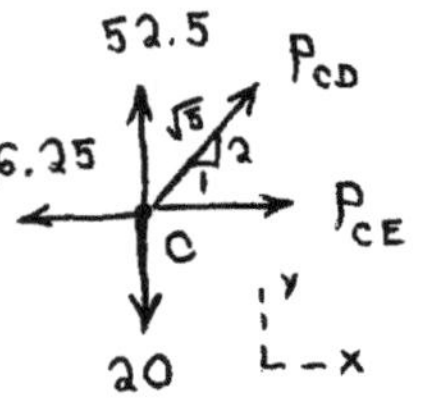

⑬ $\uparrow + \sum F_y = 0: \quad 52.5 - 20 + \frac{2}{\sqrt{5}} P_{CD} = 0$

⑭ $P_{CD} = -36.34$ kips $= 36.34$ kips (C)

⑮ $\overset{+}{\rightarrow}\sum F_x = 0: \; -26.25 + \frac{1}{\sqrt{5}} P_{CD} + P_{CE} = 0$

⑯ $P_{CE} = 26.25 - \frac{1}{\sqrt{5}}(-36.34)$

⑰ $P_{CE} = 42.50$ kips (T)

Problem 6.10:

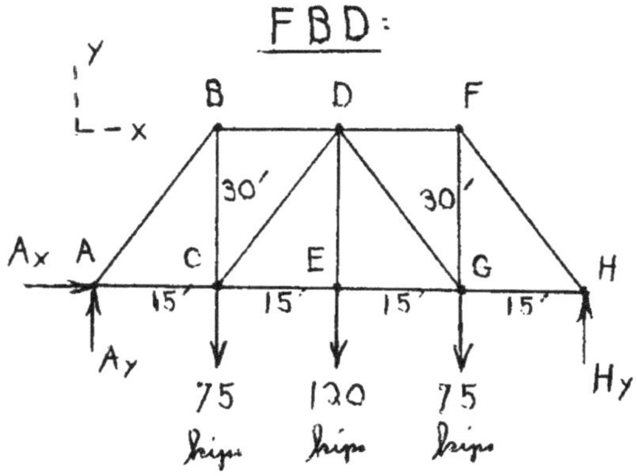

find: $P_{AC}, P_{BC}, P_{CD}, P_{CE} = ?$

* consider entire truss:

① from symmetry in geometry + loading:

$$A_x = 0 \; ; \quad A_y = H_y = \frac{270}{2}$$

② $A_y = H_y = 135$ kips

* consider joint A:

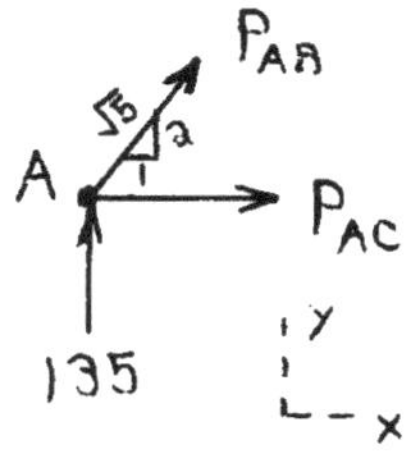

③ $+\uparrow \sum F_y = 0 : \quad \frac{2}{\sqrt{5}} P_{AB} + 135 = 0$

④ $P_{AB} = -150.9$ kips $= 150.9$ kips (C)

⑤ $\stackrel{+}{\rightarrow} \sum F_x = 0 : \quad \frac{1}{\sqrt{5}} P_{AB} + P_{AC} = 0$

⑥ $P_{AC} = -\frac{1}{\sqrt{5}}(-150.9)$

⑦ $\boxed{P_{AC} = 67.48 \text{ kips (T)}}$

* consider joint B:

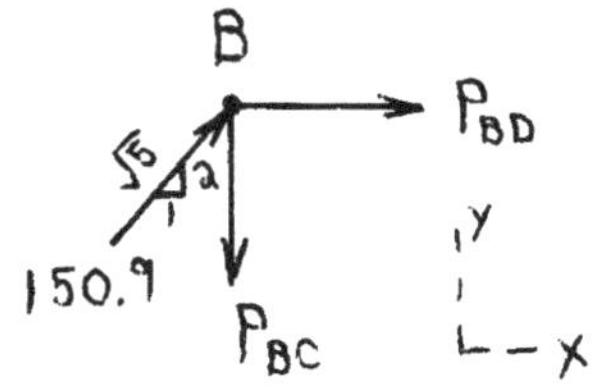

⑧ $\stackrel{+}{\rightarrow} \sum F_x = 0 : \quad \frac{1}{\sqrt{5}}(150.9) + P_{BD} = 0$

⑨ $P_{BD} = -67.48$ kips $= 67.48$ kips (C)

⑩ $+\uparrow \sum F_y = 0 : \quad \frac{2}{\sqrt{5}}(150.9) - P_{BC} = 0$

⑪ $\boxed{P_{BC} = 135.0 \text{ kips (T)}}$

* consider joint C:

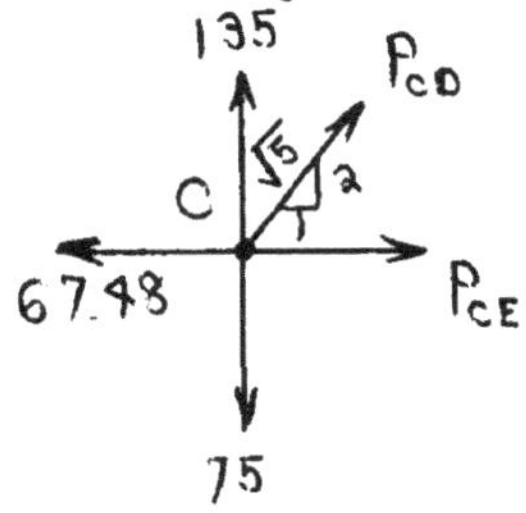

⑫ $+\uparrow \sum F_y = 0 : \quad 135 - 75 + \frac{2}{\sqrt{5}} P_{CD} = 0$

⑬ $\boxed{P_{CD} = -67.08 \text{ kips} = 67.08 \text{ kips (C)}}$

⑭ $\stackrel{+}{\rightarrow} \sum F_x = 0 : \quad -67.48 + \frac{1}{\sqrt{5}} P_{CD} + P_{CE} = 0$

⑮ $P_{CE} = 67.48 - \frac{1}{\sqrt{5}}(-67.08)$

⑯ $\boxed{P_{CE} = 97.48 \text{ kips (T)}}$

Problem 6.12:

* refer to solnś. from # 6.8

$SF_y = 3.2$

$S_y = 30$ ksi

find: $A_{AB}, A_{AC}, A_{BD}, A_{BC} = ?$

① $\sigma = \dfrac{P}{A} = \dfrac{S_y}{SF_y} \Rightarrow A = \dfrac{P(SF_y)}{S_y}$

② $A = \dfrac{P(3.2)}{(30 \text{ ksi})}$

③ from problem # 6.8:

$P_{AB} = 58.70$ kips (C)

$P_{AC} = 26.25$ kips (T)

$P_{BD} = 26.25$ kips (C)

$P_{BC} = 52.5$ kips (T)

④ subing each P from ③ into ② + solving:

$A_{AB} = 6.261 \text{ in}^2$ $A_{BD} = 2.8 \text{ in}^2$

$A_{AC} = 2.8 \text{ in}^2$ $A_{BC} = 5.6 \text{ in}^2$

Problem 6.13:

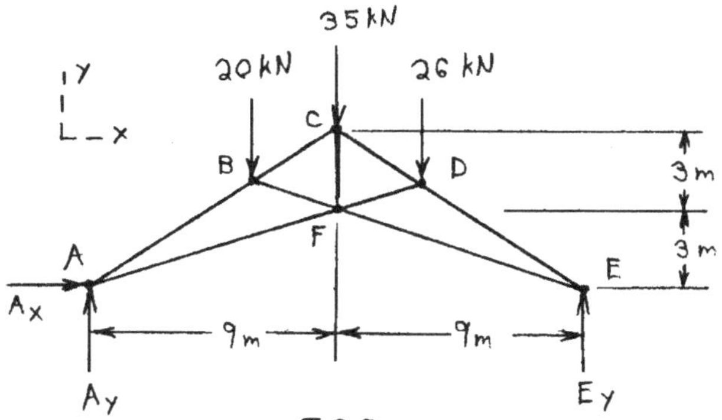

FBD:

find: $A_{AB,\ AF,\ BC,\ BF} = ?$

* 1018 steel:

$S_y = 221$ MPa

$SF_y = 3.0$

* consider geometry:

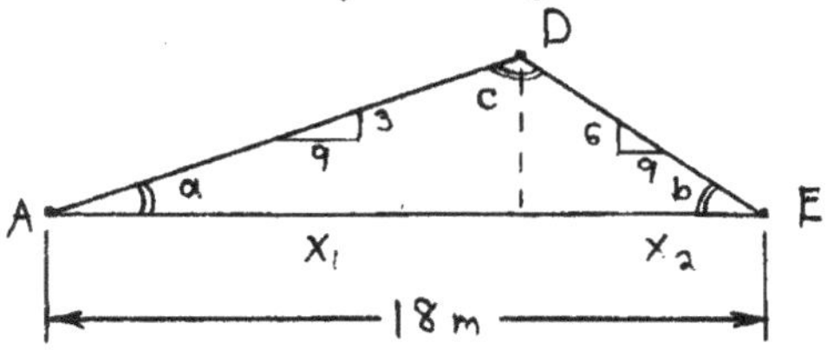

① $a = \tan^{-1}\left(\frac{3}{9}\right) = 18.43°$

② $b = \tan^{-1}\left(\frac{6}{9}\right) = 33.69°$

③ $c = 180° - 18.43° - 33.69°$

④ $c = 127.88°$

⑤ $\frac{\sin(127.88°)}{18} = \frac{\sin(18.43°)}{DE}$

⑥ $DE = (18)\frac{\sin(18.43°)}{\sin(127.88°)} = 7.210$ m

⑦ $X_2 = (7.210)\cos(33.69°) = 6.0$ m

⑧ $X_1 = 18 - 6 = 12$ m

* consider entire truss:

⑨ $\overset{+}{\rightarrow}\sum F_x = 0: \quad A_x = 0$

⑩ $\circlearrowright + \sum M_A = 0: \quad -(20)(6) - (35)(9) - (26)(12) + E_y(18) = 0$

⑪ $E_y = 41.5$ kN

⑫ $\uparrow + \sum F_y = 0: \quad A_y + (41.5) - 20 - 35 - 26 = 0$

⑬ $A_y = 39.5$ kN

* consider joint A:

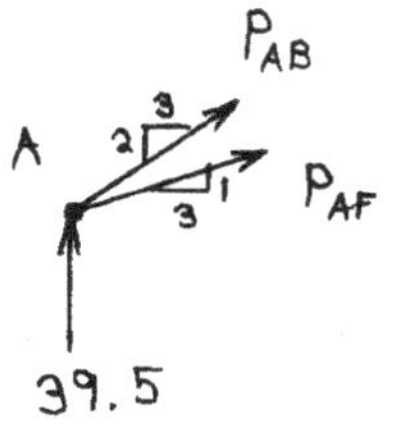

⑭ $\overset{+}{\rightarrow}\sum F_x = 0:$

$P_{AF}\left(\frac{3}{\sqrt{10}}\right) + P_{AB}\left(\frac{3}{\sqrt{13}}\right) = 0$

⑮ $P_{AF} = -P_{AB}\left(\frac{\sqrt{10}}{\sqrt{13}}\right)$

Problem 6.13: (con't)

(16) $+\uparrow\sum F_y = 0: \quad P_{AB}\left(\frac{2}{\sqrt{13}}\right) + P_{AF}\left(\frac{1}{\sqrt{10}}\right) + 39.5 = 0$

(17) subing (15) in (16):

$$P_{AB}\left(\frac{2}{\sqrt{13}}\right) + P_{AB}\left(-\frac{1}{\sqrt{13}}\right) = -39.5$$

(18) $\boxed{P_{AB} = -142.4 \text{ kN} = 142.4 \text{ kN (C)}}$.

(19) from (15): $P_{AF} = -(-142.4)\left(\frac{\sqrt{10}}{\sqrt{13}}\right) \Rightarrow \boxed{P_{AF} = 124.9 \text{ kN (T)}}$.

* consider joint B:

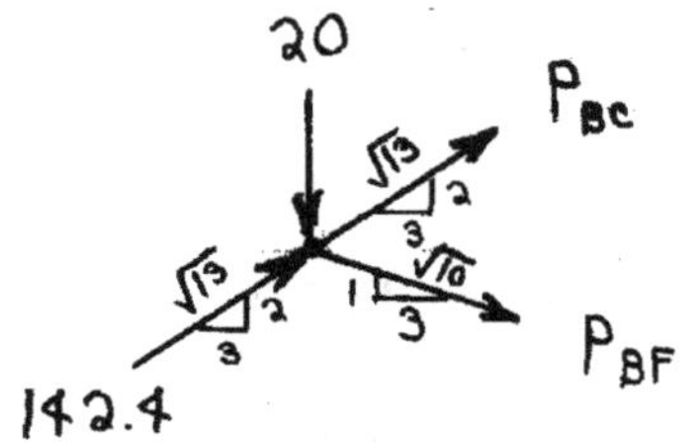

(20) $\overset{+}{\rightarrow}\sum F_x = 0:$

$$(142.4)\frac{3}{\sqrt{13}} + P_{BC}\left(\frac{3}{\sqrt{13}}\right) + P_{BF}\left(\frac{3}{\sqrt{10}}\right) = 0$$

(21) $P_{BF} = -\frac{\sqrt{10}}{\sqrt{13}}\left(P_{BC} + 142.4\right)$

(22) $+\uparrow\sum F_y = 0: \quad (142.4)\frac{2}{\sqrt{13}} - 20 + P_{BC}\left(\frac{2}{\sqrt{13}}\right) - P_{BF}\left(\frac{1}{\sqrt{10}}\right) = 0$

(23) subing (21) in (22):

$$P_{BC}\left(\frac{2}{\sqrt{13}}\right) + \frac{1}{\sqrt{13}}\left(P_{BC} + 142.4\right) = -58.99$$

(24) $P_{BC}\left(\frac{3}{\sqrt{13}}\right) = -98.48$

(25) $\boxed{P_{BC} = -118.4 \text{ kN} = 118.4 \text{ kN (C)}}$.

(26) using (25) in (21): $\boxed{P_{BF} = -21.05 \text{ kN} = 21.05 \text{ kN (C)}}$.

(27) $SF_y = \frac{S_y}{\sigma} = \frac{S_y A}{P}$

(28) $A = \frac{P(SF_y)}{S_y} = P\left(\frac{3}{221 \text{ MPa}}\right)$

(29) use values of $P_{AB,\ AF,\ BC,\ BF}$ in (28) to solve for A:

$$\boxed{\begin{aligned} A_{AB} &= 1933 \text{ mm}^2 \\ A_{AF} &= 1694 \text{ mm}^2 \\ A_{BC} &= 1607 \text{ mm}^2 \\ A_{BF} &= 285.7 \text{ mm}^2 \end{aligned}}$$.

Problem 6.17:

FBD:

find: $\boxed{P_{DF},\ P_{DB},\ P_{DE} = ?}$

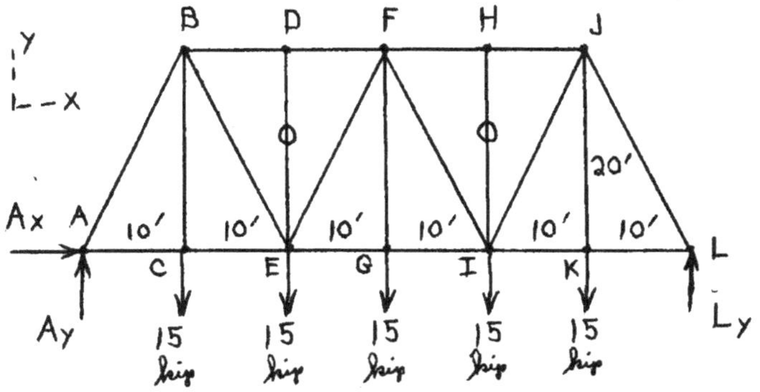

* consider entire truss:

① from symmetry in geometry & loading: $A_y = L_y = \dfrac{(15)5}{2} = 37.5$ kip

$A_x = 0$

② by inspection, DE & HI are ZFM's $\Rightarrow$ $\boxed{P_{DE} = 0}$.

$P_{DB} = P_{DF}$

③ cut truss thru members DF, EF, & EG and keep left side:

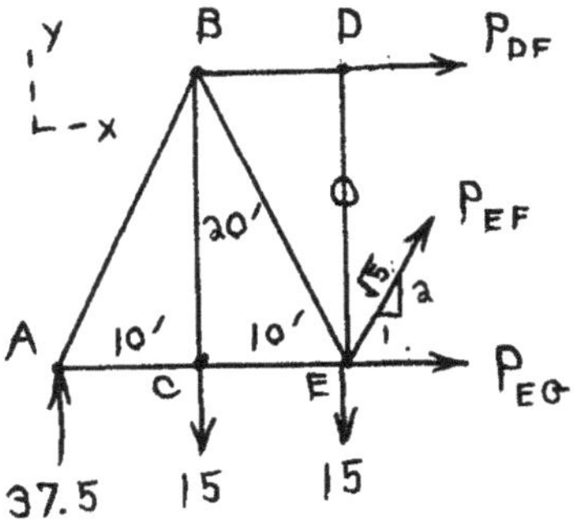

④ $\circlearrowleft + \sum M_E = 0$:

$(15)(10) - (37.5)(20) - P_{DF}(20) = 0$

⑤ $\boxed{P_{DF} = -30 \text{ kip} = 30 \text{ kip (C)} = P_{DB}}$

Problem 6.20:

FBD:

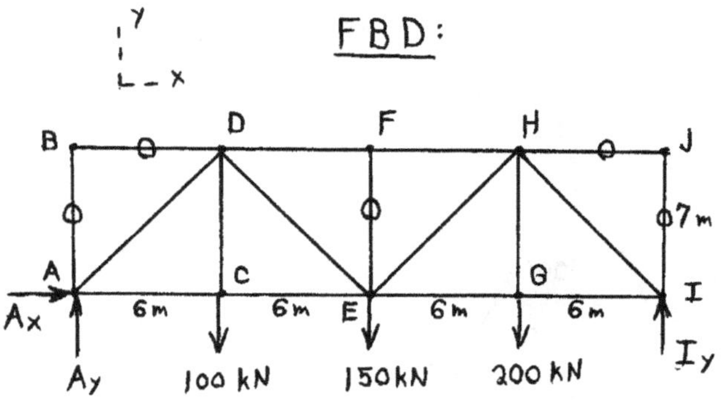

find: $\boxed{P_{DC},\ P_{DE},\ P_{DF} = ?}$.

* consider entire truss:

① $\rightarrow\!+ \sum F_x = 0: \quad A_x = 0$

② $\circlearrowleft\!+ \sum M_A = 0:$

$-(100)(6) - (150)(12) - (200)(18) + I_y(24) = 0$

③ $I_y = 250 \text{ kN}$

④ $\uparrow\!+ \sum F_y = 0: \quad A_y + (250) - 450 = 0 \Rightarrow A_y = 200 \text{ kN}$

⑤ by inspection: AB, BD, IJ, HJ, + EF are ZFM's

* consider joint A:

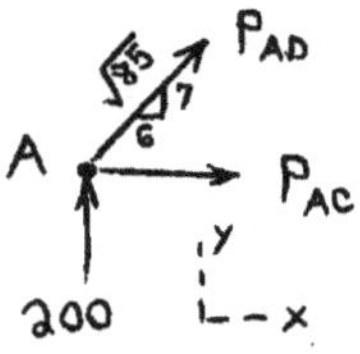

⑥ $+\uparrow \sum F_y = 0: \quad \frac{7}{\sqrt{85}} P_{AD} + 200 = 0$

⑦ $P_{AD} = -263.4 \text{ kN} = 263.4 \text{ kN (C)}$

⑧ $\rightarrow\!+ \sum F_x = 0: \quad \frac{6}{\sqrt{85}} P_{AD} + P_{AC} = 0$

⑨ $P_{AC} = -\frac{6}{\sqrt{85}}(-263.4) = 171.4 \text{ kN (T)}$

* consider joint C:

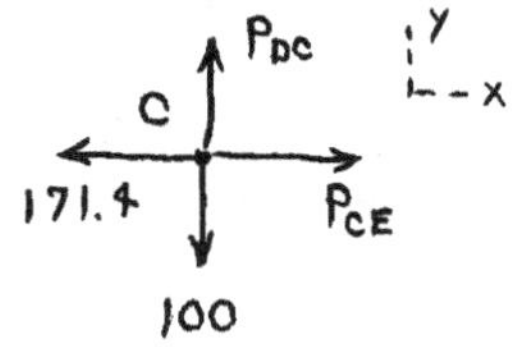

⑩ $\rightarrow\!+ \sum F_x = 0: \quad P_{CE} = 171.4 \text{ kN (T)}$

⑪ $\uparrow\!+ \sum F_y = 0: \quad \boxed{P_{DC} = 100 \text{ kN (T)}}$.

* consider joint D:

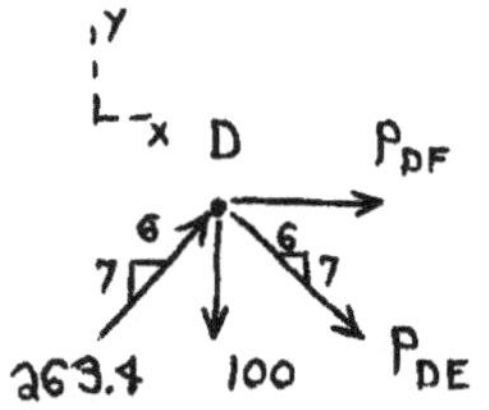

⑫ $\uparrow\!+ \sum F_y = 0: \quad \frac{7}{\sqrt{85}}(263.4) - 100 \quad \frac{7}{\sqrt{85}} P_{DE} = 0$

⑬ $\boxed{P_{DE} = 131.7 \text{ kN (T)}}$.

⑭ $\rightarrow\!+ \sum F_x = 0: \quad \frac{6}{\sqrt{85}}(263.4) + \frac{6}{\sqrt{85}} P_{DE} + P_{DF} = 0$

⑮ $P_{DF} = -\frac{6}{\sqrt{85}}(263.4 + 131.7)$

⑯ $\boxed{P_{DF} = -257.1 \text{ kN} = 257.1 \text{ kN (C)}}$.

Problem 6.23:

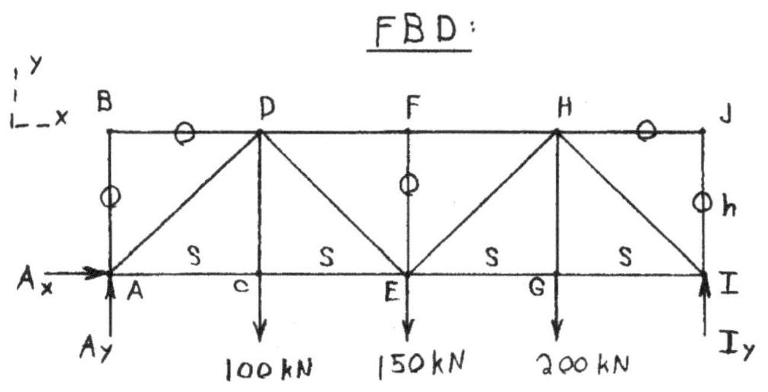

$0.5 \le \frac{s}{h} \le 2.0$

find: $\boxed{P_{AC},\ P_{AD} = ?}$

* consider entire truss:

① $\overset{+}{\rightarrow}\sum F_x = 0: \quad A_x = 0$

② $+\uparrow \sum F_y = 0: \quad A_y + I_y - 100 - 150 - 200 = 0$

③ $\circlearrowleft + \sum M_A = 0: \quad -(100)(s) - (150)(2s) - (200)(3s) + I_y(4s) = 0$

④ $I_y = 250 \text{ kN}$

⑤ using I_y in ②: $A_y + (250) - 450 = 0$

⑥ $A_y = 200 \text{ kN}$

⑦ by inspection, ZFM are: AB, BD, EF, HJ, IJ

* consider joint at A:

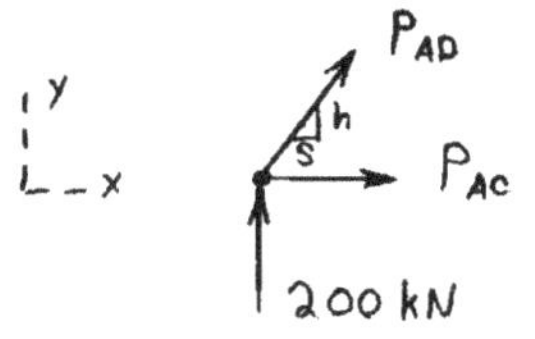

⑧ $\overset{+}{\rightarrow}\sum F_x = 0:$

$$P_{AC} + \left(\frac{s}{\sqrt{s^2+h^2}}\right) P_{AD} = 0$$

⑨ $+\uparrow \sum F_y = 0:$

$$200 + \left(\frac{h}{\sqrt{s^2+h^2}}\right) P_{AD} = 0$$

⑩ $P_{AD} = -(200)\dfrac{\sqrt{s^2+h^2}}{h}$

⑪ $P_{AD} = -\dfrac{(200)}{h}\sqrt{h^2\left(\dfrac{s^2}{h^2}+1\right)} := \boxed{-(200)\sqrt{\left(\frac{s}{h}\right)^2+1} = P_{AD}}$

⑫ using ⑩ in ⑧:

$$P_{AC} + \left(\frac{s}{\sqrt{s^2+h^2}}\right)\left[-(200)\frac{\sqrt{s^2+h^2}}{h}\right] = 0$$

⑬ $\boxed{P_{AC} = (200)\frac{s}{h}}$

⑭ use an Excel spreadsheet to plot P_{AC} + P_{AD} as functs. of $\frac{s}{h}$ from 0.5 to 2.0

Problem 6.23: (con't)

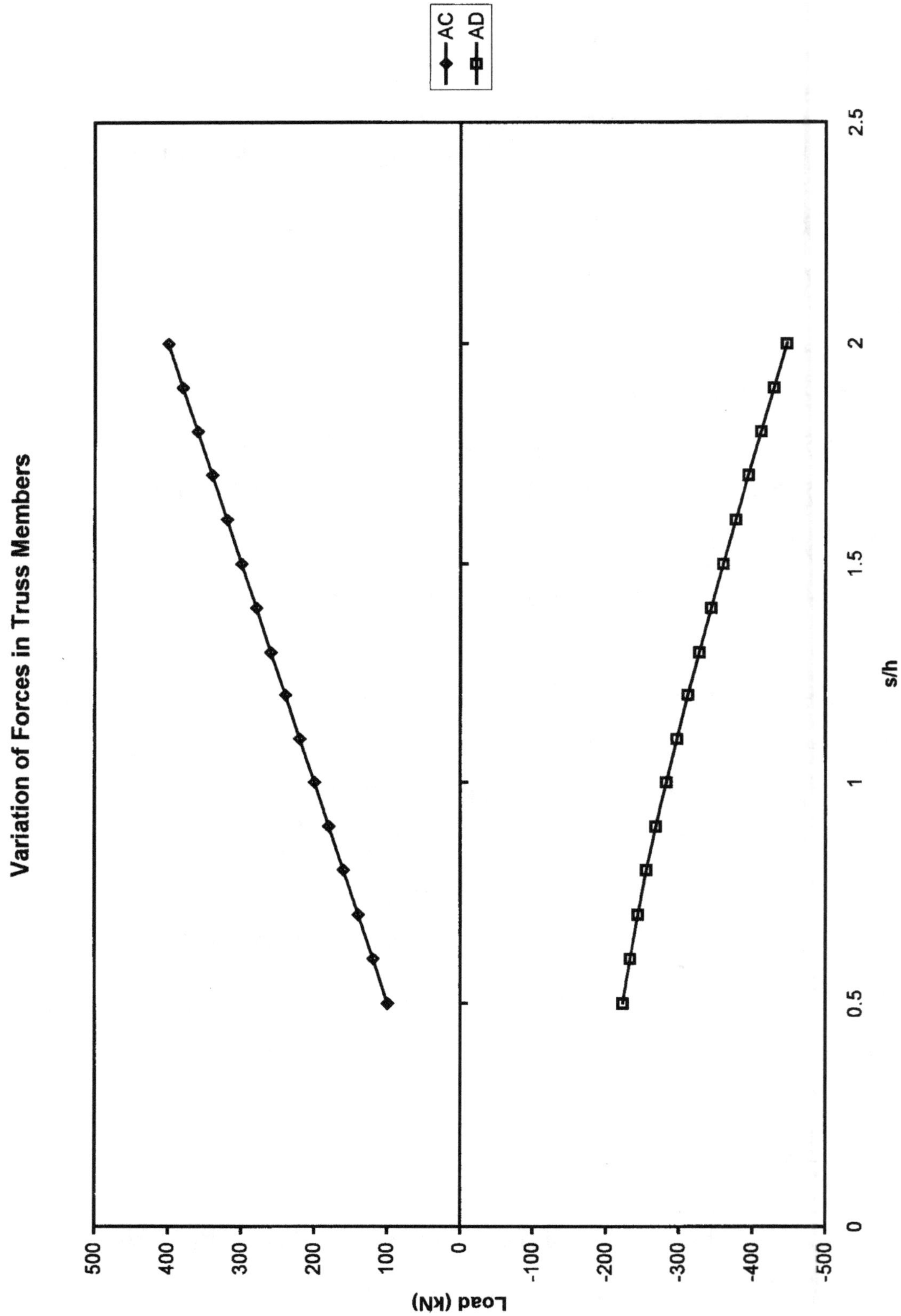

Problem 6.25:

FBD:

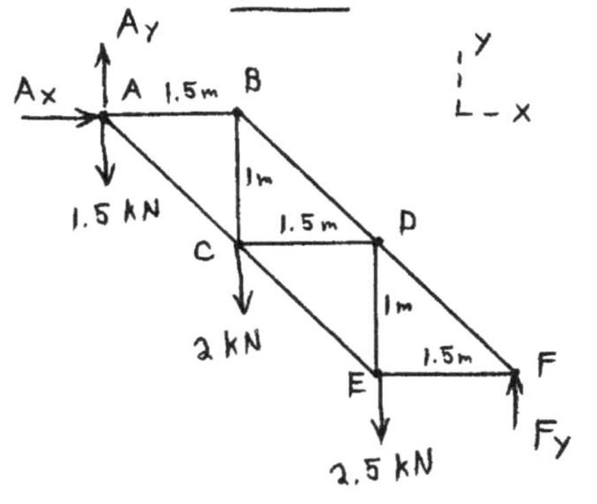

find: $\boxed{P_{AC},\ P_{CD},\ P_{CE},\ P_{DF} = ?}$

* consider entire truss:

① $\overset{+}{\rightarrow}\sum F_x = 0: \quad A_x = 0$

② $\circlearrowleft\!+\sum M_A = 0:$

$-(2)(1.5) - (2.5)(3) + F_y(4.5) = 0$

③ $F_y = 2.333 \text{ kN}$

④ $\uparrow\!+\sum F_y = 0: \quad A_y + (2.333) - 1.5 - 2 - 2.5 = 0$

⑤ $A_y = 3.667 \text{ kN}$

⑥ cut truss thru members of interest + keep rt. side:

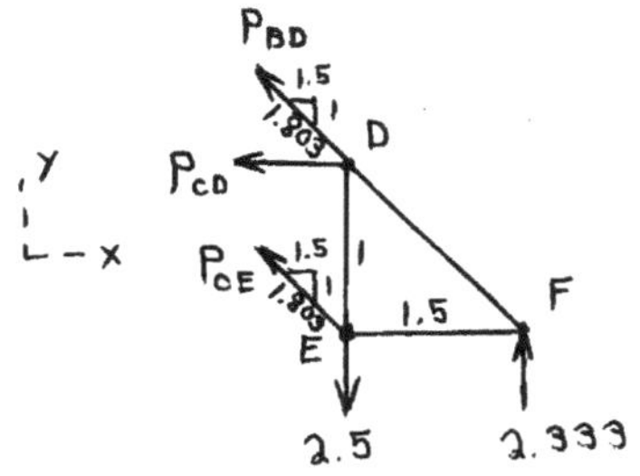

⑦ $\circlearrowleft\!+\sum M_D = 0:$

$-\left(\left(\frac{1.5}{1.803}\right)P_{CE}\right)(1) + (2.333)(1.5) = 0$

⑧ $\boxed{P_{CE} = 4.206 \text{ kN (T)}}$

⑨ $\uparrow\!+\sum F_y = 0:$

$\left(\frac{1}{1.803}\right)P_{BD} + \left(\frac{1}{1.803}\right)P_{CE} - 2.5 + 2.333 = 0$

⑩ $P_{BD} = (1.803)(2.5 - 2.333) - (4.206)$

⑪ $P_{BD} = -3.905 \text{ kN} = 3.905 \text{ kN (c)}$

⑫ $\overset{+}{\rightarrow}\sum F_x = 0: \quad -\left(\frac{1.5}{1.803}\right)(P_{BD} + P_{CE}) - P_{CD} = 0$

⑬ $P_{CD} = -\left(\frac{1.5}{1.803}\right)(-3.905 + 4.206)$

⑭ $\boxed{P_{CD} = -0.2504 \text{ kN} = 0.2504 \text{ kN (c)}}$

* consider joint A:

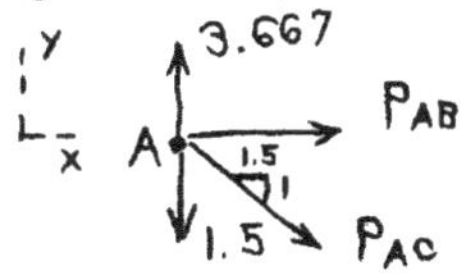

⑮ $\uparrow\!+\sum F_y = 0: \quad 3.667 - 1.5 - \left(\frac{1}{1.803}\right)P_{AC} = 0$

⑯ $\boxed{P_{AC} = 3.907 \text{ kN (T)}}$

* consider joint F:

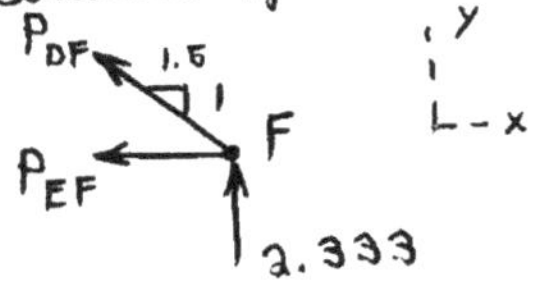

⑰ $\uparrow\!+\sum F_y = 0: \quad 2.333 + \left(\frac{1}{1.803}\right)P_{DF} = 0$

⑱ $\boxed{P_{DF} = -4.206 \text{ kN} = 4.206 \text{ kN (c)}}$

Problem 6.26:

find: P_{CD}, P_{CJ}, P_{KJ} = ?

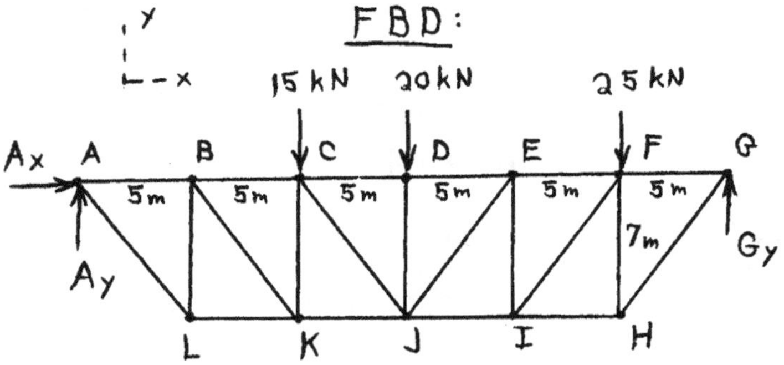

* consider entire truss:

① $\overset{\rightarrow}{+}\Sigma F_x = 0$: $A_x = 0$

② $\circlearrowleft + \Sigma M_A = 0$: $-(15)(10) - (20)(15) - (25)(25) + G_y(30) = 0$

③ $G_y = 35.83$ kN

④ $\uparrow + \Sigma F_y = 0$: $A_y + (35.83) - 15 - 20 - 25 = 0$

⑤ $A_y = 24.17$ kN

⑥ cut truss thru members of interest + keep left side:

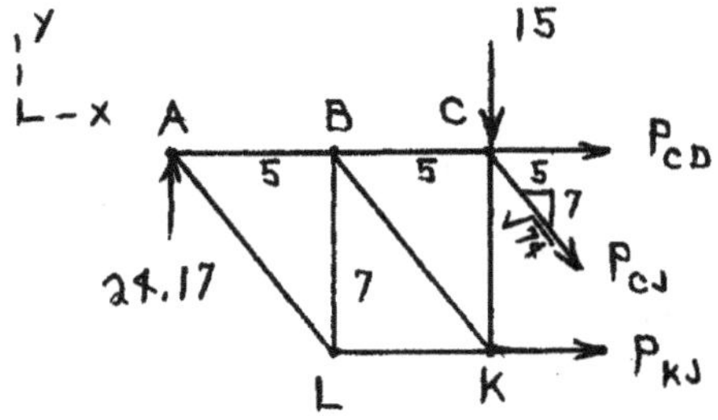

⑦ $\circlearrowleft + \Sigma M_C = 0$:

$-(24.17)(10) + P_{KJ}(7) = 0$

⑧ $P_{KJ} = 34.53$ kN (T)

⑨ $\uparrow + \Sigma F_y = 0$:

$24.17 - 15 - \frac{7}{\sqrt{74}} P_{CJ} = 0$

⑩ $P_{CJ} = 11.27$ kN (T)

⑪ $\overset{\rightarrow}{+}\Sigma F_x = 0$: $P_{CD} + \frac{5}{\sqrt{74}} P_{CJ} + P_{KJ} = 0$

⑫ $P_{CD} = -\frac{5}{\sqrt{74}}(11.27) - (34.53)$

⑬ $P_{CD} = -41.08$ kN $= 41.08$ kN (C)

Problem 6.30:

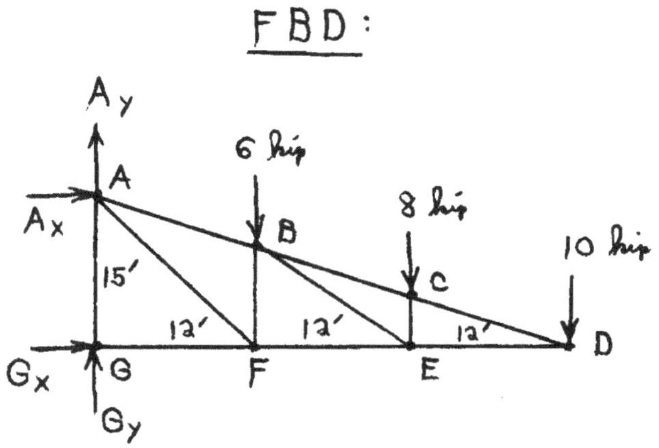

$A_{BC} = 2.5 \text{ in}^2$

1020 HR steel

find: σ_{BC}, $SF_{BC} = ?$.

* consider entire truss:

① 4 unknowns, 3 equil. eqns. ⇒ statically indeterminate problem (i.e. cannot solve for all 4 reactions)

② cut truss thru BC + keep rt. side:

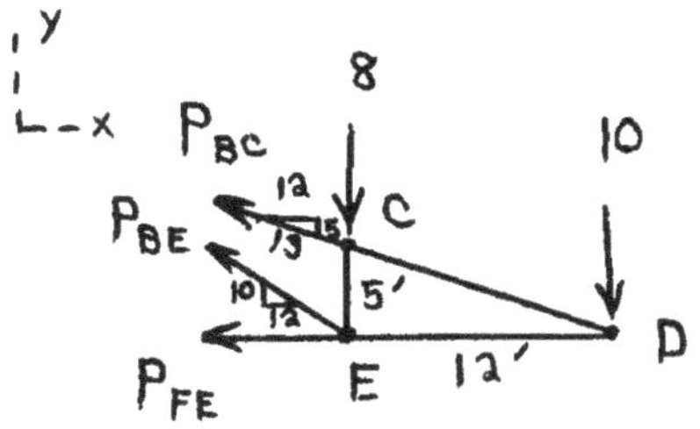

③ $\circlearrowleft + \sum M_E = 0$:

$$\left(\tfrac{12}{13} P_{BC}\right)(5) - (10)(12) = 0$$

④ $P_{BC} = 26$ kip (T)

⑤ from Appendix B-2: $S_y = 42$ ksi

⑥ $\sigma_{BC} = \dfrac{P_{BC}}{A_{BC}} = \dfrac{26}{2.5} \Rightarrow$ $\boxed{\sigma_{BC} = 10.4 \text{ ksi (T)}}$.

⑦ $SF_{BC} = \dfrac{S_y}{\sigma_{BC}} = \dfrac{42}{10.4} \Rightarrow$ $\boxed{SF_{BC} = 4.038}$.

Problem 6.35:

FBD:

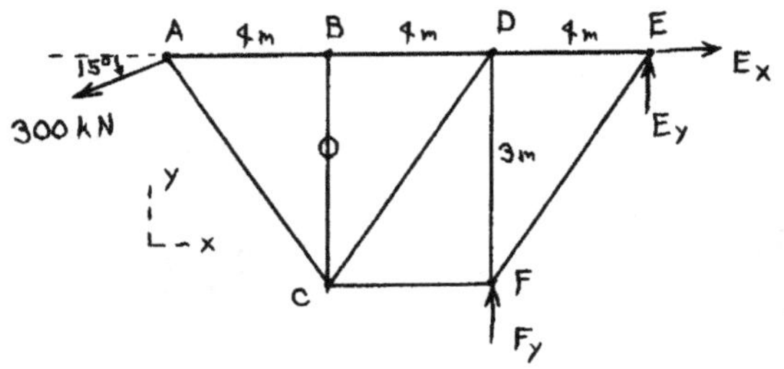

1020 HR steel $SF_y = 3.2$

find: A_{AB}, A_{BD}, A_{DE}, A_{EF}, A_{CF} = ?

① by inspection, BC is a ZFM

* consider entire truss:

② $\overset{+}{\rightarrow}\sum F_x = 0: \quad E_x - (300)\cos(15°) = 0 \Rightarrow E_x = 289.8 \text{ kN}$

③ $\circlearrowleft + \sum M_E = 0: \quad ((300)\sin(15°))(12) - F_y(4) = 0 \Rightarrow F_y = 232.9 \text{ kN}$

④ $\uparrow + \sum F_y = 0: \quad -(300)\sin(15°) + (232.9) + E_y = 0 \Rightarrow E_y = -155.3 \text{ kN}$

* consider joint A:

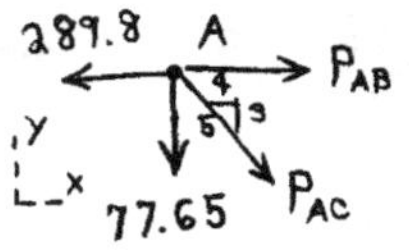

⑤ $\uparrow + \sum F_y = 0: \quad -77.65 - \frac{3}{5} P_{AC} = 0$

⑥ $P_{AC} = -129.4 \text{ kN} = 129.4 \text{ kN (C)}$

⑦ $\overset{+}{\rightarrow}\sum F_x = 0: \quad -289.8 + P_{AB} + \frac{4}{5} P_{AC} = 0$

⑧ $P_{AB} = 289.8 - \frac{4}{5}(-129.4) = 393.3 \text{ kN (T)}$

⑨ since BC is a ZFM, $P_{BD} = P_{AB} = 393.3 \text{ kN (T)}$

* consider joint E:

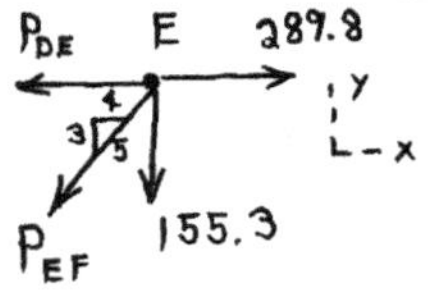

⑩ $\uparrow + \sum F_y = 0: \quad -\frac{3}{5} P_{EF} - 155.3 = 0$

⑪ $P_{EF} = -258.8 \text{ kN} = 258.8 \text{ kN (C)}$

⑫ $\overset{+}{\rightarrow}\sum F_x = 0: \quad 289.8 - P_{DE} - \frac{4}{5} P_{EF} = 0$

⑬ $P_{DE} = 289.8 - \frac{4}{5}(-258.8) = 496.8 \text{ kN (T)}$

* consider joint F:

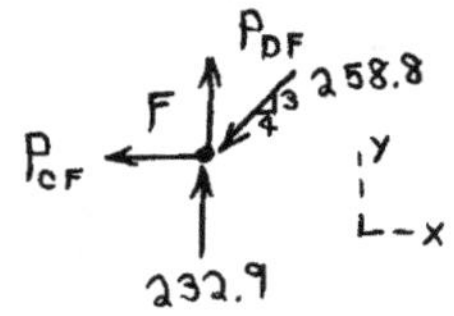

⑭ $\overset{+}{\rightarrow}\sum F_x = 0: \quad -P_{CF} - \frac{4}{5}(258.8) = 0$

⑮ $P_{CF} = -207.0 \text{ kN} = 207.0 \text{ kN (C)}$

⑯ from Appendix B-2: $S_y = 290 \text{ MPa}$

⑰ $\sigma = \frac{P}{A} = \frac{S_y}{SF_y} = \frac{290}{3.2} = 90.63 \text{ MPa}$

⑱ $A = \frac{P}{90.63 \text{ MPa}}$

⑲ subing forces P into ⑱ + solving for A:

$A_{AB} = A_{BD} = 4340 \text{ mm}^2$	$A_{EF} = 2856 \text{ mm}^2$
$A_{DE} = 5482 \text{ mm}^2$	$A_{CF} = 2284 \text{ mm}^2$

CHAPTER 7 PROBLEMS

7.1 The Patcenter in Princeton, NJ is a large open structure with a cable-stayed roof. In the figure presented below, one of the nine tubular steel masts that are uniformly spaced at 9 m intervals to support the roof structure is shown. In this design, the roof hangs from cables because the vertical side members are used only to prevent wind uplift. The tubular mast, 15 m high, is supported with a 9 m wide by 6 m high rectangular steel frame. Determine the force in the primary rod stay and the tubular steel masts if the uniformly distributed load on the roof is 70 kN/m. Assume the roof is pinned to the steel mast at point C. Hint: Refer to the Example 7.3 in the text for additional dimensions.

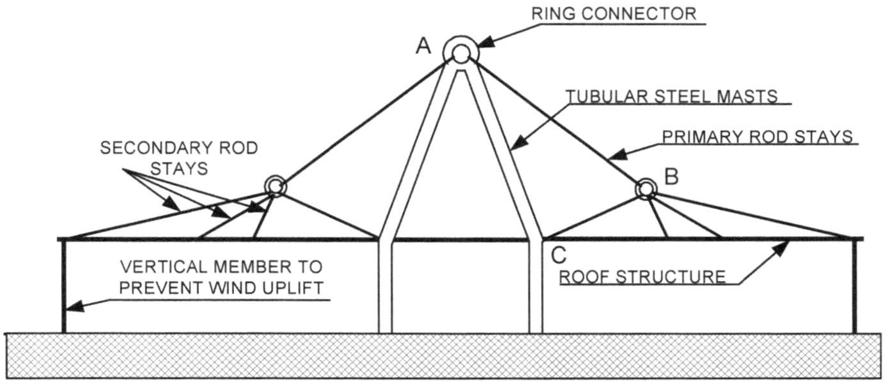

7.2 A new building similar to the Patcenter is under consideration by an architectural firm. They propose increasing the height of the mast from 15 m to 24 m to make the appearance of the structure more dramatic. If all of the other parameters are the same as given in Problem 7.1, prepare an analysis indicating the effect of this change.

7.3 A new building similar to the Patcenter is under consideration by an architectural firm. They propose a structure with the dimensions presented in the figure to the right. Determine the force in the primary rod stay AB and the tubular steel masts AC. Also find the reaction forces at point C.

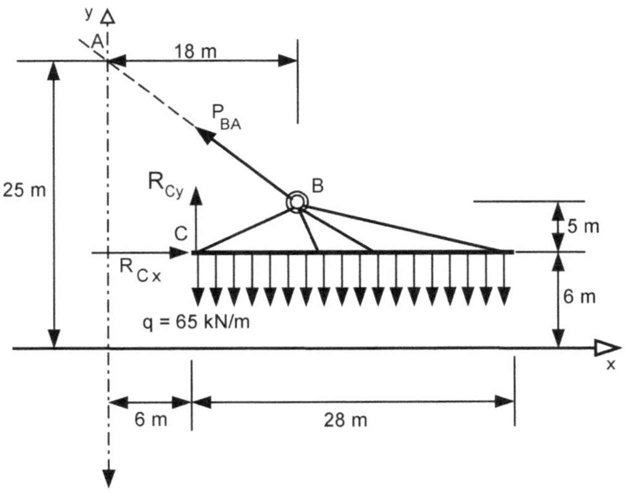

7.4 A space truss, shown in the figure to the right, is supported at point A with a ball and socket joint and with cables anchored at points B and D. A force F is applied to joint E. If F is represented in vector format as shown in the figure, determine the internal forces in the three members that intersect at joint E. Also specify the diameter of the rods, if the truss is fabricated from a 1018 A steel alloy. A safety factor of SF = 3.0 based on yield strength is specified by the building code. The truss dimensions are given in ft.

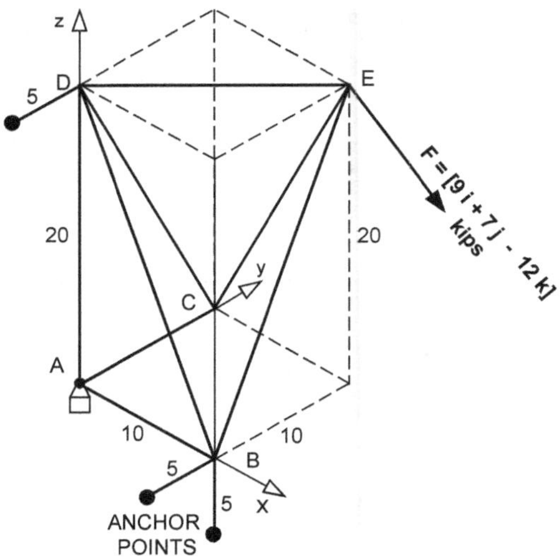

7.5 Determine the internal forces in members AB, DB and EB of the truss described in Problem 7.4.

7.6 Determine the internal forces in members AC, DC and EC of the truss described in Problem 7.4.

7.7 For the communications tower illustrated in the figure to the right, a wind force of 4,500 lb from the West is applied at a position z = 300 ft. The tower has a height H = 400 ft and a dead weight W = 50 kips. The anchor points on the ground plane are defined by: x_1 = 100ft, x_2 = 115 ft, y_1 = 130 ft and y_2 = 145 ft. The cables have a breaking strength of 26,600 lb and a ball and socket joint is used to support the tower at point O. Determine the loads in the four cables and the reaction forces at the base of the tower. Also determine the minimum safety factor.

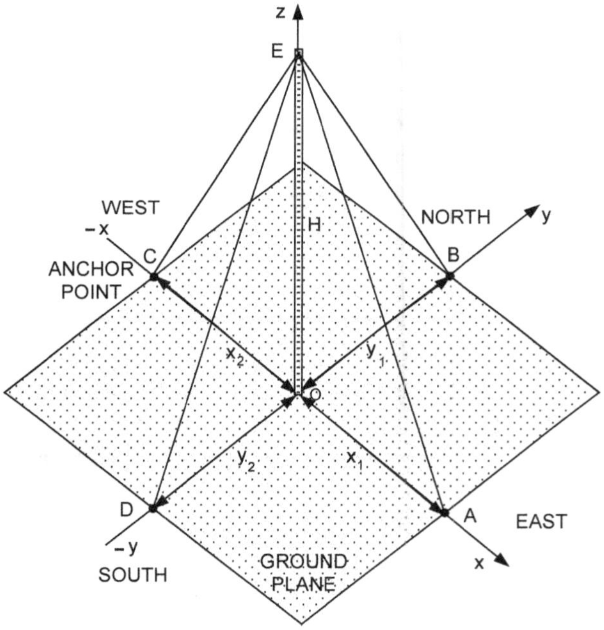

7.8 The anchor points on the ground plane for the communications tower shown in the figure to the right are changed to: x_1 = 60 m, x_2 = 70 m, y_1 = 80 m, y_2 = 90 m. The tower has a height of 120 m, a dead weight of 200kN and is subjected to a wind force of 12.6 kN from the East, that is applied at a position z = 80 m. The cables have a breaking strength of 130 kN, and the tower is supported by a ball and socket joint at point O. Determine the loads on the four cables and the reaction forces at the ball and socket joint. Also find the minimum safety factor.

7.9 If the steel used in fabricating the tower described in Problem 7.7 has a yield strength of 50 ksi, determine the cross sectional area required at the base of the tower. The safety factor for the tower is specified as 4.0. Does the wind loading condition affect the result?

7.10 If the steel used in fabricating the tower in Problem 7.8 has a yield strength of 360 MPa, determine the cross sectional area required at the base of the tower. The safety factor for the tower is specified as 5.0. Does the wind loading condition affect the result?

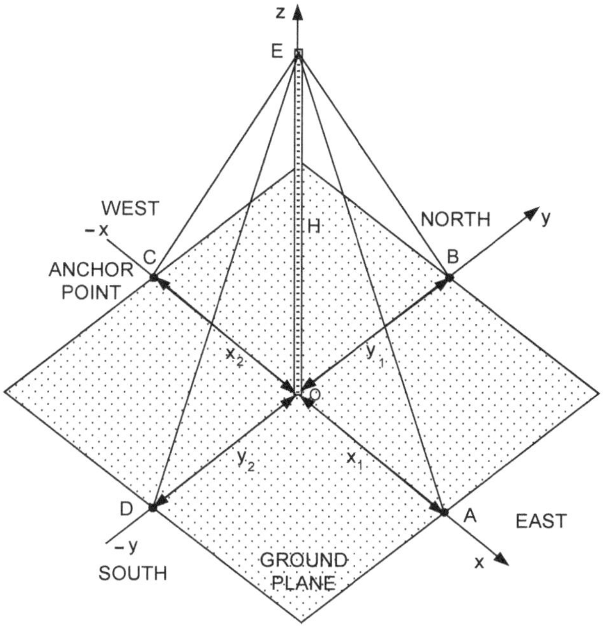

7.11 The safety factors for the communications tower described in Problems 7.7 and 7.8 are very large. Develop arguments supporting the use of relatively large safety factors in the design of communications towers. Also develop arguments for redesign with smaller diameter cable, smaller footprint and smaller cross sectional area at the base of the tower structure.

7.12 A severe ice storm strikes the communications tower coating all of the members with a thick layer of ice. The dead weight of the tower is increased from 200 kN to 320 kN. Determine the decrease in the safety factor for the conditions of Problem 7.8.

7.13 For the derrick shown in the figure below, determine the margin of safety for the three cables if the force F acting on the boom is 4.0 ton. The cable diameter is 0.5 in. for all three cables. The anchor points P and Q for two of the cables lie in the x-y plane with coordinates P= (12, –9, 0) ft and Q = (–12, – 9, 0) ft. The derrick pole is 15 ft high and the boom is 20 ft long. The derrick pole and the boom are separate components. The boom is attached to the pole at point R with a sleeve type bearing that enables the boom to rotate about the pole. However, the boom is constrained from sliding up or down the pole. The derrick pole is supported by a ball and socket joint at point O. Details of the bearing arrangement are also shown in the figure below. The cables have a specified breaking strength of 180 ksi.

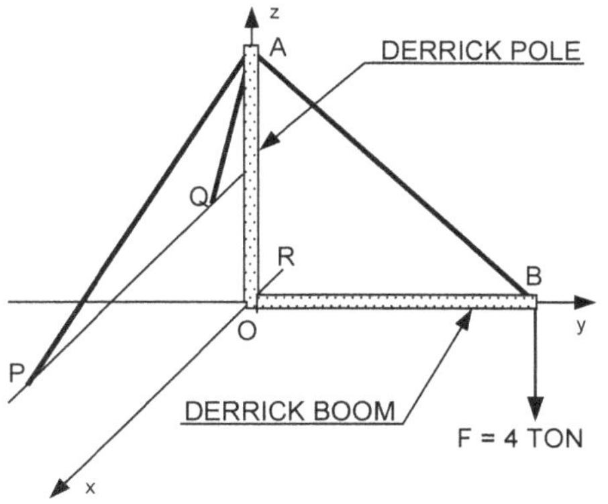

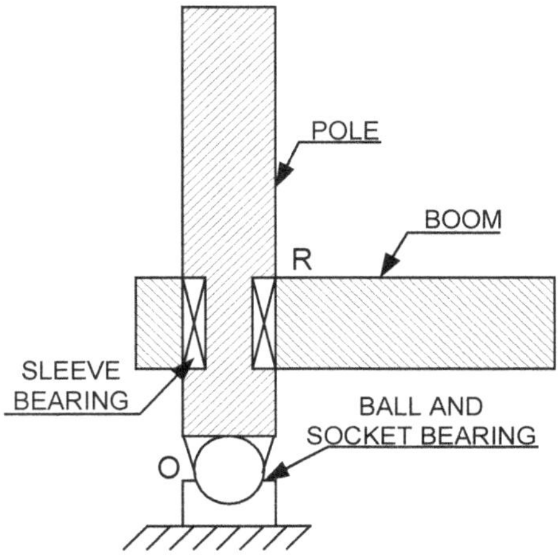

7.14 A tetrahedral space truss, shown in the figure to the right, supports a massive scoreboard and several sets of remotely controlled spotlights in an amphitheater. The base ABC of the space truss lies in the x-y plane (horizontal). The base triangle is connected to anchors in the roof beams by long cables. Determine the size of the solid round rods that are required to fabricate members AB, AC and AD of the space truss. The safety factor based on yield strength is specified as 6.0. 1020 HR steel is employed for all of the members and the weight of the scoreboard and spotlights is 20.0 kN.

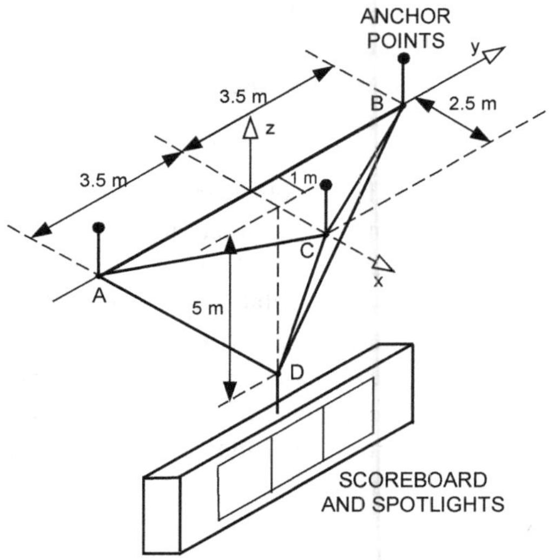

7.15 For the long horizontal boom of a construction crane, illustrated in the figure shown below, determine the internal forces in the members AB, BF, BK, FG, KL and FL. The load applied to the boom is F = 10 kip. Hint: See Example 7.8 in the text for additional details.

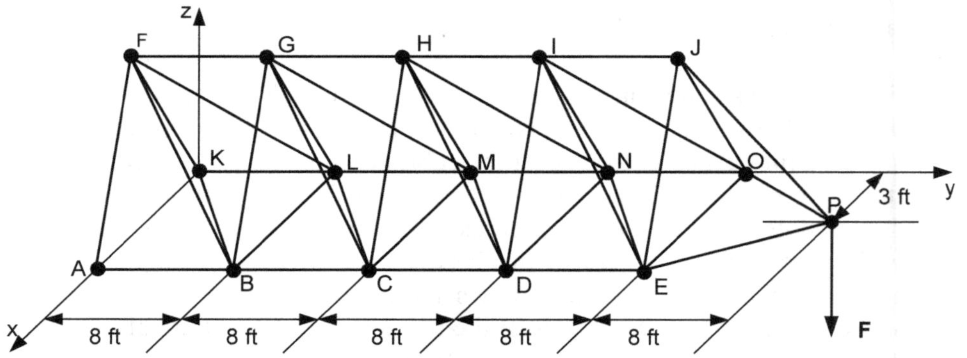

7.16 Repeat Problem 7.15 if the loading on the long horizontal boom is increased from 10 kip to 14 kip.

7.17 A hanging light assembly is positioned near the corner of a gymnasium as shown in the figure to the right. Determine the gage of the stainless steel wire required for the support of a light assembly weighing 1,100 N if a safety factor of 5.0 is specified. The wire is fabricated from 302A stainless steel. Points B and C are anchors that lie in the x-y plane and point D is an anchor that lies along the z-axis. Point A is not anchored; however it lies in the x-y plane.

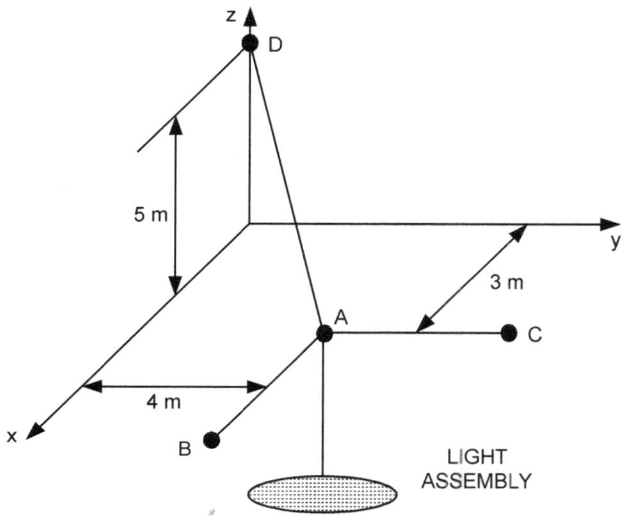

7.18 A hot air balloon, shown in the figure to the right, is moored to the ground with three cables that are anchored at points A, B and C. The coordinates (x, y, z) of the anchor points on the ground and on the basket are given in the figure. Determine the force exerted by each of the cables if the upward lift of the balloon is 950 lb. Assume that wind forces are negligible.

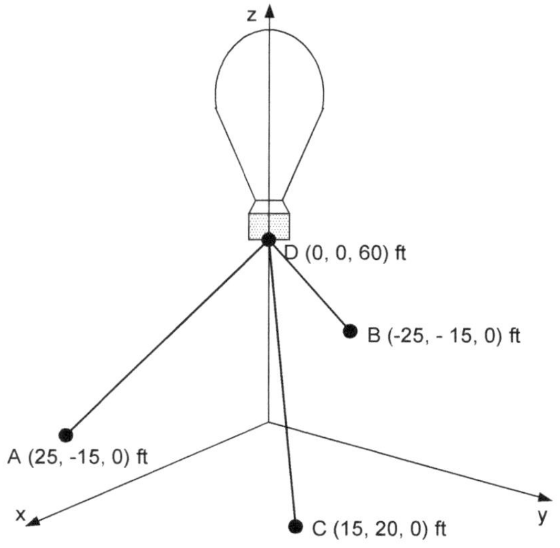

7.19 A circus cage, displayed in a large high ceiling auditorium, is supported above ground level by the three wires illustrated in the figure to the left. Determine the gage of the stainless steel wire required to support the cage weighing 2,500 lb if a safety factor of 3.5 is specified. The wire is fabricated from 4340 HR steel. The geometric parameters defining the assembly are listed in the table below. Points A, B and D locate the anchors for the cables.

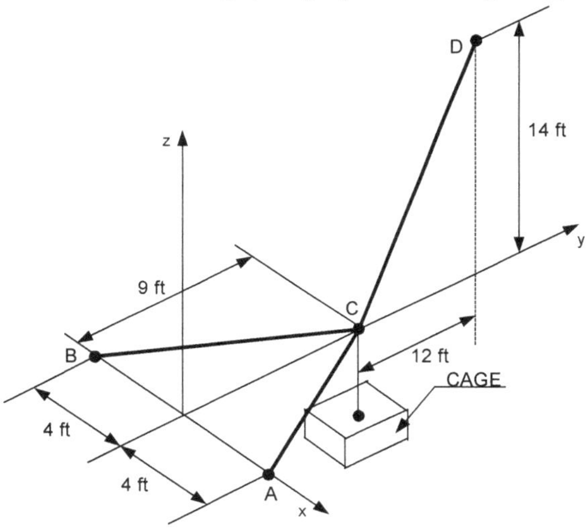

7.20 A local firm has constructed a small crane consisting of a boom supported by two steel wires BC and DE as shown in the figure to the right. The boom is fixed to the supporting wall with a ball and socket joint at point A. The wires are anchored into the wall at points C and E. Each wire has a diameter of 0.125 in. and an ultimate tensile strength of 150 ksi. Determine the maximum weight that can be supported by the boom, the support reactions at point A and the forces in wires BC and DE.

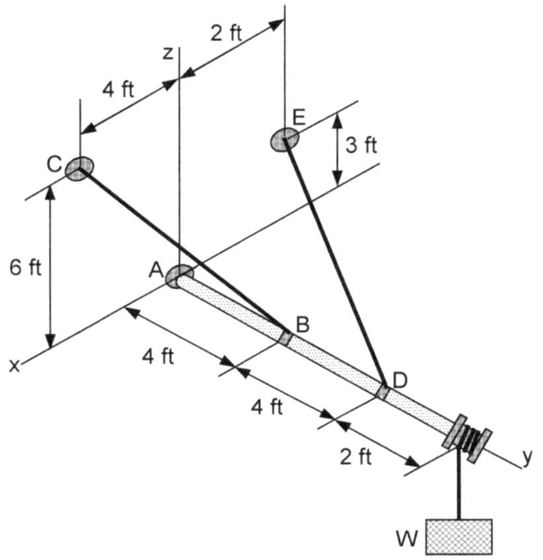

7.21 A hand operated lifting mechanism called a windless utilizes a crank to rotate a drum as shown in the figure to the right. The shaft of the mechanism is supported by a wall mounted ball and socket joint at point A and a smooth journal bearing at point B. The arm and handle of the crank in the position shown is in the y-z plane. For this position of the crank handle, determine the force F required to hold a weight of 500 kN in equilibrium. Also determine the reactions at the ball and socket joint and the journal bearing.

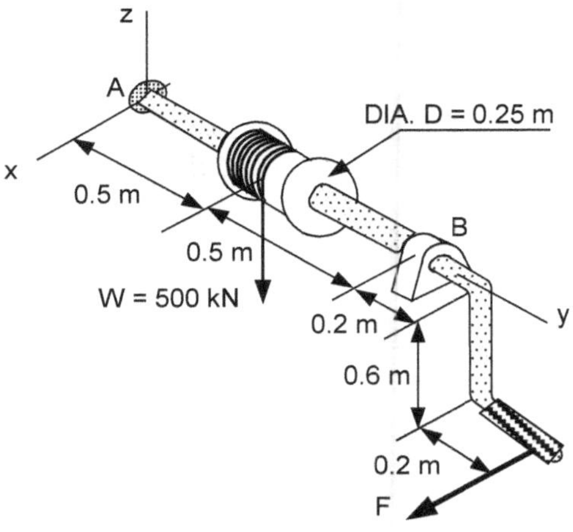

7.22 Repeat Problem 7.21 if the crank is rotated clockwise 90 degrees so that the arm and handle of the crank is in the x-y plane and the force F is acting in the positive z direction. For this position of the crank handle, determine the force F required to hold the weight of 500 kN in equilibrium. Also determine the reactions at the ball and socket joint and the journal bearing.

7.23 Repeat Problem 7.21 if the crank is rotated clockwise 180 degrees so that the handle of the crank is in the y-z plane and the force F is acting in the negative x direction. For this position of the crank handle determine, the force F required to hold the weight of 500 kN in equilibrium. Also determine the reactions at the ball and socket joint and the journal bearing.

7.24 A cross-like base and a short pole, as shown in the figure to the right, support a tabletop with a uniform thickness. Determine the largest weight W that can be applied to the table if it is placed at points A, B or C. The tabletop weighs 120 N/m^2. The dimensions of the tabletop and the base are given in the figure.

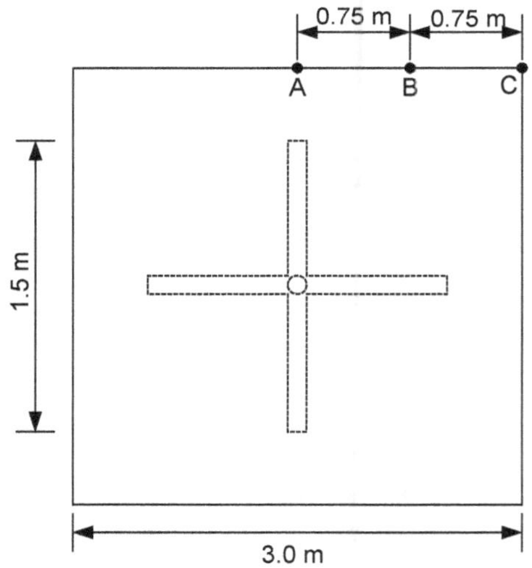

Problem 7.4:

FBD:

* 1018 A steel

$SF_y = 3.0$

find: P_{EB}, P_{EC}, P_{ED}, d_{EB}, d_{EC}, d_{ED} = ?

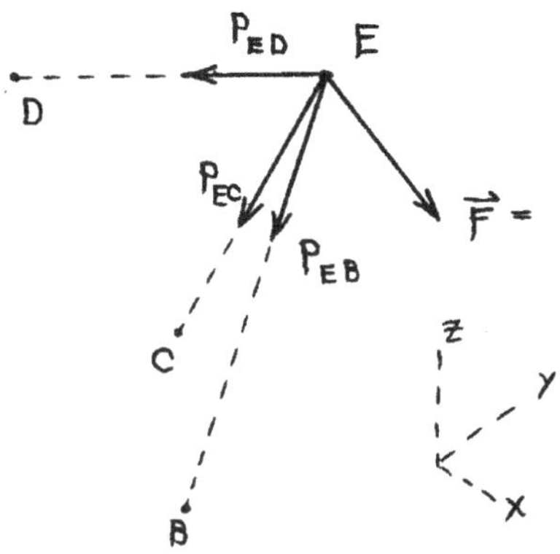

$\vec{F} = (9\vec{i} + 7\vec{j} - 12\vec{k})$ kips

① get coords. of pts. B, C, D, + E from figure in text:

$B = (10, 0, 0)$ ft $\quad D = (0, 0, 20)$ ft

$C = (0, 10, 0)$ ft $\quad E = (10, 10, 20)$ ft

② $$\vec{P}_{EB} = P_{EB}\frac{\vec{r}_{EB}}{r_{EB}} = P_{EB}\left(\frac{-10\vec{j} - 20\vec{k}}{\sqrt{500}}\right) = P_{EB}\left(-\frac{1}{\sqrt{5}}\vec{j} - \frac{2}{\sqrt{5}}\vec{k}\right)$$

③ $$\vec{P}_{EC} = P_{EC}\frac{\vec{r}_{EC}}{r_{EC}} = P_{EC}\left(\frac{-10\vec{i} - 20\vec{k}}{\sqrt{500}}\right) = P_{EC}\left(-\frac{1}{\sqrt{5}}\vec{i} - \frac{2}{\sqrt{5}}\vec{k}\right)$$

④ $$\vec{P}_{ED} = P_{ED}\frac{\vec{r}_{ED}}{r_{ED}} = P_{ED}\left(\frac{-10\vec{i} - 10\vec{j}}{\sqrt{200}}\right) = P_{ED}\left(-\frac{1}{\sqrt{2}}\vec{i} - \frac{1}{\sqrt{2}}\vec{j}\right)$$

⑤ $+\searrow \sum F_x = 0: \quad 9 - \frac{1}{\sqrt{5}}P_{EC} - \frac{1}{\sqrt{2}}P_{ED} = 0$

⑥ $+\nearrow \sum F_y = 0: \quad 7 - \frac{1}{\sqrt{5}}P_{EB} - \frac{1}{\sqrt{2}}P_{ED} = 0$

⑦ $\uparrow + \sum F_z = 0: \quad -12 - \frac{2}{\sqrt{5}}P_{EB} - \frac{2}{\sqrt{5}}P_{EC} = 0$

⑧ solving ⑤ – ⑦ simultaneously:

$P_{EB} = -8.944$ kips; $\quad P_{EC} = -4.472$ kips; $\quad P_{ED} = 15.56$ kips

⑨ from Appendix B-2: $S_y = 32$ ksi

⑩ $$\sigma_{allow} = \frac{S_y}{SF_y} = \frac{32}{3} = 10.67 \text{ ksi}$$

⑪ $$A = \frac{P}{\sigma_{allow}} = \frac{\pi}{4}d^2 \Rightarrow d = \sqrt{\frac{4P}{\pi\,\sigma_{allow}}}$$

⑫ $$d_{EB} = \sqrt{\frac{4(8.944)}{\pi(10.67)}} \Rightarrow \quad d_{EB} = 1.033 \text{ in}$$

$$d_{EC} = \sqrt{\frac{4(4.472)}{\pi(10.67)}} \Rightarrow \quad d_{EC} = 0.7305 \text{ in}$$

$$d_{ED} = \sqrt{\frac{4(15.56)}{\pi(10.67)}} \Rightarrow \quad d_{ED} = 1.363 \text{ in}$$

Problem 7.13:

FBD's:

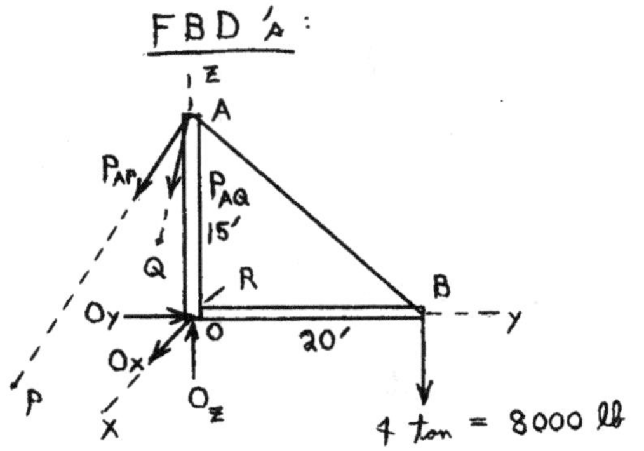

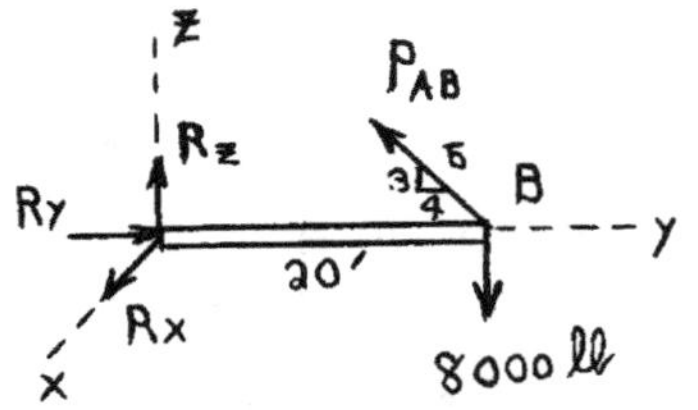

$P = (12, -9, 0)$ ft

$Q = (-12, -9, 0)$ ft

* ball + socket @ O
sleeve bearing @ R

$S_b = \sigma_{max} = 180$ ksi

$d = 0.5''$

find : $\boxed{MOS_{AB,\ AP,\ AQ} = ?}$.

① $\vec{P}_{AP} = P_{AP}\dfrac{\vec{r}_{AP}}{r_{AP}} = P_{AP}\dfrac{(12\vec{i} - 9\vec{j} - 15\vec{k})}{\sqrt{450}} = P_{AP}\left(\dfrac{4}{5\sqrt{2}}\vec{i} - \dfrac{3}{5\sqrt{2}}\vec{j} - \dfrac{1}{\sqrt{2}}\vec{k}\right)$

② $\vec{P}_{AQ} = P_{AQ}\dfrac{\vec{r}_{AQ}}{r_{AQ}} = P_{AQ}\dfrac{(-12\vec{i} - 9\vec{j} - 15\vec{k})}{\sqrt{450}} = P_{AQ}\left(-\dfrac{4}{5\sqrt{2}}\vec{i} - \dfrac{3}{5\sqrt{2}}\vec{j} - \dfrac{1}{\sqrt{2}}\vec{k}\right)$

③ $\vec{F}_B = -(8000\text{ lb})\vec{k}$

④ $\vec{O} = O_x\vec{i} + O_y\vec{j} + O_z\vec{k}$; $\vec{R} = R_x\vec{i} + R_y\vec{j} + R_z\vec{k}$

⑤ $\vec{P}_{AB} = P_{AB}\dfrac{\vec{r}_{BA}}{r_{BA}} = P_{AB}\dfrac{(-20\vec{j} + 15\vec{k})}{25} = P_{AB}\left(-\dfrac{4}{5}\vec{j} + \dfrac{3}{5}\vec{k}\right)$

* consider entire structure:

⑥ $\swarrow{+}\ \Sigma F_x = 0: \quad O_x + \dfrac{4}{5\sqrt{2}}P_{AP} - \dfrac{4}{5\sqrt{2}}P_{AQ} = 0$

⑦ $\overset{+}{\rightarrow}\ \Sigma F_y = 0: \quad O_y - \dfrac{3}{5\sqrt{2}}P_{AP} - \dfrac{3}{5\sqrt{2}}P_{AQ} = 0$

⑧ $+\uparrow\ \Sigma F_z = 0: \quad O_z - \dfrac{1}{\sqrt{2}}P_{AP} - \dfrac{1}{\sqrt{2}}P_{AQ} - 8000 = 0$

* scalar method for moments:

⑨ $\Sigma M_x = 0: \quad \left(\dfrac{3}{5\sqrt{2}}P_{AP}\right)(15) + \left(\dfrac{3}{5\sqrt{2}}P_{AQ}\right)(15) - (8000)(20) = 0$

⑩ $\Sigma M_y = 0: \quad \left(\dfrac{4}{5\sqrt{2}}P_{AP}\right)(15) - \left(\dfrac{4}{5\sqrt{2}}P_{AQ}\right)(15) = 0$

⑪ $\Sigma M_z = 0: \quad 0 = 0$

⑫ from ⑩ : $P_{AP} = P_{AQ}$

⑬ using ⑫ in ⑨ : $2\left(\dfrac{9}{\sqrt{2}}\right)P_{AP} = 160{,}000$

⑭ $P_{AP} = P_{AQ} = 12{,}571$ lb

Problem 7.13: (con't)

(15) using (14) in (6) − (8) + solving:

$O_x = 0$; $O_y = 10{,}667$ lb ; $O_z = 25{,}778$ lb

* vector method for moments:

(9) $\sum \vec{M}_o = 0 = \vec{r}_{OA} \times \vec{P}_{AP} + \vec{r}_{OA} \times \vec{P}_{AQ} + \vec{r}_{OB} \times \vec{F}_B$

(10) $$0 = \begin{vmatrix} \vec{i} & \vec{j} & \vec{k} \\ 0 & 0 & 15 \\ \frac{4}{5} & -\frac{3}{5} & -1 \end{vmatrix} \frac{P_{AP}}{\sqrt{2}} + \begin{vmatrix} \vec{i} & \vec{j} & \vec{k} \\ 0 & 0 & 15 \\ -\frac{4}{5} & -\frac{3}{5} & -1 \end{vmatrix} \frac{P_{AQ}}{\sqrt{2}} + \begin{vmatrix} \vec{i} & \vec{j} & \vec{k} \\ 0 & 20 & 0 \\ 0 & 0 & -8000 \end{vmatrix}$$

(11) $$0 = \left[\left(\frac{12}{\sqrt{2}}\right) P_{AP}\,\vec{j} + \left(\frac{9}{\sqrt{2}}\right) P_{AP}\,\vec{i}\right] + \left[-\left(\frac{12}{\sqrt{2}}\right) P_{AQ}\,\vec{j} + \left(\frac{9}{\sqrt{2}}\right) P_{AQ}\,\vec{i}\right] + \left[-160{,}000\,\vec{i}\right]$$

(12) collect like terms + set each coefficient to zero:

$\vec{i}: \quad \frac{9}{\sqrt{2}} P_{AP} + \frac{9}{\sqrt{2}} P_{AQ} - 160{,}000 = 0$

$\vec{j}: \quad \frac{12}{\sqrt{2}} P_{AP} - \frac{12}{\sqrt{2}} P_{AQ} = 0$

∴ solve for P_{AP} + P_{AQ} as previously

* consider boom RB:

(16) $\stackrel{+}{\circlearrowleft} \sum M_x = 0: \quad \left(\frac{3}{5} P_{AB}\right)(20) - (8000)(20) = 0$

(17) $P_{AB} = 13{,}333$ lb

(18) $A = \frac{\pi}{4} d^2 = \frac{\pi}{4}(0.5)^2 = 0.19635 \text{ in}^2$

(19) $\sigma = \frac{P}{A} = \frac{\sigma_{max}}{SF} \Rightarrow SF = \frac{\sigma_{max} A}{P}$

(20) $MOS = (SF - 1)(100\%) = \left(\frac{\sigma_{max} A}{P} - 1\right)(100\%)$

(21) $MOS_{AB} = \left(\frac{(180{,}000)(0.19635)}{13{,}333} - 1\right)(100\%) \Rightarrow \boxed{MOS_{AB} = 165.1\%}$.

(22) $MOS_{AP} = MOS_{AQ} = \left(\frac{(180{,}000)(0.19635)}{12{,}571} - 1\right)(100\%)$

(23) $\boxed{MOS_{AP} = MOS_{AQ} = 181.1\%}$.

Problem 7.14:

FBD:

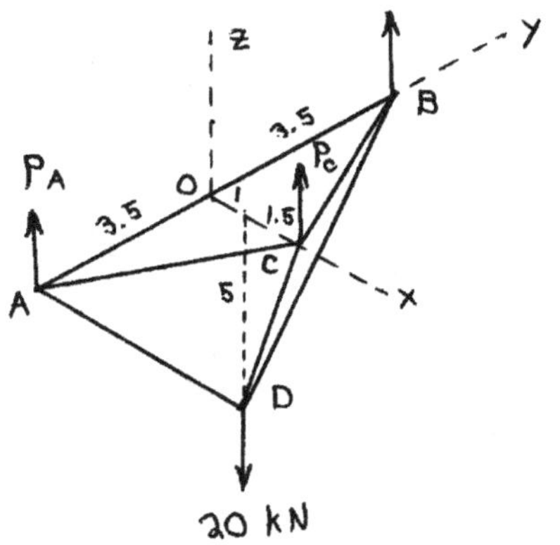

$A = (0, -3.5, 0)\ m$

$B = (0, 3.5, 0)\ m$

$C = (2.5, 0, 0)\ m$

$D = (1, 0, -5)\ m$

* 1020 HR steel $\qquad SF_y = 6.0$

find: $\boxed{d_{AB,\ AC,\ AD} = ?}$.

* consider entire truss:

① $\uparrow + \sum F_z = 0: \quad P_A + P_B + P_c - 20 = 0$

* scalar method for moments:

② $\sum M_x = 0: \quad -P_A(3.5) + P_B(3.5) = 0 \quad \Rightarrow \quad P_A = P_B$

③ $\sum M_y = 0: \quad (20)(1) - P_c(2.5) = 0 \quad \Rightarrow \quad P_c = 8\ kN$

④ using ② + ③ in ①: $\quad P_A = P_B = \frac{1}{2}(20 - 8) = 6\ kN$

* vector method for moments:

② $\sum \vec{M}_o = \vec{r}_{OA} \times \vec{P}_A + \vec{r}_{OB} \times \vec{P}_B + \vec{r}_{OC} \times \vec{P}_c + \vec{r}_{OD} \times \vec{F}_D = 0$

③ $0 = \begin{vmatrix} \vec{i} & \vec{j} & \vec{k} \\ 0 & -3.5 & 0 \\ 0 & 0 & P_A \end{vmatrix} + \begin{vmatrix} \vec{i} & \vec{j} & \vec{k} \\ 0 & 3.5 & 0 \\ 0 & 0 & P_B \end{vmatrix} + \begin{vmatrix} \vec{i} & \vec{j} & \vec{k} \\ 2.5 & 0 & 0 \\ 0 & 0 & P_c \end{vmatrix} + \begin{vmatrix} \vec{i} & \vec{j} & \vec{k} \\ 1 & 0 & -5 \\ 0 & 0 & -20 \end{vmatrix}$

④ $0 = [(-3.5 P_A)\vec{i}] + [(3.5 P_B)\vec{i}] + [(-2.5 P_c)\vec{j}] + [(-20)\vec{j}]$

⑤ collect like terms + set each coef. to zero:

$\vec{i}: \quad -3.5 P_A + 3.5 P = 0$

$\vec{j}: \quad -2.5 P_c + 20 = 0$

$\therefore$ solve for P_c, P_A, + P_B as previously

* consider joint A:

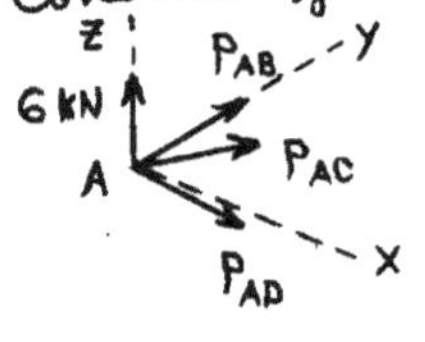

⑤ $\vec{P}_{AB} = P_{AB}\,\vec{j}$

⑥ $\vec{P}_{AC} = P_{AC}\dfrac{\vec{r}_{AC}}{r_{AC}} = P_{AC}\dfrac{(2.5\vec{i} + 3.5\vec{j})}{4.301}$

⑦ $\vec{P}_{AD} = P_{AD}\dfrac{\vec{r}_{AD}}{r_{AD}} = P_{AD}\dfrac{(1\vec{i} + 3.5\vec{j} - 5\vec{k})}{6.185}$

Problem 7.14: (con't)

⑧ $\nearrow\!+ \sum F_x = 0: \quad \frac{2.5}{4.301} P_{AC} + \frac{1}{6.185} P_{AD} = 0$

⑨ $\nearrow\!+ \sum F_y = 0: \quad P_{AB} + \frac{3.5}{4.301} P_{AC} + \frac{3.5}{6.185} P_{AD} = 0$

⑩ $\uparrow\!+ \sum F_z = 0: \quad -\frac{5}{6.185} P_{AD} + 6 = 0 \quad \Rightarrow \quad P_{AD} = 7.422 \text{ kN}$

⑪ using P_{AD} in ⑧: $\quad P_{AC} = -\frac{(4.301)}{(2.5)(6.185)}(7.422) = -2.064 \text{ kN}$

⑫ using P_{AD} + P_{AC} in ⑨:

$$P_{AB} = -\frac{3.5}{4.301}(-2.064) - \frac{3.5}{6.185}(7.422) = -2.520 \text{ kN}$$

⑬ from Appendix B-2: $\quad S_y = 290 \text{ MPa}$

⑭ $\sigma = \frac{P}{A} = \frac{S_y}{SF_y} \quad \Rightarrow \quad A = \frac{P(SF_y)}{S_y} = \frac{\pi}{4} d^2$

⑮ $d = \sqrt{\frac{4P(SF_y)}{\pi S_y}}$

⑯ $d_{AB} = \sqrt{\frac{4(2.520)(6)}{\pi(290)}} \quad \Rightarrow \quad d_{AB} = 8.148 \text{ mm}$

$d_{AC} = \sqrt{\frac{4(2.064)(6)}{\pi(290)}} \quad \Rightarrow \quad d_{AC} = 7.374 \text{ mm}$

$d_{AD} = \sqrt{\frac{4(7.422)(6)}{\pi(290)}} \quad \Rightarrow \quad d_{AD} = 13.98 \text{ mm}$

Problem 7.17:

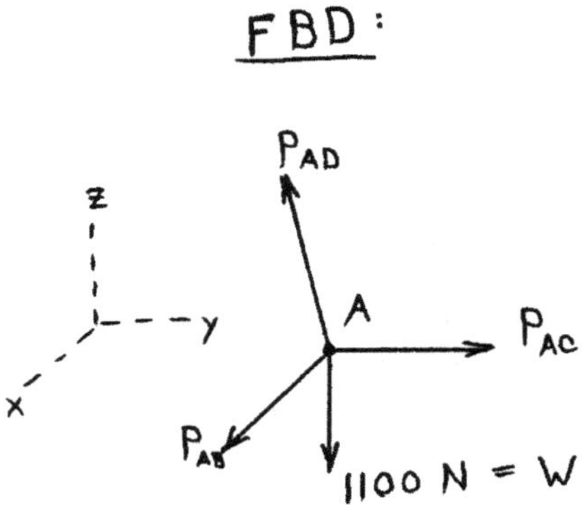

$A = (3, 4, 0)$ m

$D = (0, 0, 5)$ m

AB lies in +x direction

AC lies in +y direction

* 302A stainless steel

$SF_y = 5.0$

find: gage no. = ?

① $\vec{P}_{AB} = P_{AB}\,\vec{i}$

② $\vec{P}_{AC} = P_{AC}\,\vec{j}$

③ $\vec{P}_{AD} = P_{AD}\dfrac{\vec{r}_{AD}}{r_{AD}} = P_{AD}\dfrac{(-3\vec{i} - 4\vec{j} + 5\vec{k})}{\sqrt{50}} = P_{AD}\dfrac{(-3\vec{i} - 4\vec{j} + 5\vec{k})}{7.071}$

④ $\vec{P}_{AD} = P_{AD}(-0.4243\vec{i} - 0.5657\vec{j} + 0.7071\vec{k})$

⑤ $\vec{W} = (-1100\vec{k})$ N

⑥ $\overset{+}{\swarrow}\sum F_x = 0: \quad P_{AB} - 0.4243\,P_{AD} = 0$

⑦ $\overset{+}{\rightarrow}\sum F_y = 0: \quad P_{AC} - 0.5657\,P_{AD} = 0$

⑧ $+\uparrow\sum F_z = 0: \quad 0.7071\,P_{AD} - 1100 = 0 \quad \Rightarrow \quad P_{AD} = 1555.6$ N

⑨ using P_{AD} in ⑥ + ⑦: $P_{AB} = 660.0$ N; $P_{AC} = 880.0$ N

⑩ most highly stressed wire governs design ⇒ AD

⑪ from Appendix B-2: $S_y = 234$ MPa

⑫ $\sigma_{max} = \dfrac{S_y}{SF_y} = \dfrac{234}{5} = 46.8$ MPa

⑬ $A = \dfrac{P_{AD}}{\sigma_{max}} = \dfrac{\pi}{4}d^2 \Rightarrow d = \sqrt{\dfrac{4P_{AD}}{\pi\,\sigma_{max}}}$

⑭ $d = \sqrt{\dfrac{4(1555.6)}{\pi(46.8)}} = 6.506$ mm $= 0.2561$ in

⑮ using Appendix A, choose American S + W steel wire s.t.

$d \geq 0.2561$ in $\Rightarrow$ **gage no. 2 ($d = 0.2625$ in)**

Problem 7.18:

FBD:

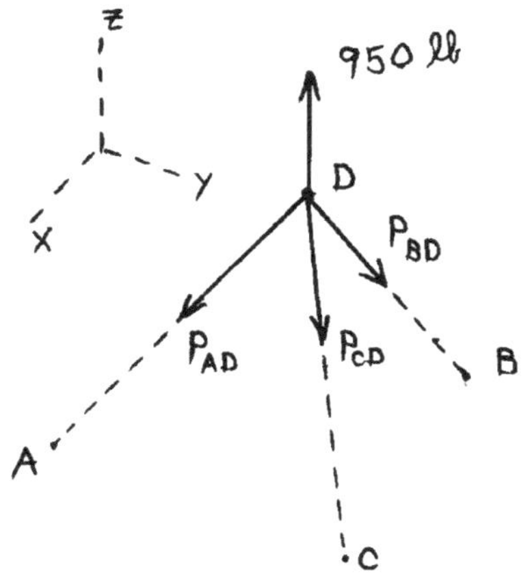

A = (25, −15, 0) ft
B = (−25, −15, 0) ft
C = (15, 20, 0) ft
D = (0, 0, 60) ft

find: $P_{AD}, P_{BD}, P_{CD} = ?$

① $\vec{P}_{AD} = P_{AD}\dfrac{\vec{r}_{DA}}{r_{DA}} = P_{AD}\dfrac{(25\vec{i} - 15\vec{j} - 60\vec{k})}{\sqrt{4450}} = P_{AD}\dfrac{(25\vec{i} - 15\vec{j} - 60\vec{k})}{66.71}$

② $\vec{P}_{AD} = P_{AD}(0.3748\vec{i} - 0.2249\vec{j} - 0.8994\vec{k})$

③ $\vec{P}_{BD} = P_{BD}\dfrac{\vec{r}_{DB}}{r_{DB}} = P_{BD}\dfrac{(-25\vec{i} - 15\vec{j} - 60\vec{k})}{66.71}$

④ $\vec{P}_{BD} = P_{BD}(-0.3748\vec{i} - 0.2249\vec{j} - 0.8994\vec{k})$

⑤ $\vec{P}_{CD} = P_{CD}\dfrac{\vec{r}_{DC}}{r_{DC}} = P_{CD}\dfrac{(15\vec{i} + 20\vec{j} - 60\vec{k})}{\sqrt{4225}} = P_{CD}\dfrac{(15\vec{i} + 20\vec{j} - 60\vec{k})}{65}$

⑥ $\vec{P}_{CD} = P_{CD}(0.2308\vec{i} + 0.3077\vec{j} - 0.9231\vec{k})$

⑦ $\vec{F}_D = (950\vec{k})$ lb

⑧ $\swarrow + \sum F_x = 0: \quad 0.3748\,P_{AD} - 0.3748\,P_{BD} + 0.2308\,P_{CD} = 0$

⑨ $\nearrow + \sum F_y = 0: \quad -0.2249\,P_{AD} - 0.2249\,P_{BD} + 0.3077\,P_{CD} = 0$

⑩ $+\uparrow \sum F_z = 0: \quad -0.8994\,P_{AD} - 0.8994\,P_{BD} - 0.9231\,P_{CD} + 950 = 0$

⑪ solving ⑧ – ⑩ simultaneously:

$P_{AD} = 165.9$ lb
$P_{BD} = 437.6$ lb
$P_{CD} = 441.1$ lb

Problem 7.19:

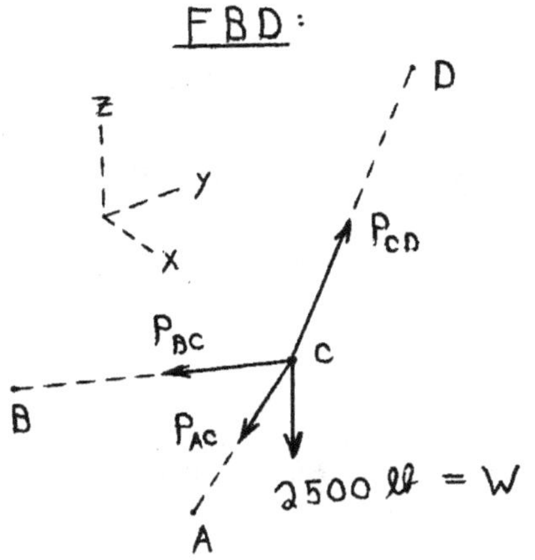

A = (4, 0, 0) ft
B = (−4, 0, 0) ft
C = (0, 9, 0) ft
D = (0, 21, 14) ft

find: gage no. = ?

* steel 4340 HR

$SF_y = 3.5$

① $\vec{P}_{AC} = P_{AC}\dfrac{\vec{r}_{CA}}{r_{CA}} = P_{AC}\dfrac{(4\vec{i} - 9\vec{j})}{\sqrt{97}} = P_{AC}(0.4061\vec{i} - 0.9138\vec{j})$

② $\vec{P}_{BC} = P_{BC}\dfrac{\vec{r}_{CB}}{r_{CB}} = P_{BC}\dfrac{(-4\vec{i} - 9\vec{j})}{\sqrt{97}} = P_{BC}(-0.4061\vec{i} - 0.9138\vec{j})$

③ $\vec{P}_{CD} = P_{CD}\dfrac{\vec{r}_{CD}}{r_{CD}} = P_{CD}\dfrac{(12\vec{j} + 14\vec{k})}{\sqrt{340}} = P_{CD}(0.6508\vec{j} + 0.7593\vec{k})$

④ $\vec{W} = (-2500\vec{k})$ lb

⑤ $+\searrow \sum F_x = 0: \quad 0.4061 P_{AC} - 0.4061 P_{BC} = 0 \Rightarrow P_{AC} = P_{BC}$

⑥ $+\nearrow \sum F_y = 0: \quad -0.9138 P_{AC} - 0.9138 P_{BC} + 0.6508 P_{CD} = 0$

⑦ $+\uparrow \sum F_z = 0: \quad 0.7593 P_{CD} - 2500 = 0 \Rightarrow P_{CD} = 3293$ lb

⑧ using P_{CD} + ⑤ in ⑥:

$2(0.9138) P_{AC} = 0.6508(3293) \Rightarrow P_{AC} = P_{BC} = 1173$ lb

⑨ most highly stressed wire governs design ⇒ CD

⑩ from Appendix B-2: $S_y = 132$ ksi

⑪ $\sigma = \dfrac{S_y}{SF_y} = \dfrac{132}{3.5} = 37.71$ ksi

⑫ $A = \dfrac{P_{CD}}{\sigma} = \dfrac{\pi}{4}d^2 \Rightarrow d = \sqrt{\dfrac{4P_{CD}}{\pi\sigma}}$

⑬ $d = \sqrt{\dfrac{4(3293)}{\pi(37{,}710)}} = 0.3334$ in

⑭ from Appendix A, American S + W steel wire:

need gage 000s (d = 0.3625 in)

Problem 7.20:

FBD:

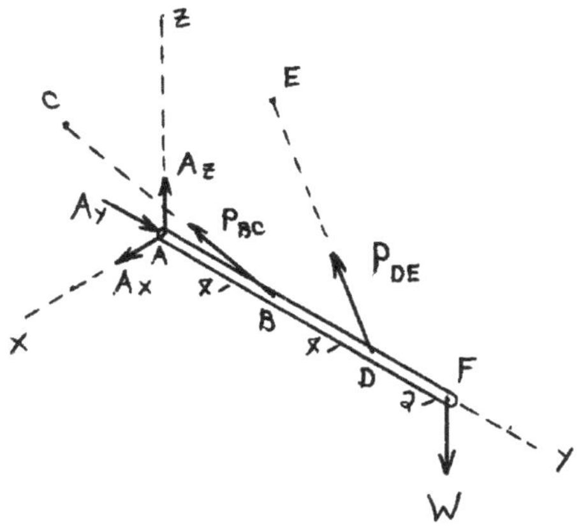

C = (4, 0, 6) ft B = (0, 4, 0) ft

E = (−2, 0, 3) ft D = (0, 8, 0) ft

* ball + socket @ A

A = (0, 0, 0) ft $D_{BC} = D_{DE} = 0.125''$

F = (0, 10, 0) ft $S_u = 150$ ksi

find: $W, A_x, A_y, A_z, P_{BC}, P_{DE} = ?$

① $\vec{P}_{BC} = P_{BC}\dfrac{\vec{r}_{BC}}{r_{BC}} = P_{BC}\dfrac{(4\vec{\imath} - 4\vec{\jmath} + 6\vec{k})}{\sqrt{68}} = P_{BC}\left(\dfrac{2}{\sqrt{17}}\vec{\imath} - \dfrac{2}{\sqrt{17}}\vec{\jmath} + \dfrac{3}{\sqrt{17}}\vec{k}\right)$

② $\vec{P}_{DE} = P_{DE}\dfrac{\vec{r}_{DE}}{r_{DE}} = P_{DE}\dfrac{(-2\vec{\imath} - 8\vec{\jmath} + 3\vec{k})}{\sqrt{77}} = P_{DE}\left(-\dfrac{2}{\sqrt{77}}\vec{\imath} - \dfrac{8}{\sqrt{77}}\vec{\jmath} + \dfrac{3}{\sqrt{77}}\vec{k}\right)$

③ $\vec{A} = A_x\vec{\imath} + A_y\vec{\jmath} + A_z\vec{k}$; $\vec{W} = -W\vec{k}$

④ $\sum F_x = 0: \quad A_x + \dfrac{2}{\sqrt{17}}P_{BC} - \dfrac{2}{\sqrt{77}}P_{DE} = 0$

⑤ $\sum F_y = 0: \quad A_y - \dfrac{2}{\sqrt{17}}P_{BC} - \dfrac{8}{\sqrt{77}}P_{DE} = 0$

⑥ $+\uparrow \sum F_z = 0: \quad A_z + \dfrac{3}{\sqrt{17}}P_{BC} + \dfrac{3}{\sqrt{77}}P_{DE} - W = 0$

* scalar method for moments:

⑦ $\sum M_x = 0: \quad \left(\dfrac{3}{\sqrt{17}}P_{BC}\right)(4) + \left(\dfrac{3}{\sqrt{77}}P_{DE}\right)(8) - W(10) = 0$

⑧ $\sum M_y = 0: \quad 0 = 0$

⑨ $\sum M_z = 0: \quad -\left(\dfrac{2}{\sqrt{17}}P_{BC}\right)(4) + \left(\dfrac{2}{\sqrt{77}}P_{DE}\right)(8) = 0$

⑩ from ⑨: $P_{DE} = 1.064\, P_{BC}$

⑪ $P_{max} = S_u A = (150{,}000)\left(\dfrac{\pi}{4}(0.125)^2\right) = 1841$ lb

⑫ since $P_{DE} > P_{BC}$, then $\boxed{P_{DE} = 1841 \text{ lb}}$ ⇒ $\boxed{P_{BC} = 1730 \text{ lb}}$

⑬ using P_{DE} + P_{BC} in ⑦ + solving: $\boxed{W = 1007 \text{ lb}}$

⑭ using P_{DE}, P_{BC}, + W in ④–⑥ + solving:

$\boxed{A_x = -419.6 \text{ lb}; \quad A_y = 2518 \text{ lb}; \quad A_z = -881.2 \text{ lb}}$

Problem 7.20: (con't)

* vector method for moments:

⑦ $\sum \vec{M}_A = 0 = \vec{r}_{AB} \times \vec{P}_{BC} + \vec{r}_{AD} \times \vec{P}_{DE} + \vec{r}_{AF} \times \vec{W}$

⑧ $$0 = \begin{vmatrix} \vec{i} & \vec{j} & \vec{k} \\ 0 & 4 & 0 \\ 2 & -2 & 3 \end{vmatrix} \frac{P_{BC}}{\sqrt{17}} + \begin{vmatrix} \vec{i} & \vec{j} & \vec{k} \\ 0 & 8 & 0 \\ -2 & -8 & 3 \end{vmatrix} \frac{P_{DE}}{\sqrt{77}} + \begin{vmatrix} \vec{i} & \vec{j} & \vec{k} \\ 0 & 10 & 0 \\ 0 & 0 & -W \end{vmatrix}$$

⑨ $$0 = \left[(12\vec{i} - 8\vec{k})\frac{P_{BC}}{\sqrt{17}}\right] + \left[(24\vec{i} + 16\vec{k})\frac{P_{DE}}{\sqrt{77}}\right] + \left[-10W\vec{i}\right]$$

⑩ collect like terms + set each coef. to zero:

$\vec{i}: \quad \frac{12}{\sqrt{17}} P_{BC} + \frac{24}{\sqrt{77}} P_{DE} - 10W = 0$

$\vec{k}: \quad -\frac{8}{\sqrt{17}} P_{BC} + \frac{16}{\sqrt{77}} P_{DE} = 0$

∴ solve for P_{BC}, P_{DE}, + W as previously

Problem 7.21:

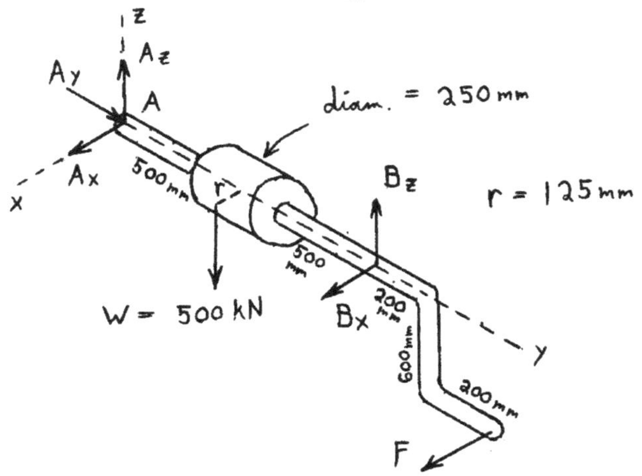

* ball + socket @ A
* smooth journal bearing @ B (no moment reactions)

find: $F, A_x, A_y, A_z, B_x, B_z = ?$

① $\vec{A} = A_x\vec{i} + A_y\vec{j} + A_z\vec{k}$; $\vec{B} = B_x\vec{i} + B_z\vec{k}$; $\vec{F} = F\vec{i}$;

$\vec{W} = -(500\text{ kN})\vec{k}$

② $\sum F_x = 0: \quad A_x + B_x + F = 0$

③ $\sum F_y = 0: \quad \boxed{A_y = 0}$

④ $\sum F_z = 0: \quad A_z + B_z - 500 = 0$

* scalar method for moments:

⑤ $\sum M_x = 0: \quad -(500)(500) + B_z(1000) = 0 \Rightarrow \boxed{B_z = 250\text{ kN}}$

⑥ $\sum M_y = 0: \quad (500)(125) - F(600) = 0 \Rightarrow \boxed{F = 104.2\text{ kN}}$

⑦ $\sum M_z = 0: \quad -B_x(1000) - F(1400) = 0$

⑧ using F in ⑦ + solving: $\boxed{B_x = -145.9\text{ kN}}$

⑨ using B_x, F, + B_z in ② + ④:

$A_x + (-145.9) + (104.2) = 0 \Rightarrow \boxed{A_x = 41.7\text{ kN}}$

$A_z + (250) - 500 = 0 \Rightarrow \boxed{A_z = 250\text{ kN}}$

* vector method for moments:

⑤ $\sum \vec{M}_A = 0 = \vec{r}_W \times \vec{W} + \vec{r}_{AB} \times \vec{B} + \vec{r}_F \times \vec{F}$

Problem 7.21: (con't)

⑥
$$0 = \begin{vmatrix} \vec{i} & \vec{j} & \vec{k} \\ 125 & 500 & 0 \\ 0 & 0 & -500 \end{vmatrix} + \begin{vmatrix} \vec{i} & \vec{j} & \vec{k} \\ 0 & 1000 & 0 \\ B_x & 0 & B_z \end{vmatrix} + \begin{vmatrix} \vec{i} & \vec{j} & \vec{k} \\ 0 & 1400 & -600 \\ F & 0 & 0 \end{vmatrix}$$

⑦
$$0 = \left[-250{,}000\,\vec{i} + 62{,}500\,\vec{j}\right] + \left[1000\,B_z\,\vec{i} - 1000\,B_x\,\vec{k}\right] + \left[-600\,F\,\vec{j} - 1400\,F\,\vec{k}\right]$$

⑧ collect like terms + set each coef. to zero:

$\vec{i}$: $-250{,}000 + B_z(1000) = 0$

$\vec{j}$: $62{,}500 - F(600) = 0$

$\vec{k}$: $-B_x(1000) - F(1400) = 0$

∴ solve for F, B_x, + B_z as previously

CHAPTER 8 PROBLEMS

8.1 For the hoist shown in the figure to the right, determine all of the external forces on member CFG as point D is moved along member AE so that the angle θ varies from 0 to 45°. We suggest that you use a spreadsheet in preparing this solution. Note the angle β = 60° and F = 580 N. The dimensions are: CF = 200 mm, FG = 320 mm and the pulley radius r = 10 mm.

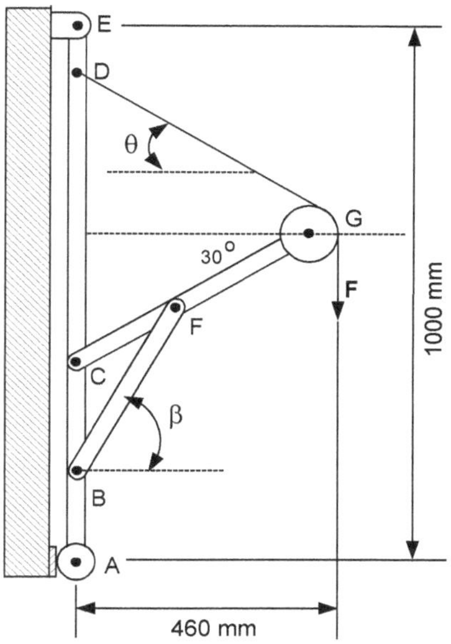

8.2 For the hoist shown in the figure to the right, determine all of the external forces on member CFG as point B is moved along member AE so that the angle β varies from 35 to 75°. We suggest that you use a spreadsheet in preparing this solution. Note the angle θ = 30° and F = 240 N. The dimensions CF = 200 mm, FG = 320 mm and the pulley radius r = 10 mm.

8.3 Prepare a FBD for the entire rectangular frame, shown in the figure to the left, if it is subjected to a concentrated force of 12.4 kN located a distance of 3.0 m from the left end of the horizontal member.

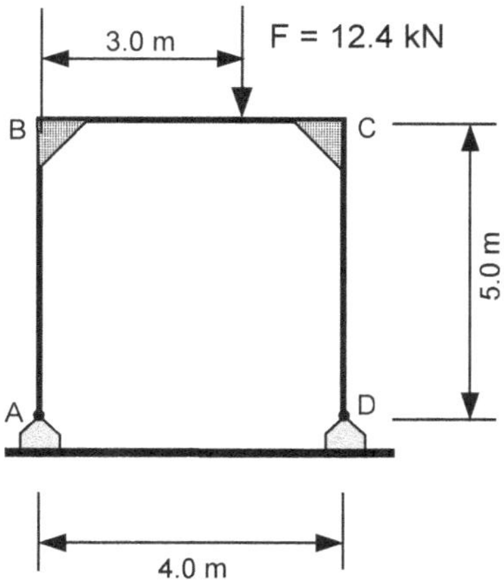

8.4 Determine the force F required to develop a pressure p = 2.1 MPa in the cylinder of the pump shown in the figure to the right. The piston area for the pump is 300 mm^2.

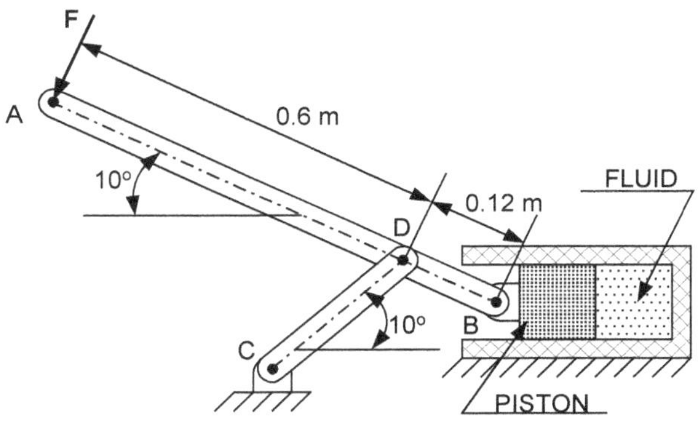

8.5 Determine the reaction force at point G and the mechanical advantage of the toggle mechanism illustrated in the figure to the right. A force F of 300 N is applied to the lever at point A. The contact at point G is made with a roller, and the links are connected with pins inserted at points B, C, D and E.

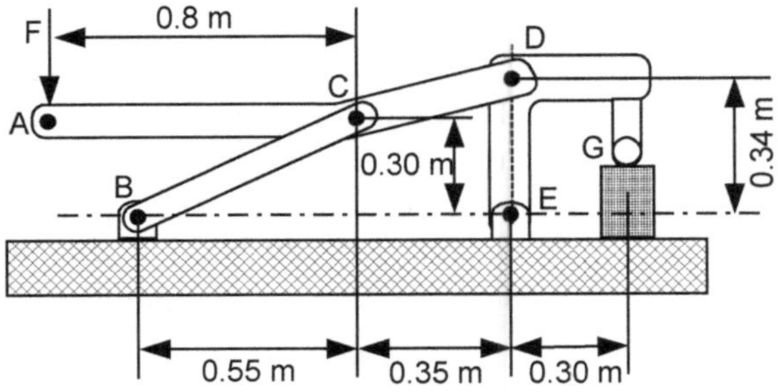

8.6 Determine the reaction force at point G and the mechanical advantage of the toggle mechanism illustrated in the figure to the right. The input force F is now 500 N. The contact at point G is made with a roller and the links are connected with pins inserted at points B, C, D and E.

8.7 Your manager believes that the dimension (e – d) for the toggle mechanism shown in the figure to the left is a critical design parameter. She asks you to determine the mechanical advantage of this mechanism if the dimension d is fixed and the dimension e is modified so that (e – d) varies from 5 mm to 50 mm. You may consider using a spreadsheet for this analysis.

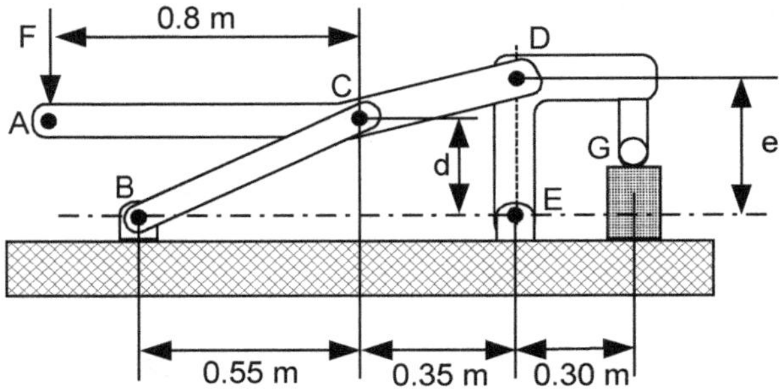

8.8 If the spring constant of the block under point G is 2.0 kN/mm, determine the vertical displacement of point G in Problem 8.5.

8.9 If the spring constant of the block under point G is 2.0 kN/mm, determine the work output of the toggle mechanism for the conditions described in Problem 8.5.

8.10 If the spring constant of the block under point G is 3.8 kN/mm, determine the vertical displacement of point G in Problem 8.6.

8.11 If the spring constant of the block under point G is 3.8 kN/mm, determine the work output of the toggle mechanism for the conditions described in Problem 8.6.

8.12 For the compaction press, shown in the figure to the right, determine the compressive force developed by the sliding platen if a force of 5 ton is applied to the toggle mechanism at point B. Also determine the mechanical advantage.

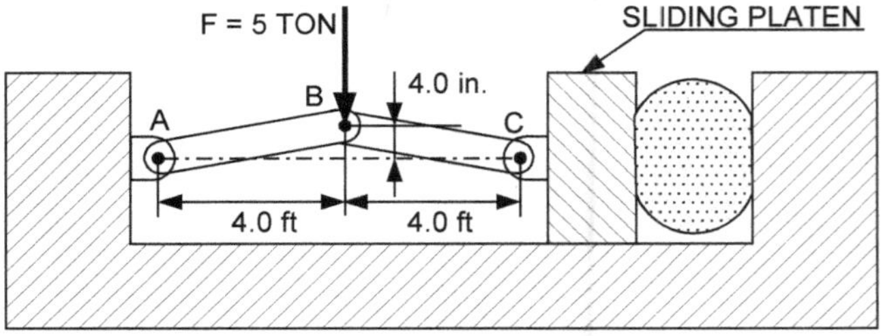

COMPACTION PRESS

8.13 A pair of pliers clamps a small diameter rubber cylinder as shown in the figure to the right. If opposing forces of 30 lb are applied to the handles of the pliers, determine the reaction forces acting on the cylinder. Also determine the work performed if each handle moves though a distance d = 0.14 in. Finally, determine the amount that the cylinder is squeezed by the application of the forces. The dimensions of the pliers are a = 1.0 in. and b = 5.5 in.

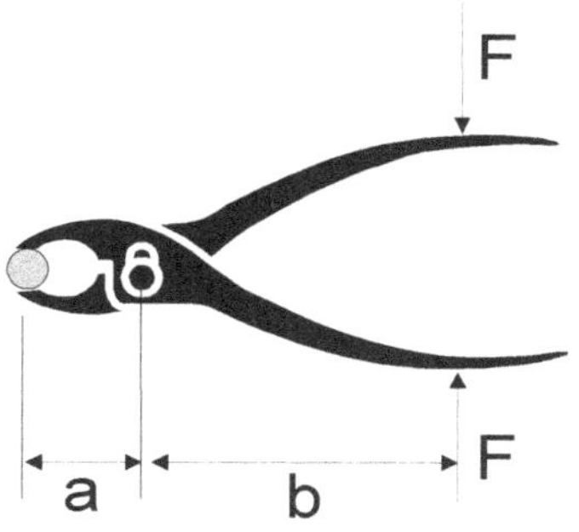

8.14 For the frame, shown in the figure to the left, determine the forces and moment at the fixed support at A, the forces at pin C and the force in link BD.

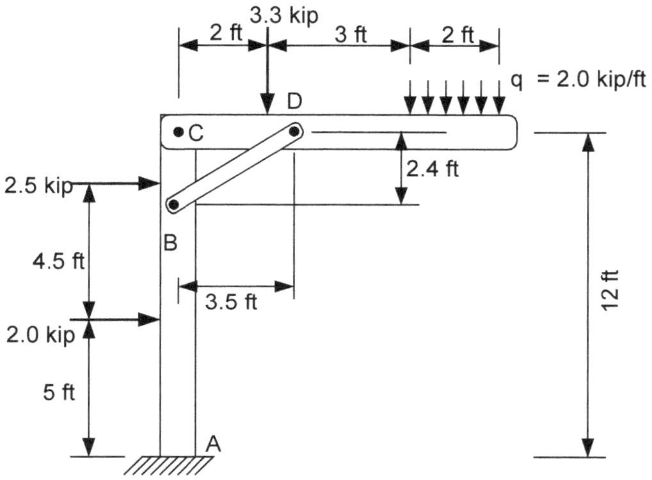

8.15 For the frame, illustrated in the figure to the right, determine all the forces acting on member ABCDE. Also determine the internal force in member AF. The attached weight W is 4.8 kN and the pulley radius is 150 mm. The dimensions in the figure are given in mm.

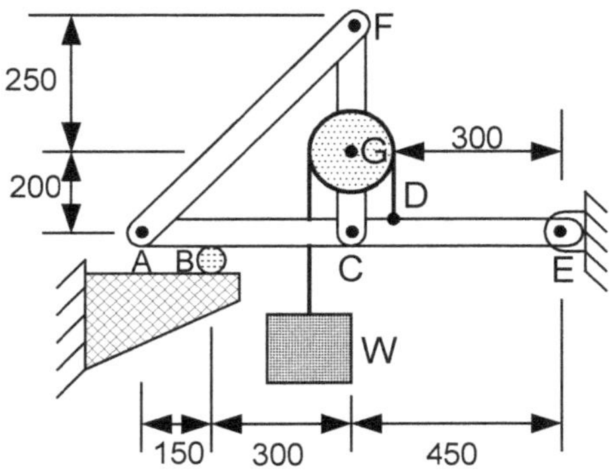

8.16 For the frame illustrated in the figure to the right, determine all the forces acting on member ABCDE. Also determine the internal force in member AF. The attached weight W is 4.8 kN and the pulley radius is 100 mm. The dimensions in the figure are given in mm.

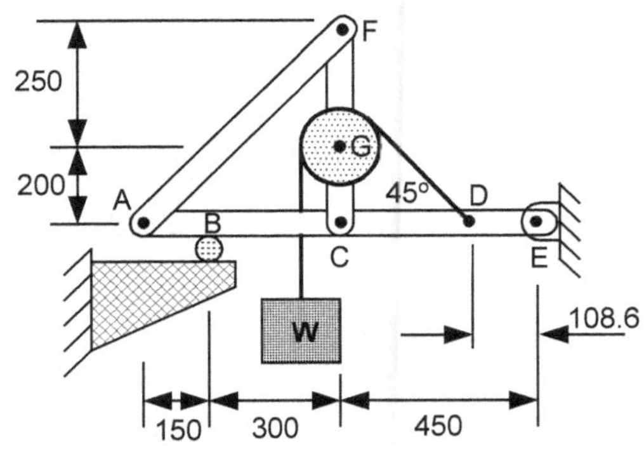

8.17 For the arm of the backhoe, shown in the figure below, determine the force that must be exerted by the hydraulic actuator AC to maintain the bucket loaded with a weight of 1.0 kN in equilibrium. Also find the forces in links BC and CE and the force acting on pin D. The bucket and the crosshatched appendage are welded together.

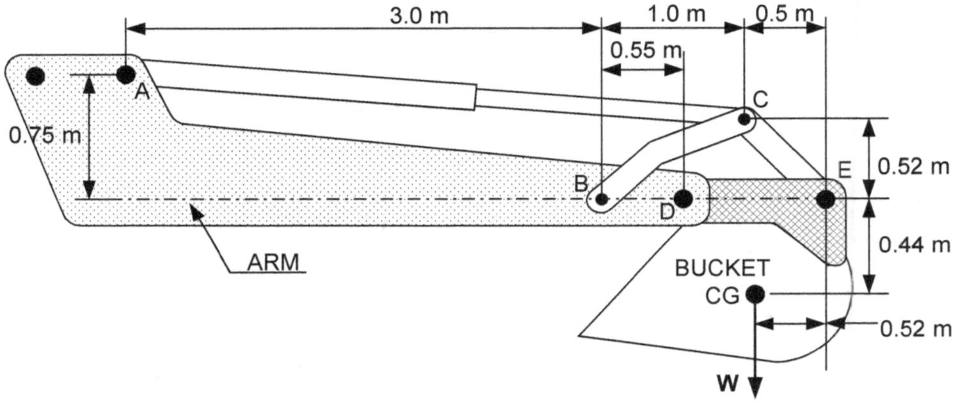

8.18 For the "Loadall" arm, presented in the figure below, determine the forces exerted by the hidden actuator and the visible actuator BC to maintain equilibrium. Also determine the forces acting at pins A and D. The arm weighs 1.5 kN and its center of gravity is located 1.8 m to the right of point A. All of the dimensions are given in meters.

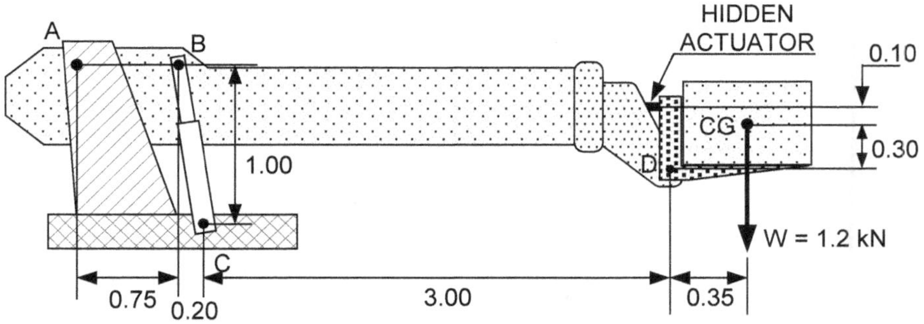

8.19 Levers are well known machines used for either amplifying or attenuating forces. When used in scales to weigh heavy objects, the levers are often arranged to compound the attenuation. Such an arrangement is illustrated in the figure below. If the pins are frictionless, show that the relation between the known scale weight w and the unknown scale weight W is:

$$W = w\left[\frac{(a+b+c+d-x)(a+b+c)b}{aec}\right]$$

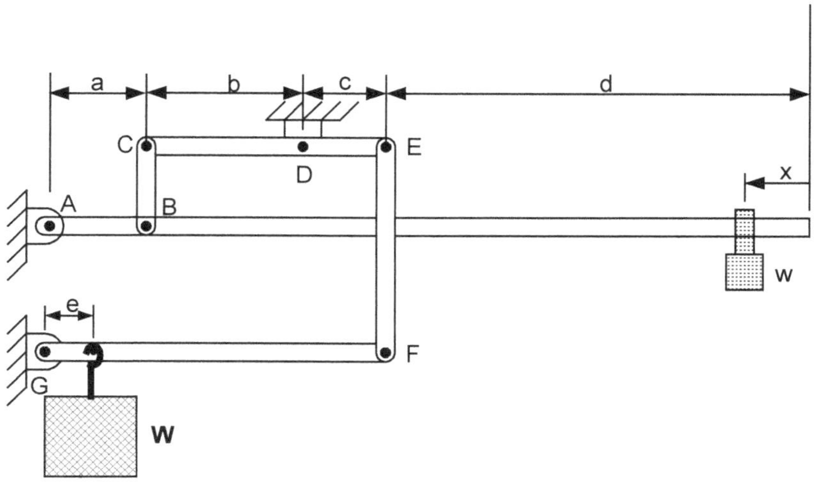

8.20 Select dimensions a through e if the scale, illustrated in the figure above, is to measure a weight W = 5.0 kN with a small sliding weight w = 10 N. The value of x = 200 mm is fixed in this design analysis.

8.21 The horizontal boom of a construction crane is counter balanced with a 15 kip weight that is centered at point B. A cable BFE anchored at points B and E supports the horizontal boom. The cable is maintained at a constant tension over its length with a small pulley located at point F. The horizontal boom is attached to the tower with a pin located at point C. The crane is lifting a load W = 10 kip from a hoist located at point D. Determine all of the forces acting on the horizontal boom, the cable tension, all of the reactions at point A and the force acting on the pin at point F. The weight of the boom is 5.0 kip and the weight of the tower member is 4.0 kip.

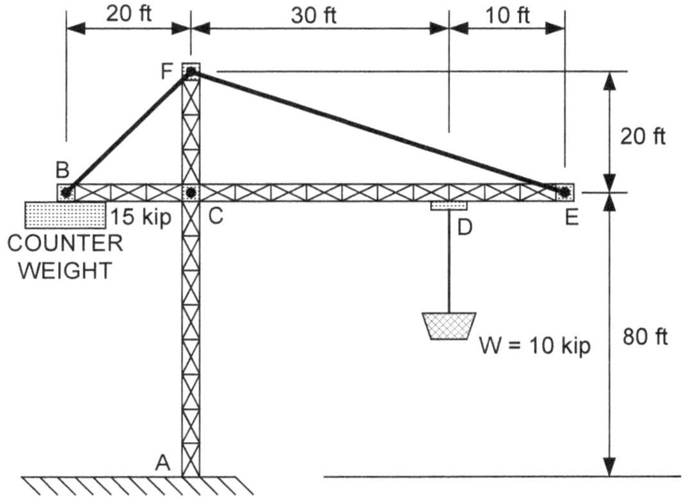

Problem 8.4:

FBD's:

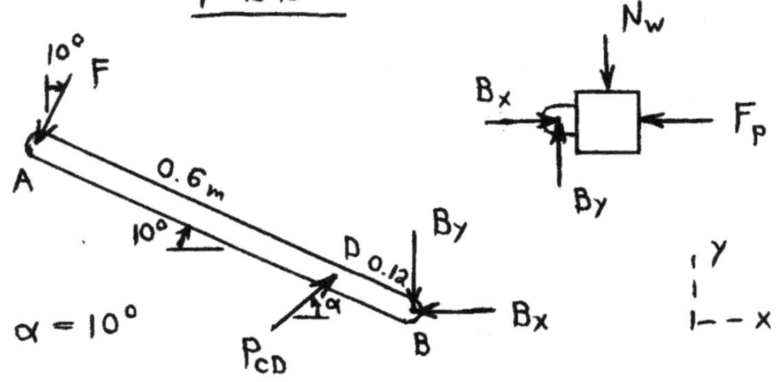

$p = 2.1$ MPa

$A = 300$ mm^2

find : $F = ?$

① $F_p = pA = (2.1)(300) = 630$ N

* consider piston :

② $\overset{+}{\rightarrow}\sum F_x = 0: \quad B_x = F_p = 630$ N

③ $+\uparrow \sum F_y = 0: \quad B_y = N_w$

* consider pump arm :

④ $\overset{+}{\rightarrow}\sum F_x = 0: \quad -F\sin(10°) + P_{CD}\cos(10°) - 630 = 0$

⑤ $+\uparrow \sum F_y = 0: \quad -F\cos(10°) + P_{CD}\sin(10°) - B_y = 0$

⑥ $\circlearrowleft + \sum M_B = 0: \quad F(0.72) - (P_{CD}\sin(20°))(0.12) = 0$

⑦ $P_{CD} = F\dfrac{(0.72)}{(\sin(20°))(0.12)} = F(17.54)$

⑧ using ⑦ in ④ :

$-F(0.17365) + (F(17.54))(0.98481) = 630$

⑨ $\boxed{F = 36.84 \text{ N}}$.

* continuing problem :

$P_{CD} = 646.2$ N

$B_y = 75.94$ N

Problem 8.5:

FBD's:

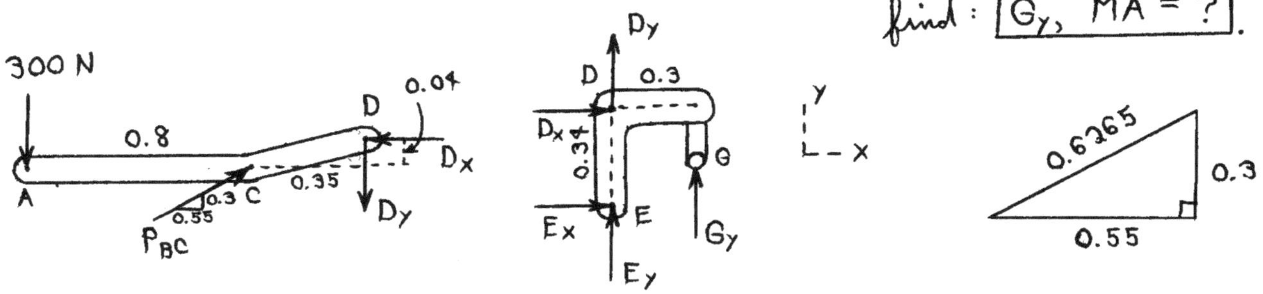

find: $\boxed{G_y,\ MA = ?}$.

* consider member ACD:

① $\circlearrowleft + \sum M_D = 0:$

$$(300)(1.15) - \left(P_{BC}\left(\frac{0.3}{0.6265}\right)\right)(0.35) + \left(P_{BC}\left(\frac{0.55}{0.6265}\right)\right)(0.04) = 0$$

② $P_{BC} = 2604\text{ N}$

③ $\overset{+}{\rightarrow} \sum F_x = 0: \quad P_{BC}\left(\frac{0.55}{0.6265}\right) - D_x = 0$

④ $D_x = (2604)\left(\frac{0.55}{0.6265}\right) \Rightarrow D_x = 2286\text{ N}$

⑤ $\uparrow + \sum F_y = 0: \quad P_{BC}\left(\frac{0.3}{0.6265}\right) - D_y - 300 = 0$

⑥ $D_y = -300 + (2604)\left(\frac{0.3}{0.6265}\right) \Rightarrow D_y = 946.9\text{ N}$

* consider member EDG:

⑦ $\circlearrowleft + \sum M_E = 0: \quad G_y(0.3) - D_x(0.34) = 0$

⑧ $G_y = (2286)\frac{(0.34)}{(0.3)} \Rightarrow \boxed{G_y = 2591\text{ N}}$.

⑨ $MA = \frac{F_{out}}{F_{in}} = \frac{2591}{300}$

⑩ $\boxed{MA = 8.637}$.

Problem 8.12:

FBD's:

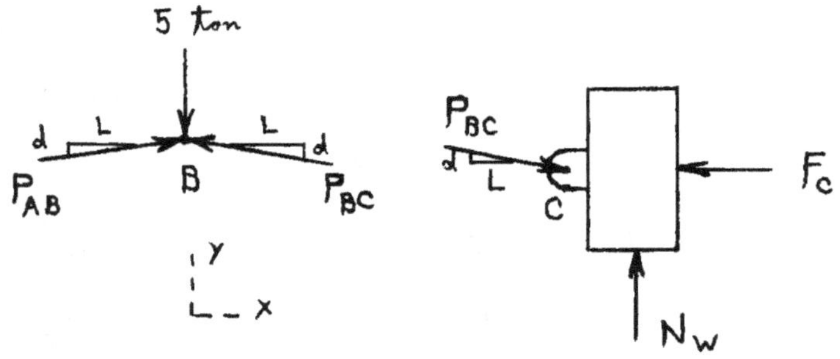

$L = 4\ ft = 48\ in$

$d = 4\ in$

find: F_c, MA = ?

① by inspection: AB + BC are 2-force members

* consider pt. B:

② $\overset{+}{\rightarrow}\sum F_x = 0: \quad P_{AB}\left(\frac{48}{48.17}\right) - P_{BC}\left(\frac{48}{48.17}\right) = 0 \Rightarrow P_{AB} = P_{BC}$

③ $+\uparrow\sum F_y = 0: \quad P_{AB}\left(\frac{4}{48.17}\right) + P_{BC}\left(\frac{4}{48.17}\right) - 5 = 0$

④ using $P_{AB} = P_{BC}$ in ③ + solving:

$P_{AB} = P_{BC} = 30.11$ ton

* consider sliding platen:

⑤ $\overset{+}{\rightarrow}\sum F_x = 0: \quad P_{BC}\left(\frac{48}{48.17}\right) - F_c = 0$

⑥ $+\uparrow\sum F_y = 0: \quad N_w - P_{BC}\left(\frac{4}{48.17}\right) = 0$

⑦ using P_{BC} in ⑤:

$F_c = (30.11)\left(\frac{48}{48.17}\right) \Rightarrow \boxed{F_c = 30\text{ ton}}$

⑧ $MA = \frac{F_{out}}{F_{in}} = \frac{30}{5}$

⑨ $\boxed{MA = 6}$

* remaining answer:

$N_w = 2.5$ ton

Problem 8.13:

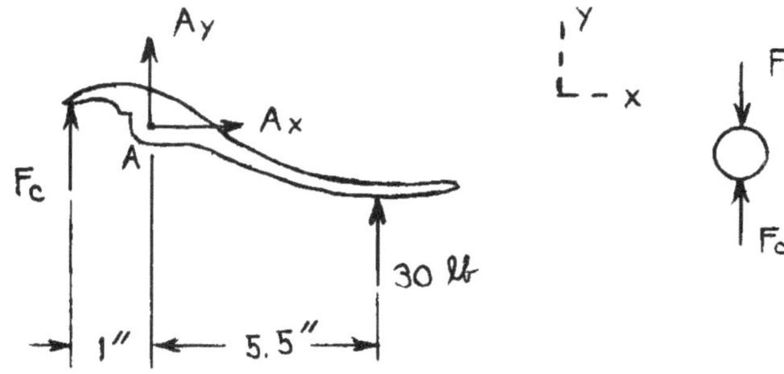

$$\underline{F\,B\,D's:}$$

find: $\boxed{F_c, \; W, \; \Delta d_c \; = \; ?}$

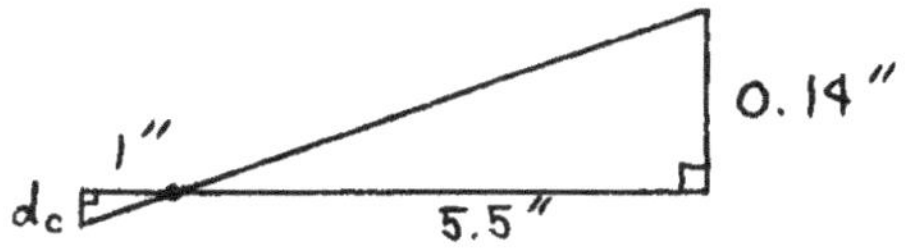

* consider 1 handle of pliers:

① $\xrightarrow{+} \sum F_x = 0:$ $A_x = 0$

② $\uparrow + \sum F_y = 0:$ $A_y + F_c + 30 = 0$

③ $\circlearrowleft + \sum M_A = 0:$ $- F_c \, (1) + (30)(5.5) = 0$

④ $\boxed{F_c = 165 \; lb}$.

⑤ $W = Fd = (30)(2(0.14)) \implies \boxed{W = 8.4 \; in\text{-}lb}$.

⑥ from similar $\Delta's:$

$$\frac{d_c}{1} = \frac{0.14}{5.5} \implies d_c = 0.02545 \; in$$

⑦ $\Delta d_c = 2(0.02545) \implies \boxed{\Delta d_c = 0.0509 \; in}$

* remaining answer:

$$A_y = -195 \; lb$$

Problem 8.14:

FBD's :

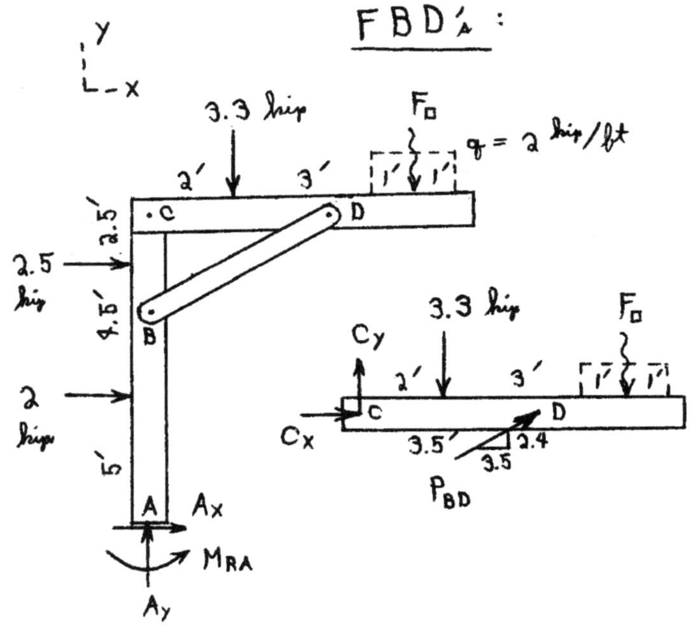

find : $A_x,\ A_y,\ M_{RA},\ C_x,\ C_y,\ P_{BD} = ?$

① $F_\square = qL = (2)(2)$

② $F_\square = 4$ kip

* consider entire frame :

③ $\overset{+}{\rightarrow}\sum F_x = 0$:

$2.5 + 2 + A_x = 0$

④ $\boxed{A_x = -4.5 \text{ kip}}$

⑤ $+\uparrow\sum F_y = 0$: $A_y - 3.3 - 4 = 0 \Rightarrow \boxed{A_y = 7.3 \text{ kip}}$

⑥ $\circlearrowleft + \sum M_A = 0$: $M_{RA} - (2)(5) - (2.5)(9.5) - (3.3)(2) - (4)(6) = 0$

⑦ $\boxed{M_{RA} = 64.35 \text{ kip-ft}}$

* consider member CD :

⑧ $\circlearrowleft + \sum M_c = 0$: $-(3.3)(2) - (4)(6) + \left(P_{BD}\left(\frac{2.4}{4.244}\right)\right)(3.5) = 0$

⑨ $\boxed{P_{BD} = 15.46 \text{ kip}}$

⑩ $+\uparrow\sum F_y = 0$: $C_y + (15.46)\left(\frac{2.4}{4.244}\right) - 3.3 - 4 = 0$

⑪ $\boxed{C_y = -1.443 \text{ kip}}$

⑫ $\overset{+}{\rightarrow}\sum F_x = 0$: $C_x + (15.46)\left(\frac{3.5}{4.244}\right) = 0$

⑬ $\boxed{C_x = -12.75 \text{ kip}}$

Problem 8.15:

FBD's: find: $\boxed{B_y, E_x, E_y, C_x, C_y, P_{AF} = ?}$

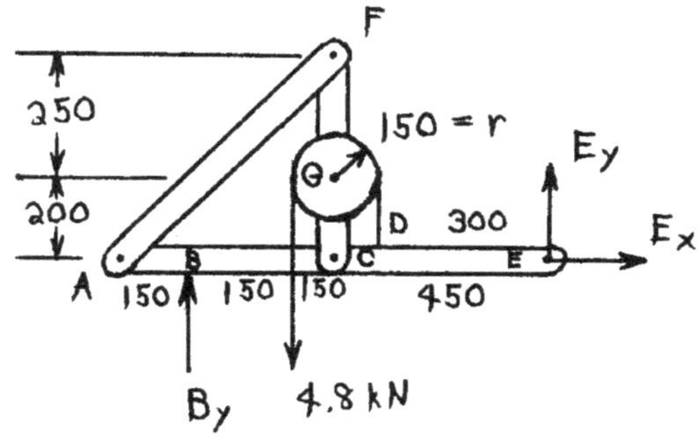

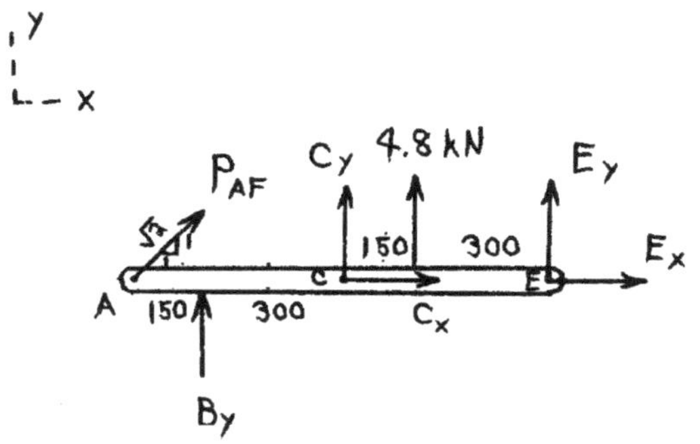

* consider entire frame:

① $\overset{+}{\rightarrow}\sum F_x = 0: \quad \boxed{E_x = 0}$

② $\circlearrowleft + \sum M_E = 0: \quad (4.8)(600) - B_y(750) = 0 \quad \Rightarrow \quad \boxed{B_y = 3.84 \text{ kN}}$

③ $+\uparrow \sum F_y = 0: \quad (3.84) + E_y - 4.8 = 0 \quad \Rightarrow \quad \boxed{E_y = 0.960 \text{ kN}}$

* consider member ABCDE:

④ $\circlearrowleft + \sum M_A = 0: \quad (3.84)(150) + C_y(450) + (4.8)(600) + (0.96)(900) = 0$

⑤ $\boxed{C_y = -9.6 \text{ kN}}$

⑥ $+\uparrow \sum F_y = 0: \quad 3.84 + \frac{1}{\sqrt{2}} P_{AF} + (-9.6) + 4.8 + 0.96 = 0$

⑦ $\boxed{P_{AF} = 0}$

⑧ $\overset{+}{\rightarrow}\sum F_x = 0: \quad \frac{1}{\sqrt{2}}(0) + 0 + C_x = 0 \quad \Rightarrow \quad \boxed{C_x = 0}$

Problem 8.17:

$\underline{FBD}$:

find: $\boxed{P_{AC}, \ P_{BC}, \ P_{CE}, \ D = ?}$

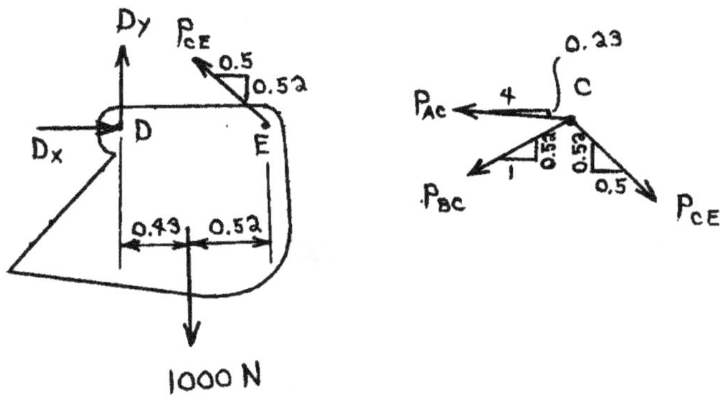

① by inspection: AC, BC, & CE are 2-force members

* consider bucket:

② $\circlearrowright + \sum M_E = 0$: $(1000)(0.52) - D_y(0.95) = 0 \implies D_y = 547.4 \text{ N}$

③ $\uparrow + \sum F_y = 0$: $D_y - 1000 + P_{CE}\left(\frac{0.52}{0.7214}\right) = 0$

④ $P_{CE} = \left(\frac{0.7214}{0.52}\right)(1000 - 547.4) \implies \boxed{P_{CE} = 627.9 \text{ N}}$

⑤ $\rightarrow + \sum F_x = 0$: $D_x - P_{CE}\left(\frac{0.5}{0.7214}\right) = 0$

⑥ $D_x = (627.9)\left(\frac{0.5}{0.7214}\right) = 435.2 \text{ N}$

⑦ $D = \sqrt{D_x^2 + D_y^2} = \sqrt{(435.2)^2 + (547.4)^2} \implies \boxed{D = 699.3 \text{ N}}$

* consider pin C:

⑧ $\rightarrow + \sum F_x = 0$: $P_{CE}\left(\frac{0.5}{0.7214}\right) - P_{BC}\left(\frac{1}{1.127}\right) - P_{AC}\left(\frac{4}{4.007}\right) = 0$

⑨ $+\uparrow \sum F_y = 0$: $-P_{CE}\left(\frac{0.52}{0.7214}\right) - P_{BC}\left(\frac{0.52}{1.127}\right) + P_{AC}\left(\frac{0.23}{4.007}\right) = 0$

⑩ subbing $P_{CE} = 627.9$ in ⑧ & ⑨ and solving simultaneously:

$\boxed{\begin{array}{l} P_{BC} = -834.4 \text{ N} \\ P_{AC} = 1178 \text{ N} \end{array}}$

CHAPTER 9 PROBLEMS

9.1 Write an engineering brief describing the factors affecting friction forces acting between two bodies.

9.2 Write an engineering brief explaining the reasons for the difference between static and dynamic friction coefficients.

9.3 Derive Eq. (9.2).

9.4 Explain the conditions that must exist before you apply the equation $(F_f)_{max} = \mu N$.

9.5 Suppose you develop a fixture consisting of an inclined plane and a block of some solid material to measure the material's coefficient of friction. You measure and record the angle of the inclined plane at the instant motion is initiated. Determine the average value of the friction coefficient from the five separate tests and its range using the data given in the table below.

Problem No.	Angle	Angle	Angle	Angle	Angle
9.5a	22.0°	22.2°	24.3°	21.8°	22.5°
9.5b	24.0°	23.2°	26.3°	24.8°	23.5°
9.5c	32.0°	31.2°	33.3°	32.8°	32.7°
9.5d	37.0°	38.2°	37.9°	39.0°	39.2°

9.6 Derive Eq. (9.7).

9.7 Determine the force F required to move the sled, illustrated in the figure to the right, if the coefficient of friction between the sled and the surface is $\mu = 0.28$.

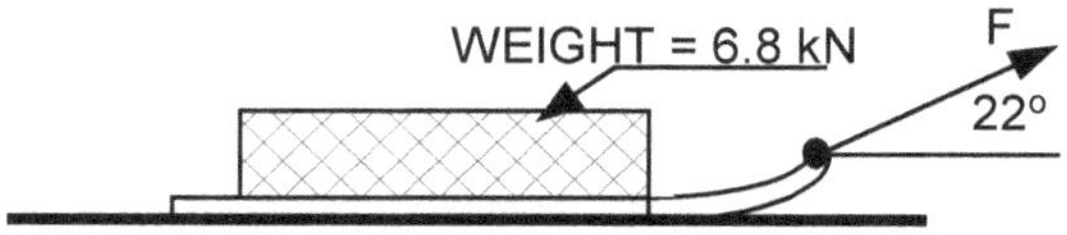

9.8 Determine the force required to slide a box, illustrated in the figure to the left, across a level floor if the force is applied at an angle $\alpha = 10°$. The coefficient of friction between the box and the floor is $\mu = 0.25$.

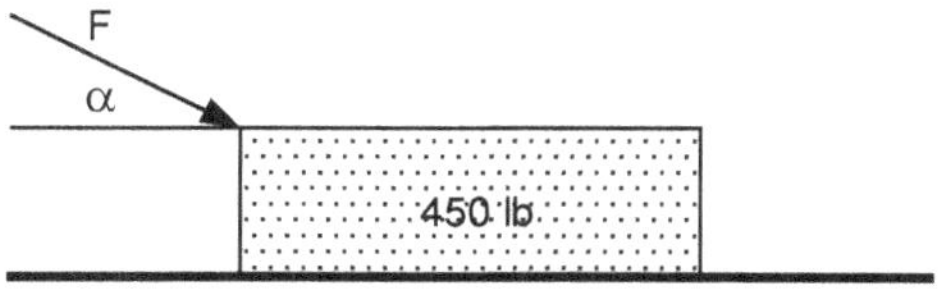

9.9 A worker attempts to slide a wooden crate along a level concrete floor in a warehouse as shown in the figure to the right. The weight of the crate is 80 lb and the worker weighs 120 lb. The coefficient of friction between the crate and the floor $\mu_c = 0.55$ and between the worker and the floor $\mu_w = 0.70$. Note that the crate is short and the worker must lean over to apply a force at an angle of 30° relative to the floor. Is it possible for the worker to move the crate? Justify your answer by showing all the relevant calculations.

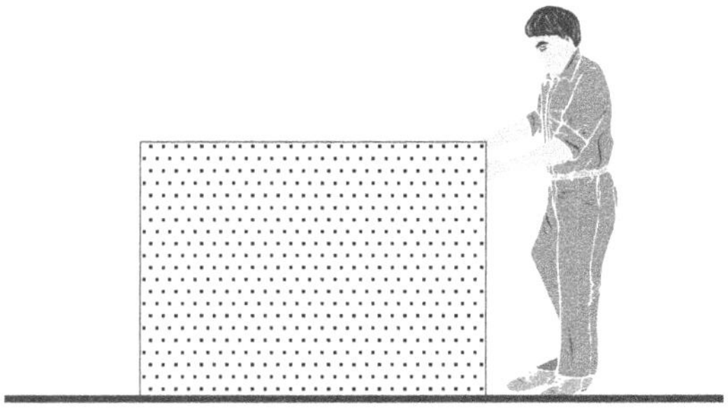

9.10 It has snowed on the day of your final examination for this course, and you find the roads covered with a coating of ice and snow. In your attempt to drive to the University, you cannot negotiate a hill with a grade of 4°. Estimate the coefficient of friction between the road and the tire for your instructor, as you explain the reason for missing the final examination.

9.11 Two blocks rest on a surface as illustrated in the figure to the right. If F_2 = 400 lb, determine the maximum force F_1 that can be applied before one of the blocks moves. The coefficient of friction between the two blocks is 0.28 and between block B and the surface it is 0.22. Block A weighs 1,200 lb and Block B weighs 2,000 lb.

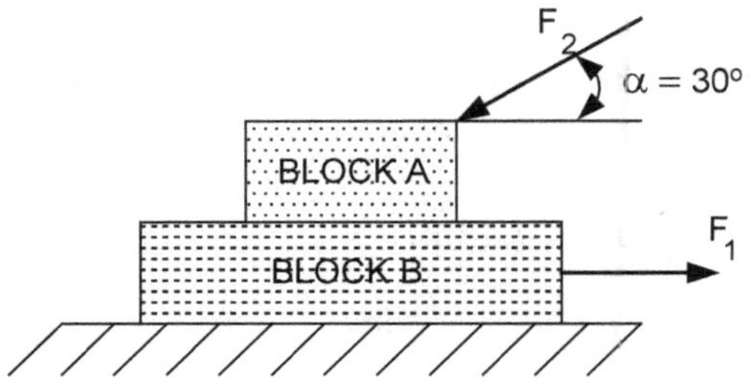

9.12 Determine if the ladder, shown in the figure to the left, is stable or not. The man on the ladder weighs 145 lb and the ladder weighs 52 lb. The coefficient of friction at the base of the ladder is 0.30 and the coefficient of friction between the ladder and the wall is 0.15.

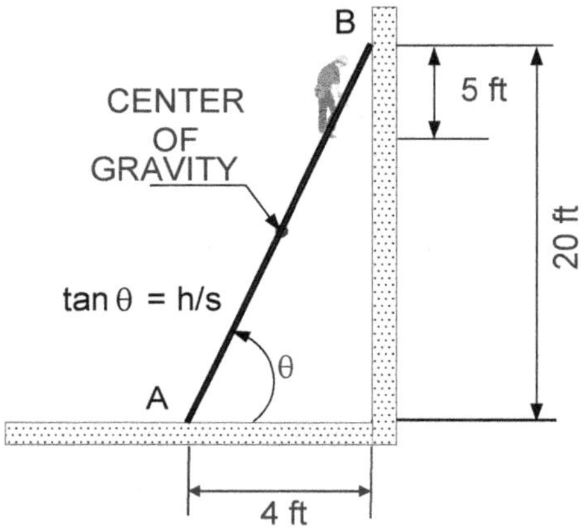

9.13 Determine the force required to move the crate, shown in the figure to the right, and indicate whether it will slide or tip. The crate weighs 720N and the coefficient of friction between it and the floor is μ = 0.42.

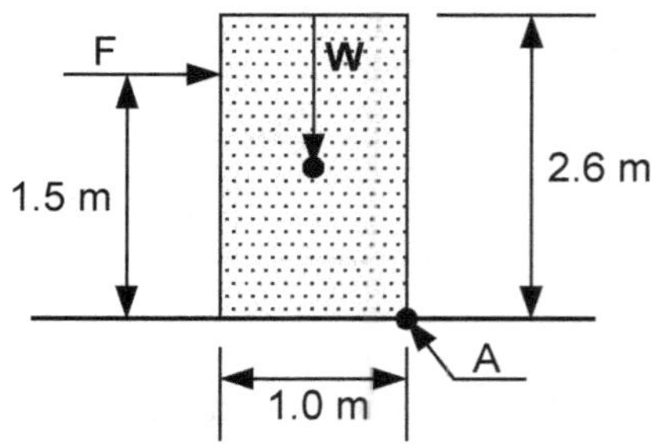

9.14 A cylindrical assembly is pulled by a force F to roll up and over a step as shown in the figure to the left. The cylinder weighs 420 N, and the coefficient of friction between the cylinder and the surfaces at points A and B is 0.43. The radii r_1 = 0.5 m and r_2 = 1.0 m. Determine if the cylinder will roll up and over the step or slip and remain in the corner. Show all of your work and justify your answer.

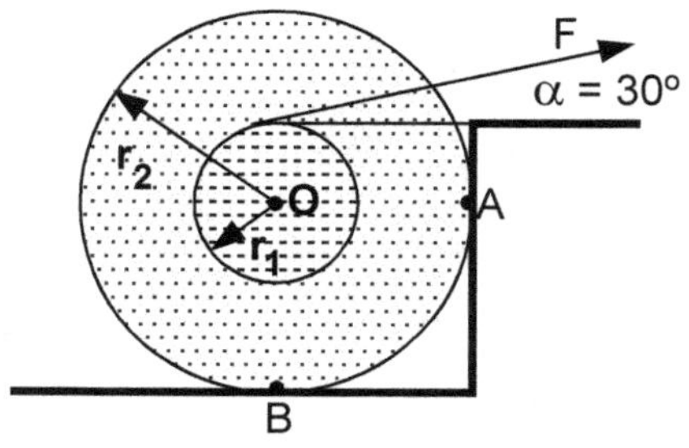

9.15 A wedge inserted at point A is used to lift one end of the machine bed shown in the figure to the right. If the wedge has an included angle $\alpha = 12°$, determine the force necessary to drive the wedge if the coefficient of friction with both surfaces $\mu = 0.35$, and the weight per unit area of the machine bed is 4 kN/m^2.

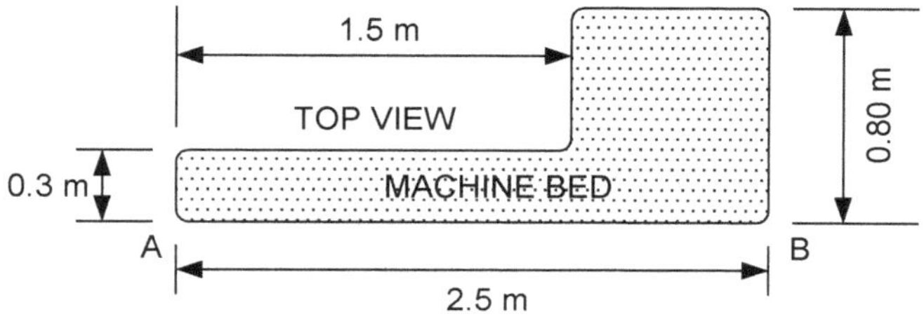

9.16 Reconsider Problem 9.15; however, the wedge is to lift the machine bed from point B and the angle of the wedge is reduced from 12° to 6°.

9.17 Determine the machine coefficient for the wedge, which measures its effectiveness as a simple lifting machine for the results obtained in Problem 9.15. Discuss the parameters leading to the best machine coefficient. Recall, work is defined as **F • d**.

9.18 If the machine bed in Problem 9.15 is lifted through a distance of 60 mm, determine the work expended in driving the wedge. Also determine the work accomplished in lifting the weight of the machine bed. Recall that work is defined as **F • d**.

9.19 If the machine bed in Problem 9.16 is lifted through a distance of 30 mm, determine the work expended in driving the wedge. Also determine the work accomplished in lifting the weight of the machine bed. Recall that work is defined as **F • d**.

9.20 For the conditions described in Problem 9.15, determine if the wedge will hold when the driving force is removed. Also determine the force F_R required to remove the wedge.

9.21 For the conditions described in Problem 9.16, determine if the wedge will hold when the driving force is removed. Also determine the force F_R required to remove the wedge.

9.22 You are lifting one corner of a car with a screw jack. The car weighs 3800lb, and the screw has a diameter of 1.0 in. and helix angle of 4°. The screw is lubricated with a thick, heavy grease and the coefficient of friction between the screw and the nut is $\mu = 0.11$. Determine the torque necessary to turn the screw.

9.23 You are lifting one corner of a heavy plate of steel with a screw jack. The plate weighs 10,000 N, and the screw has a helix angle of 3°. The screw is lubricated with a thick, heavy grease and the coefficient of friction between the screw and the nut is $\mu = 0.11$. Determine the torque necessary to turn the screw.

9.24 You are lifting one corner of a truck with a screw jack. The truck weighs 25 kN, and the screw has a helix angle of 4.5°. The screw is lubricated with a heavy grease and the coefficient of friction between the screw and the nut is $\mu = 0.11$. Determine the torque necessary to turn the screw.

9.25 Determine the locking angle α for helical threads if the coefficient of friction is 0.11.

9.26 Write an engineering brief describing the dangers associated with using a screw with a helical angle of more than 6° in the design of a jack to be used in lifting heavy weights. The screw is made from steel and the nut is made from brass.

9.27 A barrel with a diameter of 0.7 m and weighing 500 N is being rolled up an inclined ramp as shown in the figure to the right. The ramp weighs 250 N. If the coefficient of friction is 0.24 between all contacting surfaces, determine the distance x through which the barrel can be rolled before slippage occurs.

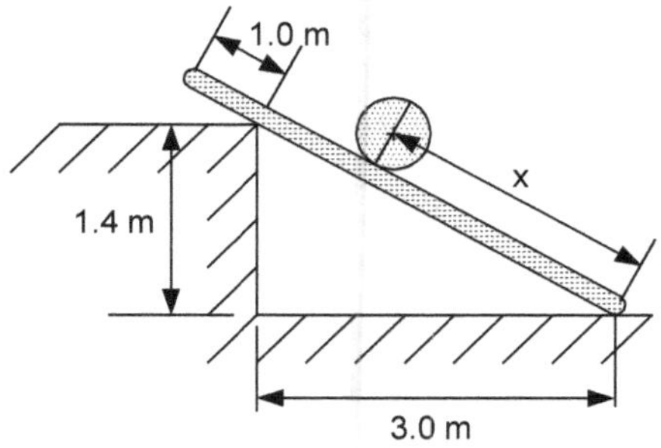

9.28 Block B with a mass m_B = 19 kg sits on top of a larger block A with a mass m_A = 32 kg. A horizontal force F is applied to block B as shown in the figure to the left. If the coefficient of friction between all of the surfaces is 0.28, determine the force F required to initiate motion. Describe the motion that occurs.

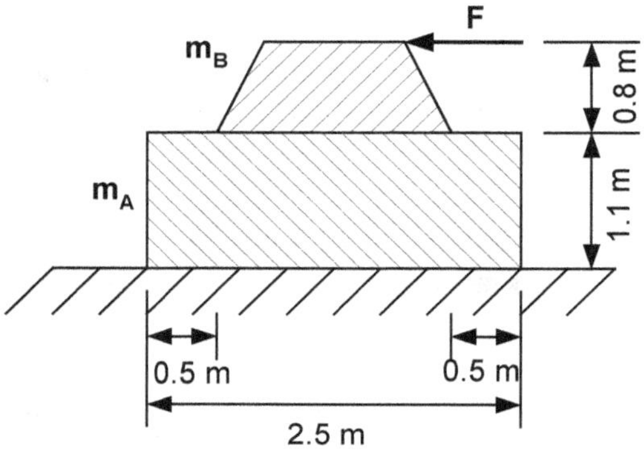

9.29 The links in the toggle mechanism, depicted in the figure to the right, are sufficiently light in weight to be neglected. The pins are frictionless. The coefficient of friction between the block and the surface is 0.34, and the block weighs 3.8 kN. Determine the angle θ that permits the application of the largest force F for which motion of the block does not occur.

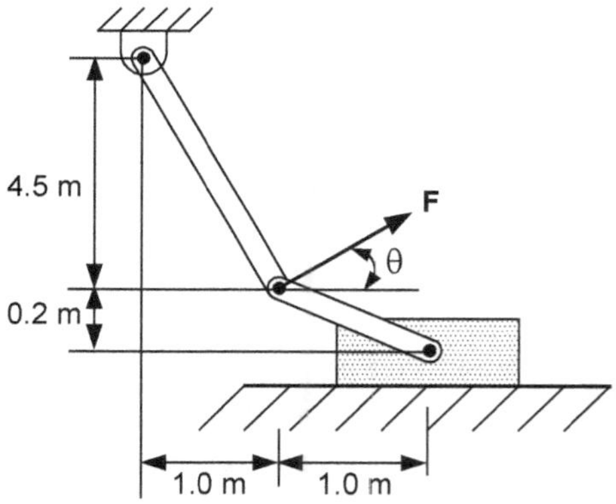

9.30 A carton weighing W_A = 85 lb has tipped and is resting against another carton weighing W_B = 125 as shown in the figure to the left. The coefficient of friction between the floor and the first (tipped) carton is μ_A = 0.45 and the coefficient of friction between the floor and the second (upright) carton is μ_B = 0.40. Determine if the boxes are in equilibrium in the position shown. Neglect the effect of friction between the cartons at the contact point.

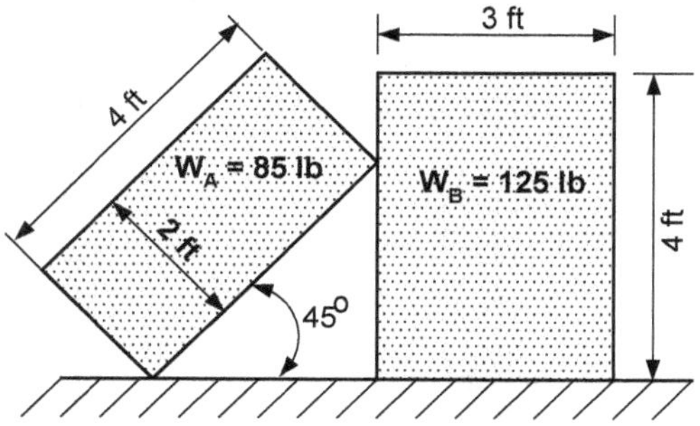

9.31 A crate with a weight W = 2.5 kN is connected to two opposing weights W_1 and W_2 by thin cables that pass over frictionless pulleys as shown in the figure to the right. Determine the minimum and maximum values for the weight W_1 so that motion does not occur. The coefficient of friction between the crate and the floor is 0.28.

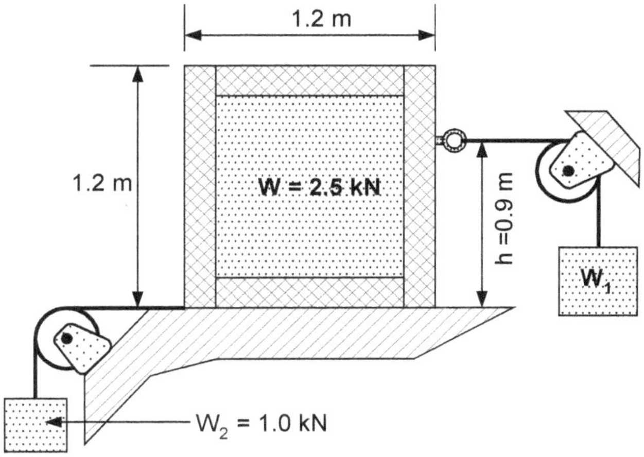

9.32 Consider the crate, shown in the figure to the right, that is subjected to the conditions given in Problem 9.31 except for the height h. Determine the product of h times the weight W_1 that produces motion by slipping and tipping simultaneously.

Problem 9.5(a):

$\theta_s = 22.0°;\ 22.2°;\ 24.3°;\ 21.8°;\ 22.5°$

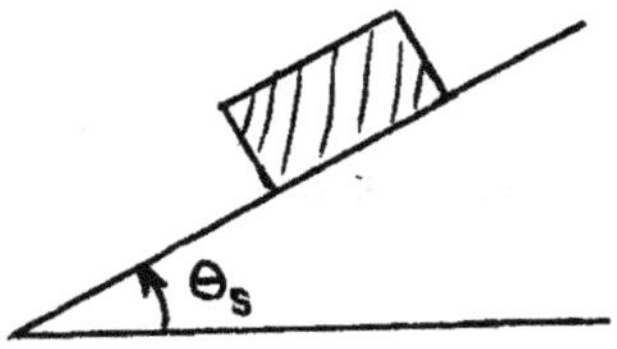

find: $\boxed{\bar{\mu},\ \mu_{range} = ?}$.

① $\mu = TAN\,\theta_s \Rightarrow \mu = 0.4040;\ 0.4081;\ 0.4515;\ 0.4000;\ 0.4142$

② $\bar{\mu} = \dfrac{\Sigma \mu_i}{5} = \dfrac{(0.4040 + 0.4081 + 0.4515 + 0.4000 + 0.4142)}{5}$

③ $\boxed{\bar{\mu} = 0.4156}$.

④ $\mu_{range} = \mu_{max} - \mu_{min} = 0.4515 - 0.4000$

⑤ $\boxed{\mu_{range} = 0.0515}$.

Problem 9.5(b):

$\theta_s = 24.0°;\ 23.2°;\ 26.3°;\ 24.8°;\ 23.5°$

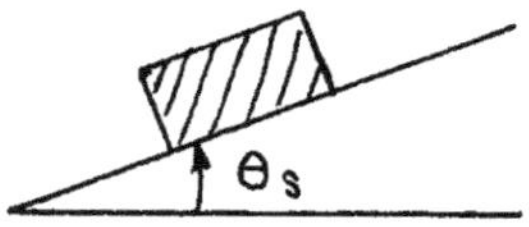

find: $\boxed{\bar{\mu},\ \mu_{range} = ?}$.

① $\mu = TAN\,\theta_s \Rightarrow \mu = 0.4452;\ 0.4286;\ 0.4942;\ 0.4621;\ 0.4348$

② $\bar{\mu} = \dfrac{\sum_{i=1}^{n} \mu_i}{n} = \dfrac{(0.4452 + 0.4286 + 0.4942 + 0.4621 + 0.4348)}{5}$

③ $\boxed{\bar{\mu} = 0.4530}$.

④ $\mu_{range} = \mu_{max} - \mu_{min} = 0.4942 - 0.4286$

⑤ $\boxed{\mu_{range} = 0.0656}$.

Problem 9.7:

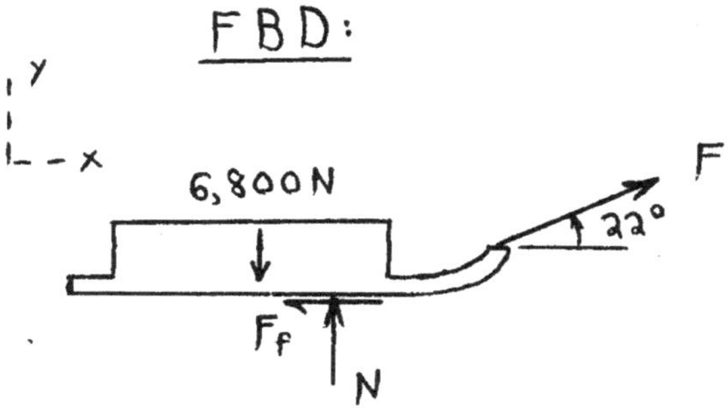

$\mu = 0.28$ * impending motion

find : $\boxed{F = ?}$.

① $\overset{+}{\rightarrow}\Sigma F_x = 0: \quad F\cos(22°) - F_f = 0 \Rightarrow F_f = F\cos(22°)$

② $\uparrow + \Sigma F_y = 0: \quad F\sin(22°) + N - 6{,}800 = 0$

③ $N = 6{,}800 - F\sin(22°)$

④ for impending motion : $F_f = \mu N$

⑤ subing ① + ③ in ④: $F\cos(22°) = (0.28)(6{,}800 - F\sin(22°))$

⑥ $F(0.92718 + 0.10489) = 1{,}904$

⑦ $\boxed{F = 1{,}845 \text{ N}}$.

* completing problem : $N = 6{,}109$ N

Problem 9.9:

FBD's:

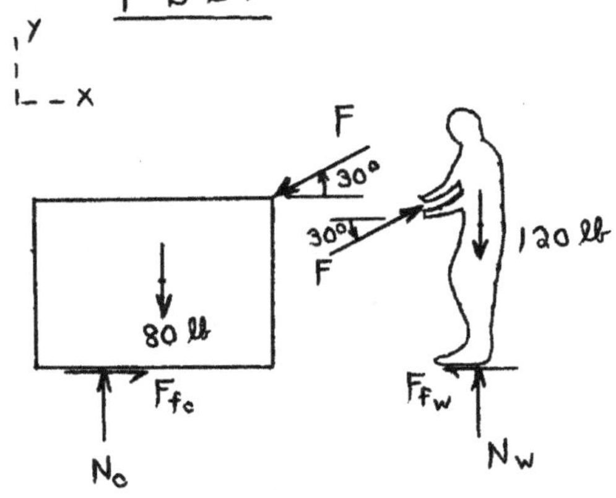

$\mu_c = 0.55 \quad \mu_w = 0.70$

find: can worker move crate?

* consider crate:

① $\overset{+}{\rightarrow}\Sigma F_x = 0: \quad F_{fc} - F\cos(30°) = 0 \Rightarrow F_{fc} = F\cos(30°)$

② $+\uparrow\Sigma F_y = 0: \quad N_c - 80 - F\sin(30°) = 0$

③ $N_c = 80 + F\sin(30°)$

④ for impending motion: $F_{fc} = \mu_c N_c$

⑤ subing ① + ③ in ④: $F\cos(30°) = (0.55)(80 + F\sin(30°))$

⑥ $F(0.86603 - 0.275) = 44$

⑦ $F = 74.45$ lb

* consider worker:

⑧ $\overset{+}{\rightarrow}\Sigma F_x = 0: \quad (74.45)\cos(30°) - F_{fw} = 0$

⑨ $F_{fw} = 64.48$ lb

⑩ $+\uparrow\Sigma F_y = 0: \quad N_w + (74.45)\sin(30°) - 120 = 0$

⑪ $N_w = 82.78$ lb

⑫ $(F_{fw})_{max} = \mu_w N_w = (0.70)(82.78) = 57.95$ lb

∴ since $F_{fw} = 64.48$ lb exceeds $(F_{fw})_{max} = 57.95$ lb, then worker __cannot__ move crate (i.e. worker slips).

Problem 9.12:

FBD:

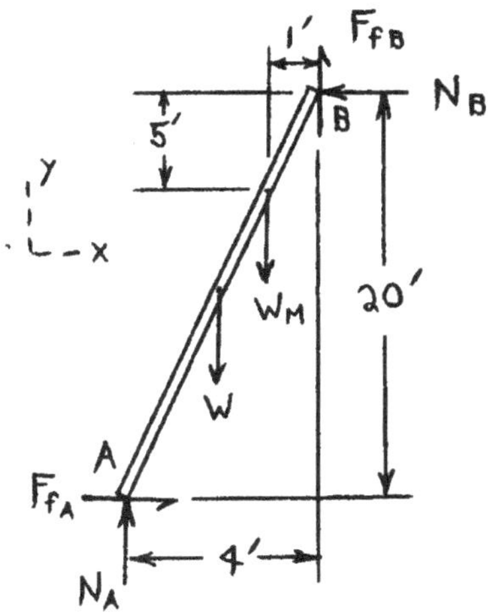

$W = 52$ lb $\qquad \mu_A = 0.30$

$W_M = 145$ lb $\qquad \mu_B = 0.15$

find: stability = ?

① $\overset{+}{\rightarrow}\sum F_x = 0: \quad F_{fA} - N_B = 0 \Rightarrow F_{fA} = N_B$

② $\uparrow + \sum F_y = 0: \quad N_A + F_{fB} - 52 - 145 = 0 \Rightarrow N_A + F_{fB} = 197$

③ $\circlearrowleft + \sum M_A = 0: \; -(52)(2) - (145)(3) + F_{fB}(4) + N_B(20) = 0$

④ $4F_{fB} + 20N_B = 539$

⑤ assume impending motion @ B: $F_{fB} = \mu_B N_B = 0.15 N_B$

⑥ subing ⑤ in ④: $4(0.15 N_B) + 20 N_B = 539$

⑦ $N_B = 26.17$ lb

⑧ $F_{fB} = 0.15(26.17) = 3.926$ lb

⑨ using F_{fB} + N_B in ① + ②:

$F_{fA} = 26.17$ lb

$N_A = 197 - (3.926) = 193.1$ lb

⑩ $(F_{fA})_{max} = \mu_A N_A = (0.3)(193.1) = 57.93$ lb

⑪ $F_{fA} = 26.17 < 57.93 = (F_{fA})_{max} \Rightarrow$ pt. A does not slip

∴ ladder is stable.

Problem 9.13:

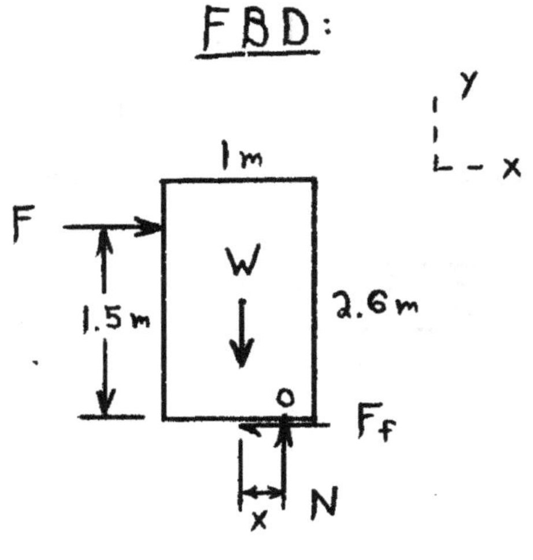

$W = 720\text{ N}$ $\mu = 0.42$

* crate begins to move

find: $F = ?$, slide or tip?

① $\xrightarrow{+}\sum F_x = 0:\quad F - F_f = 0 \;\Rightarrow\; F = F_f$

② $+\uparrow\sum F_y = 0:\quad N - 720 = 0 \;\Rightarrow\; N = 720\text{ N}$

* assume crate slides:

③ $F_f = \mu N = (0.42)(720) = 302.4\text{ N}$

④ using ③ in ①: $F_{slide} = 302.4\text{ N}$

* assume crate tips:

⑤ $x = \frac{1}{2}(1) = 0.5\text{ m}$

⑥ $\circlearrowleft + \sum M_o = 0:\quad (720)(0.5) - F(1.5) = 0$

⑦ $F_{tip} = 240\text{ N}$

⑧ choose smallest F $\Rightarrow$ $F = 240\text{ N}$, for tipping.

Problem 9.30:

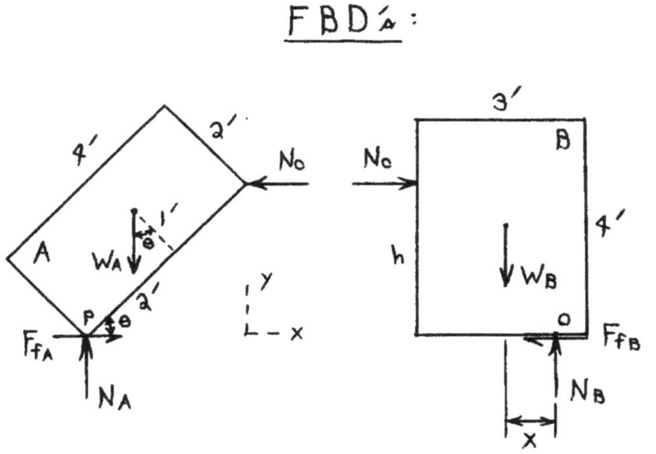

$W_A = 85$ lb

$W_B = 125$ lb

$\theta = 45°$

$\mu_A = 0.45$

$\mu_B = 0.40$

find: equilibrium ?

* consider block A:

① $\overset{+}{\rightarrow}\sum F_x = 0: \quad F_{fA} - N_c = 0 \quad \Rightarrow \quad F_{fA} = N_c$

② $+\uparrow\sum F_y = 0: \quad N_A - 85 = 0 \quad \Rightarrow \quad N_A = 85$ lb

③ $\circlearrowleft+\sum M_P = 0: \quad N_c\left((4)\sin(45°)\right) - (85)\left((2)\cos(45°) - (1)\sin(45°)\right) = 0$

④ $N_c = 21.25$ lb

⑤ from ①: $F_{fA} = N_c = 21.25$ lb

⑥ $(F_{fA})_{max} = \mu_A N_A = (0.45)(85) = 38.25$ lb

⑦ since $F_{fA} = 21.25$ lb < 38.25 lb $= (F_{fA})_{max}$, then A does not slip

* consider block B:

⑧ $\overset{+}{\rightarrow}\sum F_x = 0: \quad N_c - F_{fB} = 0 \quad \Rightarrow \quad F_{fB} = N_c = 21.25$ lb

⑨ $+\uparrow\sum F_y = 0: \quad N_B - 125 = 0 \quad \Rightarrow \quad N_B = 125$ lb

⑩ $\circlearrowleft+\sum M_o = 0: \quad N_B\, x - N_c\left((4)\sin(45°)\right) = 0$

⑪ $x = \dfrac{(21.25)\left((4)\sin(45°)\right)}{125} = 0.4808$ ft

⑫ $(F_{fB})_{max} = \mu_B N_B = (0.4)(125) = 50$ lb

⑬ since $F_{fB} = 21.25$ lb < 50 lb $= (F_{fB})_{max}$, then B does not slip

⑭ since $x = 0.4808$ ft < 1.5 ft, then B does not tip

∴ **both blocks are in equilibrium**.

Problem 9.31:

FBD's :

$\mu = 0.28$

* no motion

find : $(W_1)_{max}$, $(W_1)_{min}$ = ?

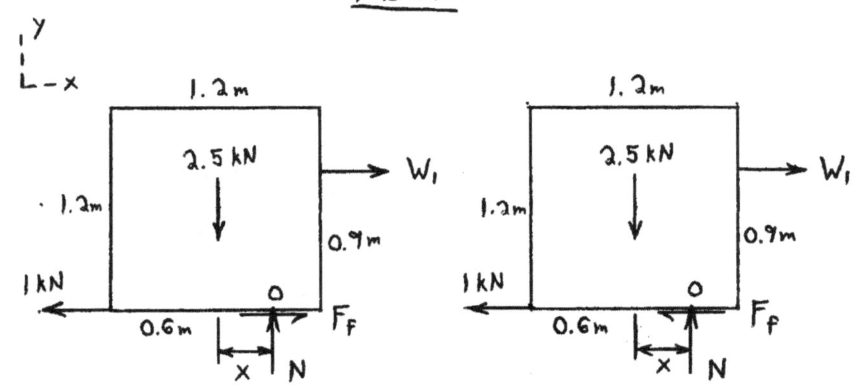

crate moves left — crate moves right

* consider sliding to left :

① $\overset{+}{\rightarrow}\Sigma F_x = 0: \; -1 + W_1 + F_f = 0 \;\Rightarrow\; W_1 = 1 - F_f$

② $+\uparrow\Sigma F_y = 0: \; N - 2.5 = 0 \;\Rightarrow\; N = 2.5 \text{ kN}$

③ $F_f = \mu N = (0.28)(2.5) = 0.7 \text{ kN}$

④ using ③ in ①: $W_1 = 1 - (0.7) = 0.3 \text{ kN}$

* consider sliding to right :

⑤ $\overset{+}{\rightarrow}\Sigma F_x = 0: \; -1 - F_f + W_1 = 0 \;\Rightarrow\; W_1 = 1 + F_f$

⑥ N + F_f are unchanged $\Rightarrow$ $N = 2.5 \text{ kN}$; $F_f = 0.7 \text{ kN}$

⑦ subing ⑥ in ⑤: $W_1 = 1 + (0.7) = 1.7 \text{ kN}$

* consider tipping to right :

⑧ $x = \frac{1}{2}(1.2) = 0.6 \text{ m}$

⑨ $\circlearrowleft + \Sigma M_o = 0: \; (2.5)(0.6) - W_1(0.9) = 0$

⑩ $W_1 = 1.667 \text{ kN}$

⑪ choose smallest loads for moving both left + right

$\Rightarrow$ $(W_1)_{min} = 0.3 \text{ kN}$

$(W_1)_{max} = 1.667 \text{ kN}$

APPENDIX C PROBLEMS

C.1 Determine the area of a right triangle if its base is 30 mm and its height is 70 mm.

C.2 Determine the area of an isosceles triangle if its base is 30 mm and its height is 70 mm.

C.3 Determine the area of a right triangle if its base is 4in. and its height is 7 in.

C.4 Determine the area of an isosceles triangle if its base is 5 in. and its height is 9.5 in.

C.5 Determine the area of a circle with a diameter of 14 mm.

C.6 Determine the area of a circle with a diameter of 3.2 in.

C.7 Determine the area of a circle with a diameter of 88 mm.

C.8 Determine the area of a circle with a diameter of 3.6 in.

C.9 Determine the area of an ellipse if the major and minor axes are defined by a = 2b = 16 mm.

C.10 Determine the area of an ellipse if the major and minor axes are defined by a = 3b = 6 in.

C.11 Determine the area of an ellipse if the major and minor axes are defined by a = 4b = 20 mm.

C.12 Determine the area of an ellipse if the major and minor axes are defined by a = (3.5)b = 7.0 in.

C.13 Draw a right triangle with a base of 30 mm and height of 70 mm and then locate and dimension its centroid.

C.14 Draw an isosceles triangle with a base of 30 mm and height of 70 mm and then locate and dimension its centroid.

C.15 Draw a right triangle with a base of 4in. and height of 7 in. and then locate and dimension its centroid.

C.16 Draw an isosceles triangle with a base of 5 in. and height of 9.5 in. and then locate and dimension its centroid.

C.17 Verify the location of the centroid for the semicircular area defined in the figure to the right.

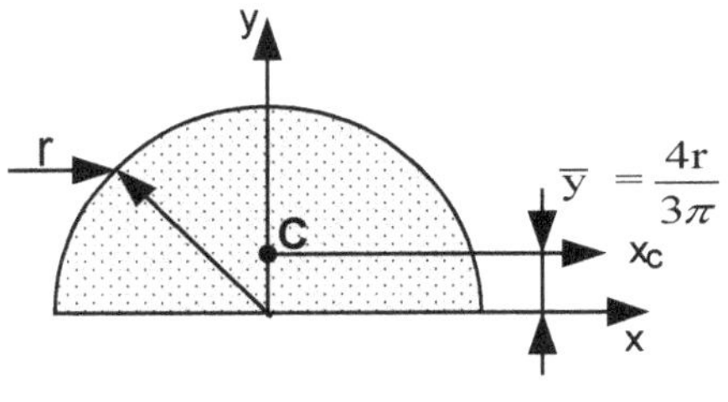

C.18 Verify the location of the centroid for the semi-elliptical area defined in the figure to the right.

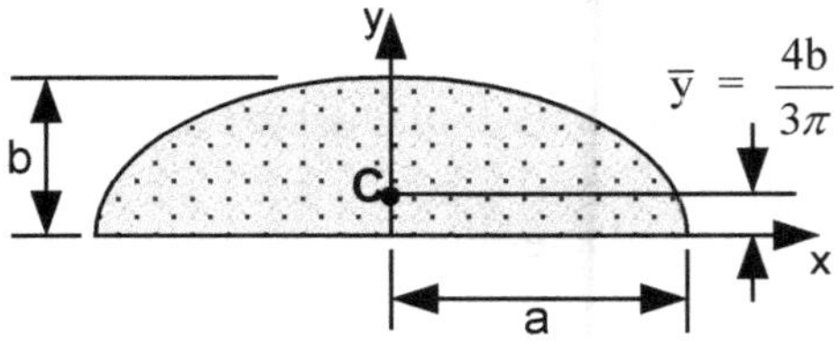

C.19 Verify the location of the centroid for the parabolic area defined in the figure to the left.

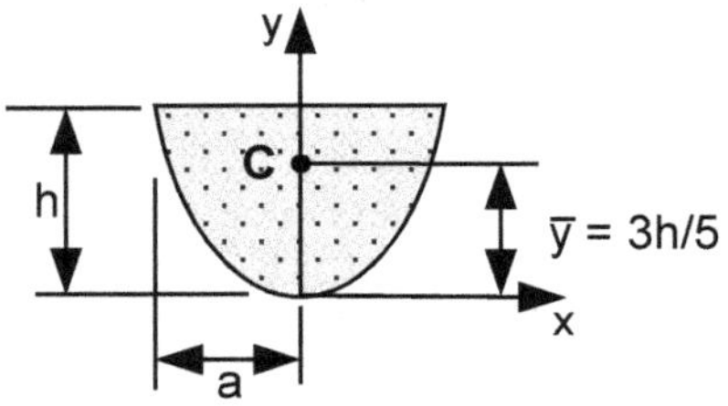

C.20 Verify the location of the centroid for the circular sector defined in the figure to the right.

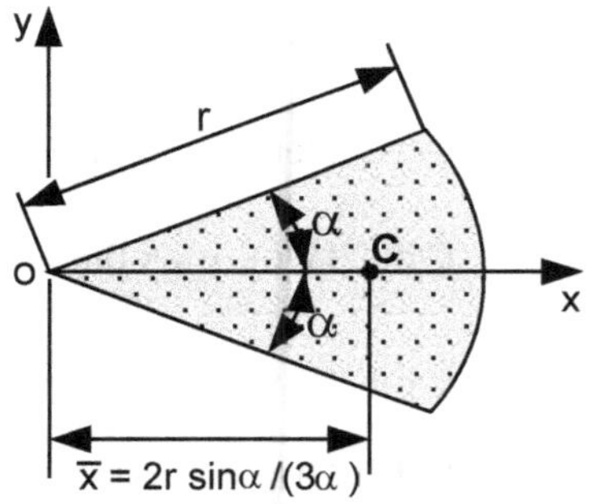

C.21 Verify the location of the centroid for the general spandrel defined in the figure to the left.

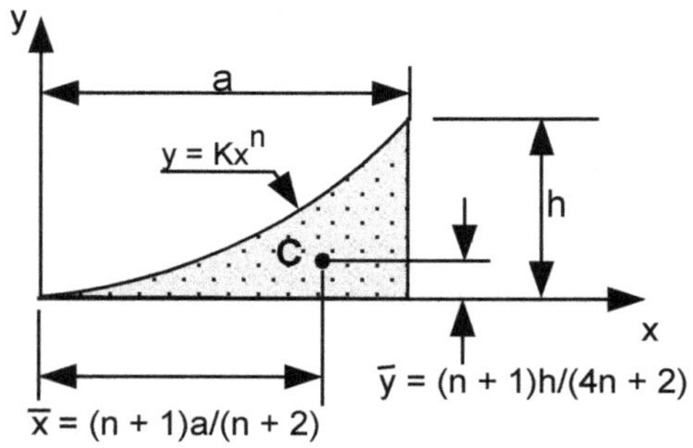

C.22 For the unusual plate shown in the figure to the right, derive an equation locating the centroid relative to point A. The elliptical hole is centered in the rectangular portion of the plate.

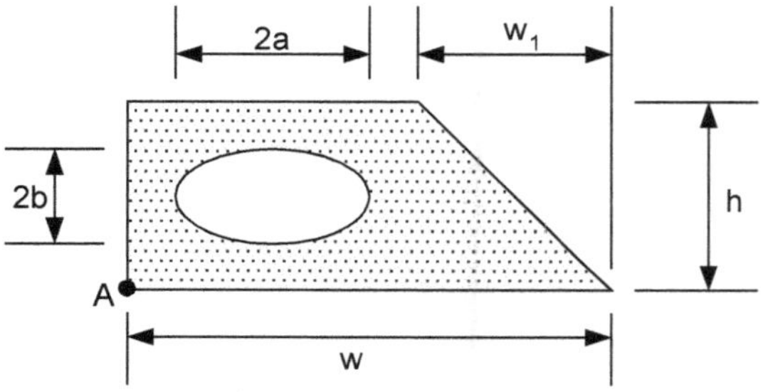

Properties of Areas

C.23 Determine the location of the centroid relative to point A for the plate defined in the figure to the right. All dimensions are given in mm, and the elliptical hole is centered in the rectangular portion of the plate.

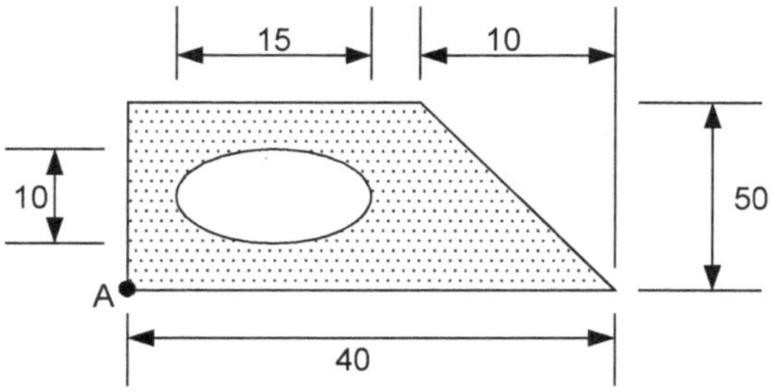

C.24 Divide the plate, illustrated in the figure to the right, into three areas (an ellipse, rectangle and triangle). Determine the moment of inertia for each area about its centroidal axis.

C.25 Verify that the moment of inertia of an ellipse about its centroidal axis is given by:

$$I_x = \pi a b^3/4 \qquad \text{and} \qquad I_y = \pi a^3 b/4$$

C.26 Determine the moment of inertia for the total area of the plate, shown in the figure to the left, about its centroidal axis.

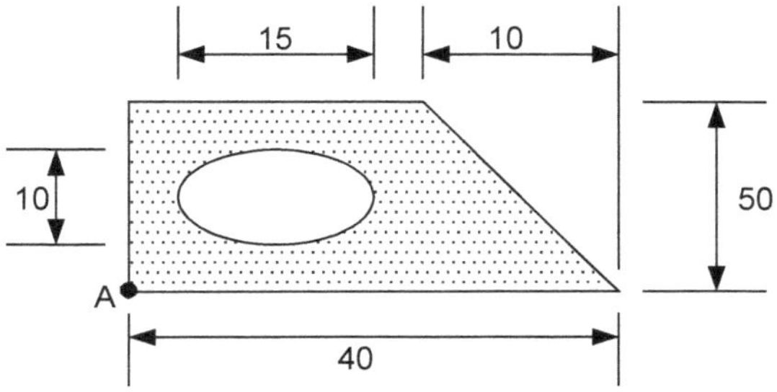

C.27 For the U shaped section, shown in the figure to the right, determine the location of the centroid $\overline{y}$ relative to an axis along its base. All dimensions are in inches.

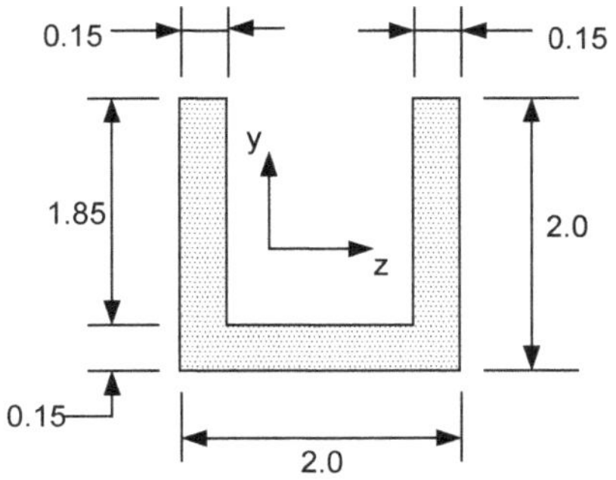

C.28 Determine the location of the centroid $\overline{y}$, relative to an axis along its base, for the U shaped section shown in the figure to the left. All dimensions are in inches.

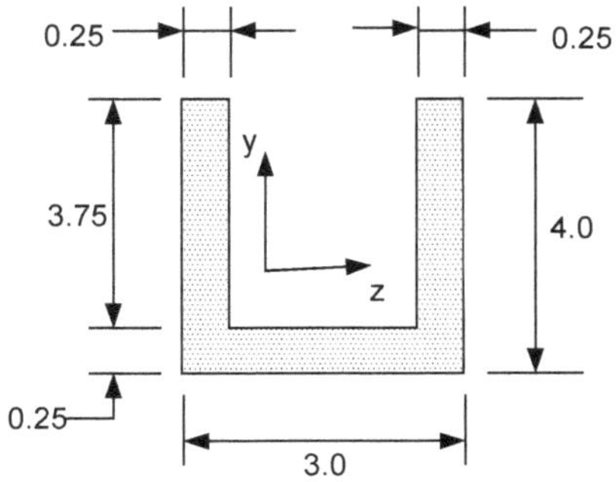

C.29 Determine the moment of inertia I_z about the centroid of the U shaped member defined in (a) Problem C.27 and (b) Problem C.28.

C.30 Verify that the polar moment of inertia of a quarter circle area with a radius R is $J_o = \pi R^4/8$.

C.31 For the uncommon shape presented in the figure to the right, determine the centroid and the moment of inertia relative to the centroidal axis. The rectangular hole is centered in the elliptical area.

C.32 For the uncommon shape presented in the figure to the right, determine the moment of inertia relative to the z-axis.

C.33 For the uncommon shape presented in the figure to the right, determine the moment of inertia relative to the y-axis.

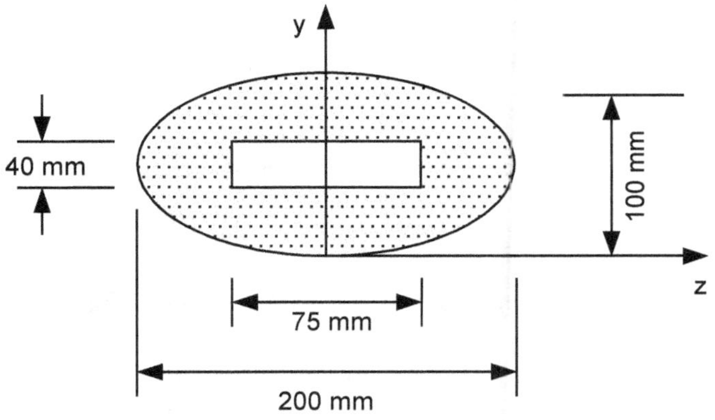

C.34 For the irregular cross sectional area, described in the figure to the left, locate the centroid relative to both the x and y axes. The dimensions are given in mm.

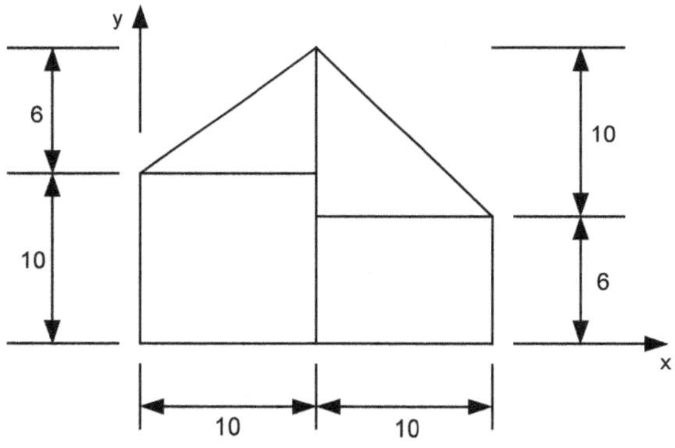

Problem C.17:

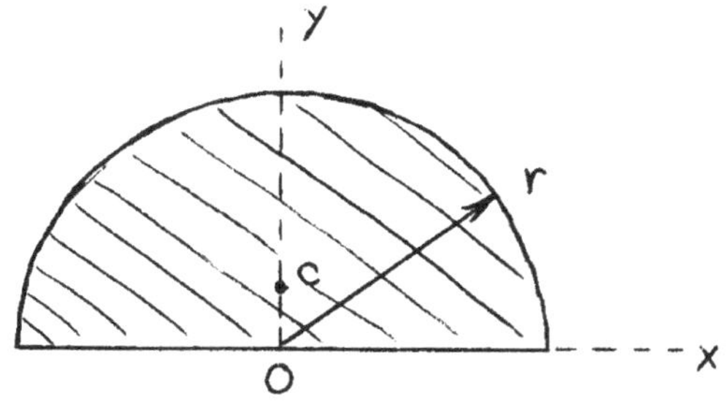

verify: location of centroid for semicircular area, by integration method

$$c = \left(0, \frac{4r}{3\pi}\right)$$

① since semicircular area is symmetric w.r.t. y-axis (i.e. in horizontal direction) $\Rightarrow$ $\boxed{\bar{x} = 0}$.

② recall eqn. of a circle, centered at O: $x^2 + y^2 = r^2$

③ $Q_x = \int_A y\,dA = A\bar{y} \Rightarrow \bar{y} = \dfrac{Q_x}{A}$

④ consider differential area dA, parallel to y-axis:

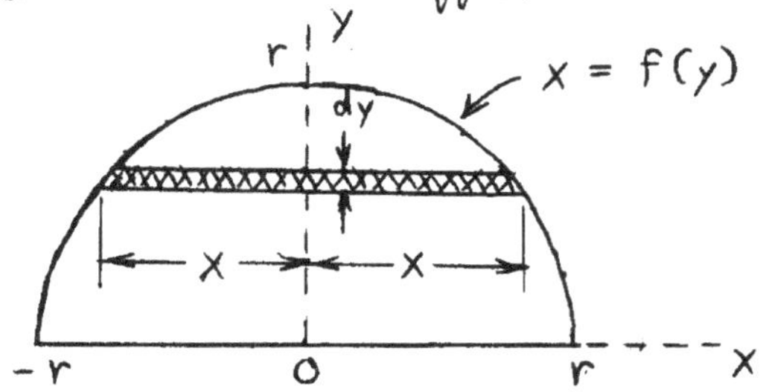

⑤ $dA = (2x)\,dy$

⑥ using ⑤ in ③:

$$Q_x = \int_A y(2x)\,dy$$

⑦ from ②: $x = (r^2 - y^2)^{1/2}$

⑧ subing ⑦ in ⑥: $Q_x = \int_0^r 2y(r^2 - y^2)^{1/2}\,dy$

⑨ let $u = r^2 - y^2 \Rightarrow du = -2y\,dy$

⑩ $Q_x = -\int_0^r u^{1/2}\,du = -\dfrac{2}{3}u^{3/2}\Big]_0^r$

⑪ $Q_x = -\dfrac{2}{3}(r^2 - y^2)^{3/2}\Big]_0^r$

⑫ $Q_x = -\dfrac{2}{3}\left[0 - r^3\right] = \dfrac{2}{3}r^3$

⑬ recall for a semicircle: $A = \dfrac{\pi r^2}{2}$

⑭ $\bar{y} = \dfrac{Q_x}{A} = \dfrac{\frac{2}{3}r^3}{\frac{\pi r^2}{2}} \Rightarrow \boxed{\bar{y} = \dfrac{4r}{3\pi}}$.

Problem C.18:

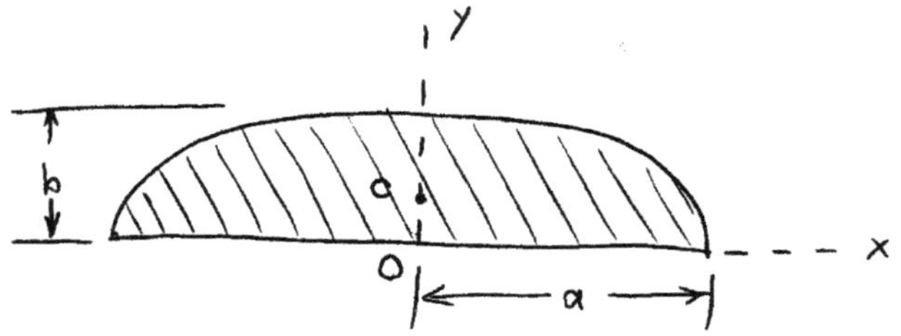

verify: location of centroid, by integration method

$C = \left(0, \frac{4b}{3\pi}\right)$

① recall eqn. of an ellipse, centered at O: $\frac{x^2}{a^2} + \frac{y^2}{b^2} = 1$

② since elliptical area is symmetric w.r.t. y-axis (i.e. in horizontal direction) $\Rightarrow$ $\boxed{\bar{x} = 0}$.

③ $Q_x = \int_A y\,dA = A\bar{y} \Rightarrow \bar{y} = \frac{Q_x}{A}$

④ from symmetry w.r.t. y-axis, only 1/2 of the ellipse must be considered:

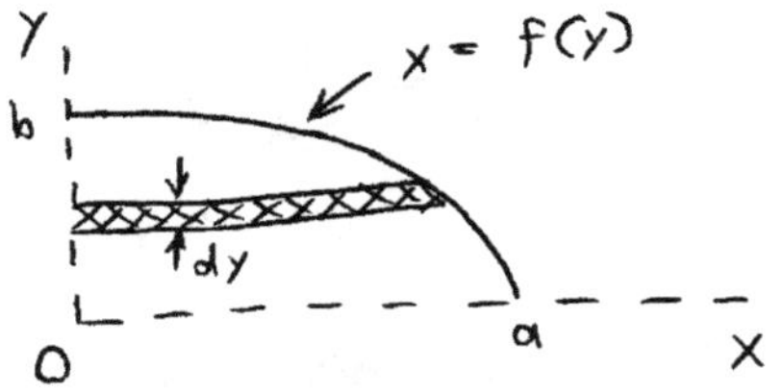

⑤ $Q_x = 2\int_0^b y(x\,dy)$

⑥ $x = a\left(1 - \frac{y^2}{b^2}\right)^{1/2}$

⑦ subing ⑥ in ⑤:

$$Q_x = 2\int_0^b a\left(1 - \frac{y^2}{b^2}\right)^{1/2} y\,dy$$

⑧ let $u = 1 - \frac{y^2}{b^2} \Rightarrow du = -\frac{2y}{b^2}\,dy$

⑨ $Q_x = -b^2 a \int_0^b u^{1/2}\,du = -b^2 a\left(\frac{2}{3}u^{3/2}\right)\Big]_0^b$

⑩ $Q_x = -b^2 a\left(\frac{2}{3}\left(1 - \frac{y^2}{b^2}\right)^{3/2}\right)\Big]_0^b$

⑪ $Q_x = -\frac{2b^2 a}{3}\left[0 - 1\right] = \frac{2}{3}ab^2$

⑫ recall for a 1/2 ellipse: $A = \frac{1}{2}\pi ab$

⑬ $\bar{y} = \frac{Q_x}{A} = \frac{2/3\,ab^2}{1/2\,\pi ab} \Rightarrow \boxed{\bar{y} = \frac{4b}{3\pi}}$.

Problem C.23:

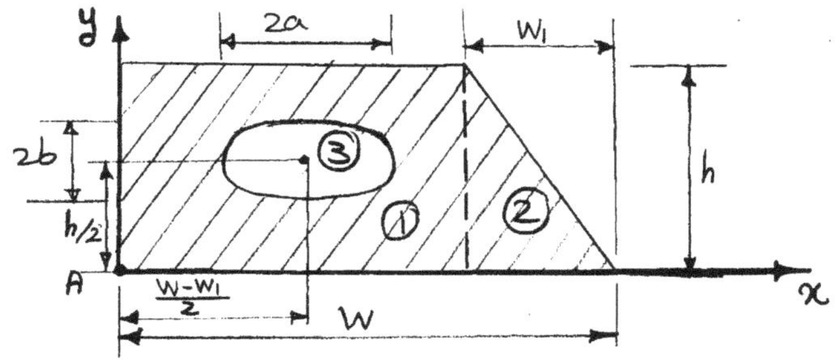

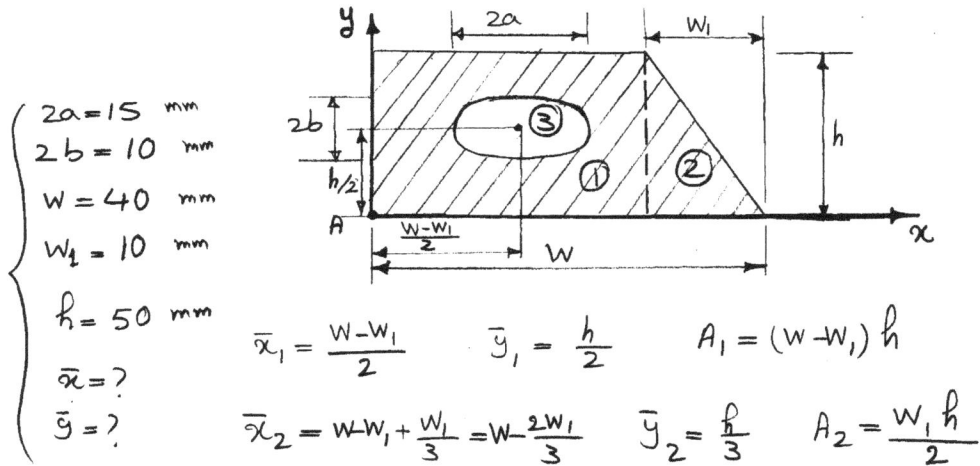

$$\begin{cases} 2a = 15 \text{ mm} \\ 2b = 10 \text{ mm} \\ W = 40 \text{ mm} \\ W_1 = 10 \text{ mm} \\ h = 50 \text{ mm} \\ \bar{x} = ? \\ \bar{y} = ? \end{cases}$$

$$\bar{x}_1 = \frac{W - W_1}{2} \qquad \bar{y}_1 = \frac{h}{2} \qquad A_1 = (W - W_1)\,h$$

$$\bar{x}_2 = W - W_1 + \frac{W_1}{3} = W - \frac{2W_1}{3} \qquad \bar{y}_2 = \frac{h}{3} \qquad A_2 = \frac{W_1 h}{2}$$

$$\bar{x}_3 = \frac{W - W_1}{2} \qquad \bar{y}_3 = \frac{h}{2} \qquad A_3 = \pi a b$$

$$\bar{x} = \frac{\sum_{n=1}^{3} \bar{x}_n A_n}{A_t}$$

$$\bar{x} = \frac{\bar{x}_1 A_1 + \bar{x}_2 A_2 - \bar{x}_3 A_3}{A_1 + A_2 - A_3}$$

$$\bar{x}_1 = 15 \qquad \bar{y}_1 = 25 \qquad A_1 = 1500$$

$$\bar{x}_2 = 33.3 \qquad \bar{y}_2 = 16.7 \qquad A_2 = 250$$

$$\bar{x}_3 = 15 \qquad \bar{y}_3 = 25 \qquad A_3 = \pi \times 7.5 \times 5 = 117.8$$

$$A_t = 1632.2$$

$$\bar{x} = \frac{15 \times 1500 + 33.3 \times 250 - 15 \times 117.8}{1632.2}$$

$$\therefore \boxed{\bar{x} = 17.8 \text{ mm}}$$

$$\bar{y} = \frac{\sum_{n=1}^{3} \bar{y}_n A_n}{A_t}$$

$$\bar{y} = \frac{\bar{y}_1 A_1 + \bar{y}_2 A_2 - \bar{y}_3 A_3}{A_1 + A_2 - A_3}$$

$$\therefore \boxed{\bar{y} = 23.7 \text{ mm}}$$

Problem C.23 - Alternate Method:

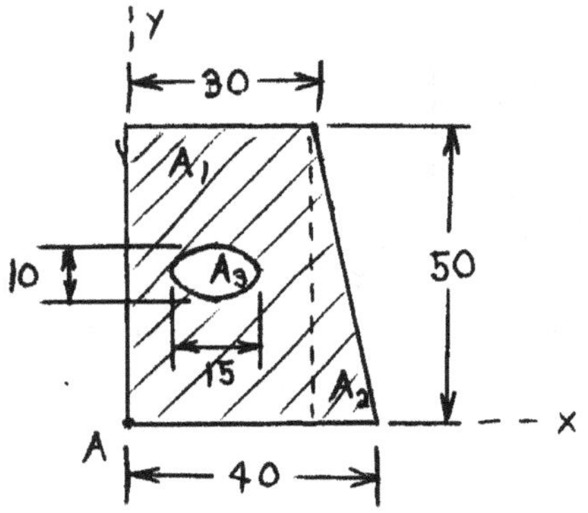

* elliptical hole centered in rectangle
* all dimensions in mm

find: $\bar{x}, \bar{y} = ?$

① break area into 3 parts: rectangle, triangle, + elliptical hole

②

	A_i (mm^2)	$\bar{x}_i$ (mm)	$A_i \bar{x}_i$ (mm^3)
A_1	$(30)(50) = 1500$	$\frac{1}{2}(30) = 15$	22,500
A_2	$\frac{1}{2}(10)(50) = 250$	$30 + \frac{(10)}{3} = 33.33$	8,333
A_3	$-\pi(7.5)(5) = -117.8$	$\frac{1}{2}(30) = 15$	$-1,767$
Σ	1,632.2		29,066

③ $\bar{x} = \dfrac{\Sigma A_i \bar{x}_i}{\Sigma A_i} = \dfrac{29,066}{1,632.2} \Rightarrow \boxed{\bar{x} = 17.81 \text{ mm}}$

④

	A_i (mm^2)	$\bar{y}_i$ (mm)	$A_i \bar{y}_i$ (mm^3)
A_1	1500	$\frac{1}{2}(50) = 25$	37,500
A_2	250	$\frac{1}{3}(50) = 16.67$	4,168
A_3	-117.8	$\frac{1}{2}(50) = 25$	$-2,945$
Σ	1,632.2		38,723

⑤ $\bar{y} = \dfrac{\Sigma A_i \bar{y}_i}{\Sigma A_i} = \dfrac{38,723}{1,632.2} \Rightarrow \boxed{\bar{y} = 23.72 \text{ mm}}$

Problem C.25:

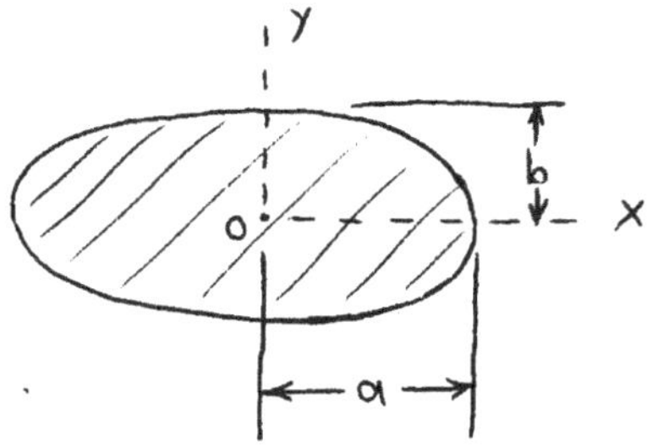

ellipse : centered at origin O

verify : $I_x = \frac{\pi}{4} a b^3$

$I_y = \frac{\pi}{4} b a^3$

① recall eqn. for ellipse centered at origin : $\frac{x^2}{a^2} + \frac{y^2}{b^2} = 1$

② $x^2 = \frac{a^2}{b^2}(b^2 - y^2) \Rightarrow x = \frac{a}{b}\sqrt{b^2 - y^2}$

③ consider ellipse :

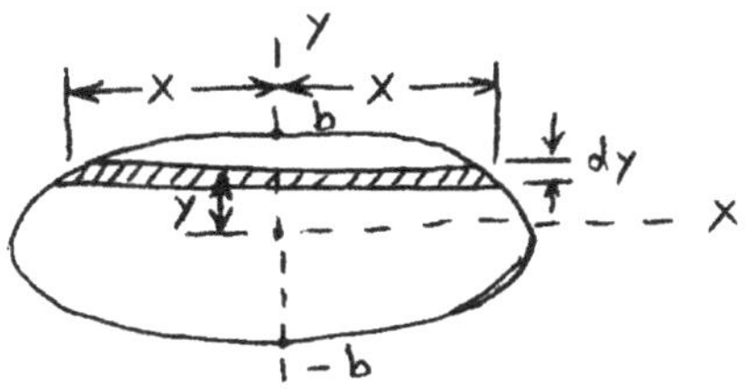

④ $I_x = \int_A y^2 \, dA$

⑤ $I_x = \int_A y^2 (2x \, dy)$

⑥ $I_x = 2\int_A y^2 x \, dy$

⑦ subing ② in ⑥ :

$$I_x = 2\frac{a}{b}\int_{-b}^{b} y^2 \sqrt{b^2 - y^2} \, dy$$

⑧ from table of integrals :

$$\int y^2 \sqrt{b^2 - y^2} \, dy = -\frac{y}{4}\sqrt{(b^2 - y^2)^3} + \frac{b^2}{8}\left(y\sqrt{b^2 - y^2} + b^2 \sin^{-1}\left(\frac{y}{b}\right)\right)$$

⑨ using ⑧ in ⑦ :

$$I_x = 2\frac{a}{b}\left\{-\frac{y}{4}\sqrt{(b^2 - y^2)^3} + \frac{b^2}{8} y\sqrt{b^2 - y^2} + \frac{b^4}{8}\sin^{-1}\left(\frac{y}{b}\right)\right\}\Bigg]_{-b}^{b}$$

⑩ $I_x = 2\frac{a}{b}\left\{-0 + 0 + \frac{b^4}{8}\left(\frac{\pi}{2}\right) + 0 - 0 - \frac{b^4}{8}\left(-\frac{\pi}{2}\right)\right\}$

⑪ $I_x = 2\frac{a}{b}\left\{\frac{\pi b^4}{8}\right\}$

⑫ $\boxed{I_x = \frac{\pi}{4} a b^3}$.

⑬ using similar analysis for I_y :

$y = \frac{b}{a}\sqrt{a^2 - x^2}$

$I_y = \int_A x^2 \, dA = 2\int_{-a}^{a} x^2 y \, dx \Rightarrow \boxed{I_y = \frac{\pi}{4} b a^3}$.

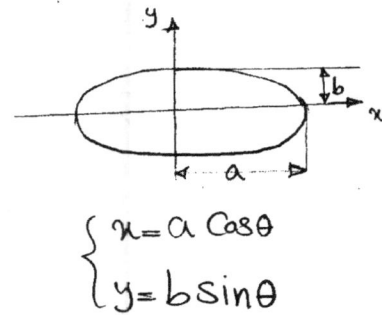

Problem C.25 - Alternate Method:

ellipse $\quad I_x = \frac{\pi a b^3}{4} \quad$ and $\quad I_y = \frac{\pi a^3 b}{4}$

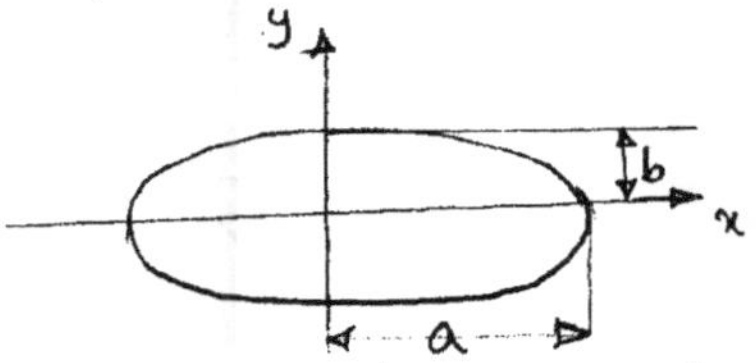

$$\begin{cases} x = a\cos\theta \\ y = b\sin\theta \end{cases}$$

$$I_x = \int y^2\, dA$$

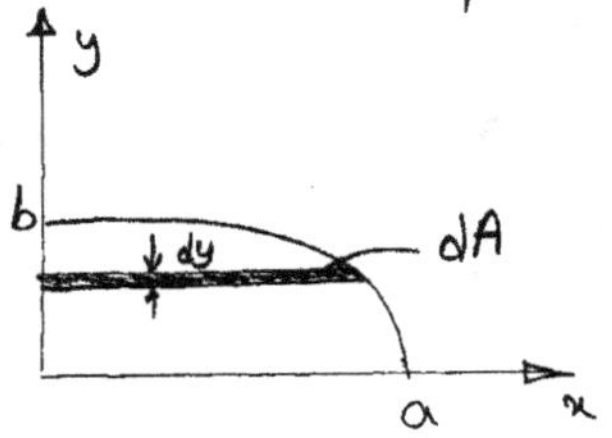

$$dA = x\,dy = (a\cos\theta)(b\cos\theta)\,d\theta$$

$$dA = ab\cos^2\theta\, d\theta$$

$$y^2 = b^2\sin^2\theta$$

$$I_x = \int_A y^2\, dA = 4\int_0^{\pi/2} b^2\sin^2\theta\,(ab)\cos^2\theta\, d\theta$$

$$I_x = 4ab^3\int_0^{\pi/2}\sin^2\theta\cos^2\theta\, d\theta$$

$$I_x = ab^3\int_0^{\pi/2}\sin^2 2\theta\, d\theta$$

$$I_x = ab^3\int_0^{\frac{\pi}{2}}\left(\frac{1-\cos 4\theta}{2}\right)d\theta$$

$$I_x = ab^3\left(\frac{\theta}{2} - \frac{1}{8}\underset{0}{\cancel{\sin 4\theta}}\right)\Bigg|_0^{\frac{\pi}{2}}$$

$$\boxed{I_x = \frac{\pi}{4}ab^3}$$

similarly, $\quad I_y = \int x^2\, dA = \frac{\pi a^3 b}{4}$

$$\boxed{I_y = \frac{\pi}{4}a^3 b}$$

Problem C.26:

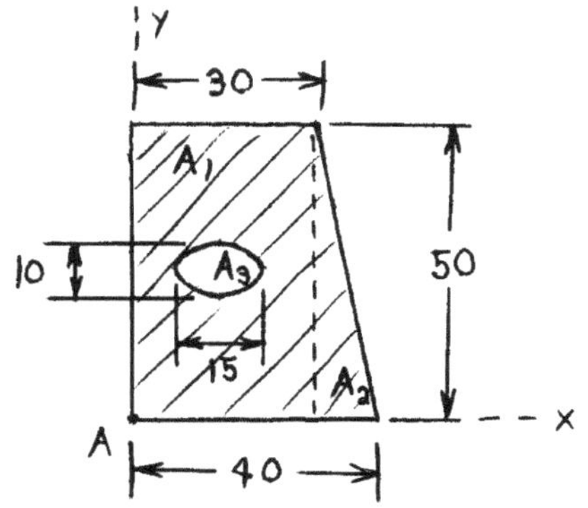

* elliptical hole centered in rectangle

* all dimensions in mm

find: $I_x, I_y = ?$ (about centroidal axes)

① break area into 3 parts: rectangle, triangle, + elliptical hole

②

	A_i (mm^2)	$\bar{X}_i$ (mm)	$A_i\bar{X}_i$ (mm^3)
A_1	$(30)(50) = 1500$	$\frac{1}{2}(30) = 15$	22,500
A_2	$\frac{1}{2}(10)(50) = 250$	$30 + \frac{(10)}{3} = 33.33$	8,333
A_3	$-\pi(7.5)(5) = -117.8$	$\frac{1}{2}(30) = 15$	−1,767
Σ	1,632.2		29,066

③ $$\bar{X} = \frac{\Sigma A_i \bar{x}_i}{\Sigma A_i} = \frac{29{,}066}{1{,}632.2} = 17.81 \text{ mm}$$

④

	A_i (mm^2)	$\bar{y}_i$ (mm)	$A_i\bar{y}_i$ (mm^3)
A_1	1500	$\frac{1}{2}(50) = 25$	37,500
A_2	250	$\frac{1}{3}(50) = 16.67$	4,168
A_3	−117.8	$\frac{1}{2}(50) = 25$	−2,945
Σ	1,632.2		38,723

⑤ $$\bar{y} = \frac{\Sigma A_i \bar{y}_i}{\Sigma A_i} = \frac{38{,}723}{1{,}632.2} = 23.72 \text{ mm}$$

Problem C.26: (con't)

⑥ now place coord. system @ centroid $C = (\bar{x}, \bar{y})$

⑦ $I_{x_1} = I_{xc_1} + A_1 d_{x_1}^2 = \frac{1}{12} b_1 h_1^3 + A_1 (\bar{x}_1 - \bar{x})^2$

⑧ $I_{x_1} = \frac{(30)(50)^3}{12} + (1500)(15 - 17.81)^2 = 324,340 \text{ mm}^4$

⑨ $I_{x_2} = I_{xc_2} + A_2 d_{x_2}^2 = \frac{1}{36} b_2 h_2^3 + A_2 (\bar{x}_2 - \bar{x})^2$

⑩ $I_{x_2} = \frac{(10)(50)^3}{36} + (250)(33.33 - 17.81)^2 = 94,940 \text{ mm}^4$

⑪ $I_{x_3} = I_{xc_3} + A_3 d_{x_3}^2 = \frac{\pi}{4} a_3 b_3^3 + A_3 (\bar{x}_3 - \bar{x})^2$

⑫ $I_{x_3} = \frac{\pi (7.5)(5)^3}{4} + (117.8)(15 - 17.81)^2 = 1666.5 \text{ mm}^4$

⑬ $I_x = I_{x_1} + I_{x_2} - I_{x_3} = 324,340 + 94,940 - 1666.5$

⑭ $\boxed{I_x = 417,610 \text{ mm}^4}$.

⑮ $I_{y_1} = I_{yc_1} + A_1 d_{y_1}^2 = \frac{1}{12} h_1 b_1^3 + A_1 (\bar{y}_1 - \bar{y})^2$

⑯ $I_{y_1} = \frac{(50)(30)^3}{12} + (1500)(25 - 23.72)^2 = 114,960 \text{ mm}^4$

⑰ $I_{y_2} = I_{yc_2} + A_2 d_{y_2}^2 = \frac{1}{36} h_2 b_2^3 + A_2 (\bar{y}_2 - \bar{y})^2$

⑱ $I_{y_2} = \frac{(50)(10)^3}{36} + (250)(16.67 - 23.72)^2 = 13,815 \text{ mm}^4$

⑲ $I_{y_3} = I_{yc_3} + A_3 d_{y_3}^2 = \frac{\pi}{4} b_3 a_3^3 + A_3 (\bar{y}_3 - \bar{y})^2$

⑳ $I_{y_3} = \frac{\pi (5)(7.5)^3}{4} + (117.8)(25 - 23.72)^2 = 1849.7 \text{ mm}^4$

㉑ $I_y = I_{y_1} + I_{y_2} - I_{y_3} = 114,960 + 13,815 - 1849.7$

㉒ $\boxed{I_y = 126,930 \text{ mm}^4}$.

Problem C.27:

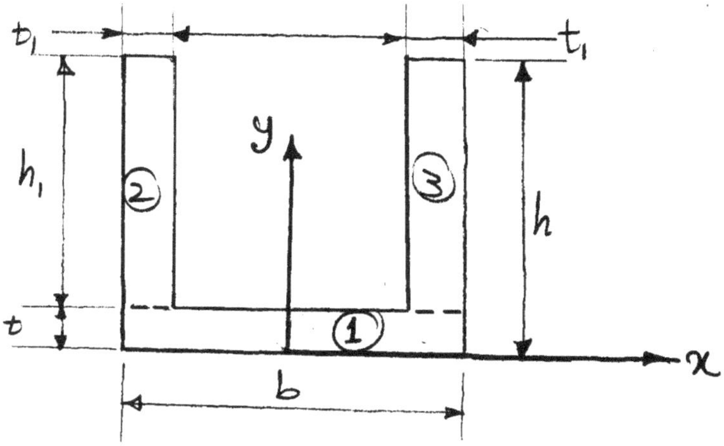

$$\begin{cases} b = 2'' \text{ (in)} \\ t = 0.15'' \\ t_1 = 0.15'' \\ h = 2.0'' \\ h_1 = 1.85'' \\ \bar{y} = ? \\ \bar{x} = 0 \end{cases}$$

$$\bar{y}_1 = \frac{t}{2} = \frac{0.15}{2} = 0.075''$$

$$\bar{y}_2 = \bar{y}_3 = t + \frac{h_1}{2} = 0.15 + \frac{1.85}{2} = 1.075''$$

$$A_1 = bt = 2 \times 0.15 = 0.3 \quad in^2$$

$$A_2 = A_3 = h_1 t_1 = 1.85 \times 0.15 = 0.2775 \quad in^2$$

$$A_t = A_1 + A_2 + A_3 = 0.855 \quad in^2$$

$$\bar{y} = \frac{\sum_{n=1}^{3} \bar{y}_n A_n}{A_t} = \frac{\bar{y}_1 A_1 + \bar{y}_2 A_2 + \bar{y}_3 A_3}{A_t}$$

$$\therefore \boxed{\bar{y} = 0.724 \text{ in}}$$

Problem C.27 - Alternate Method:

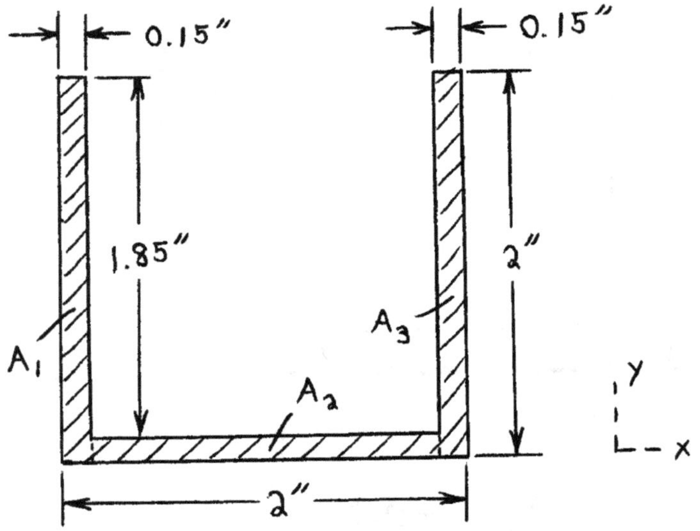

find: $\boxed{\bar{y} = ?}$.

① break area into 3 rectangles

②

	A_i (in^2)	$\bar{y}_i$ (in)	$A_i \bar{y}_i$ (in^3)
A_1	(0.15)(2) = 0.3	½ (2) = 1	0.3
A_2	(0.15)(1.7) = 0.255	½ (0.15) = 0.075	0.01913
A_3	(0.15)(2) = 0.3	½ (2) = 1	0.3
Σ	0.855		0.61913

③ $\bar{y} = \dfrac{\Sigma A_i \bar{y}_i}{\Sigma A_i} = \dfrac{0.61913}{0.855} \Rightarrow \boxed{\bar{y} = 0.7241 \text{ in}}$.

Problem C.28:

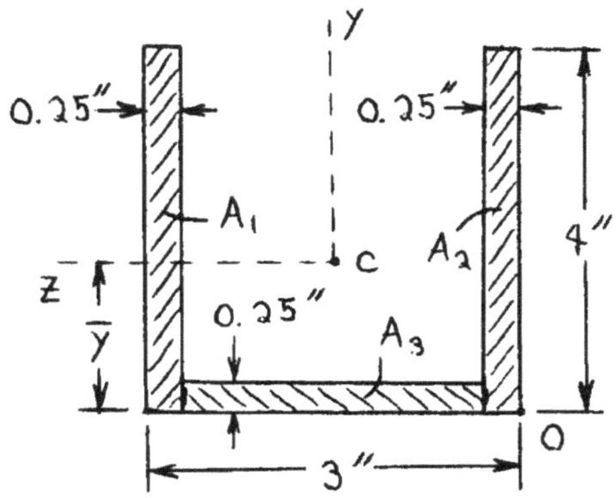

find: $\boxed{\bar{y} = ?}$ (w.r.t. base of section)

① break area into 3 rectangles: A_1, A_2, A_3

② place O @ lower rt. corner

③

	A_i (in²)	$\bar{y}_i$ (in)	
A_1	$(0.25)(4) = 1$	$\frac{1}{2}(4) = 2$	2
A_2	$(0.25)(4) = 1$	$\frac{1}{2}(4) = 2$	2
A_3	$(2.5)(0.25) = 0.625$	$\frac{1}{2}(0.25) = 0.125$	0.078125
Σ	2.625		4.078125

④ $$\bar{y} = \frac{\Sigma A_i \bar{y}_i}{\Sigma A_i} = \frac{4.078125}{2.625} \Rightarrow \boxed{\bar{y} = 1.554 \text{ in}}.$$

Problem C.29(a):

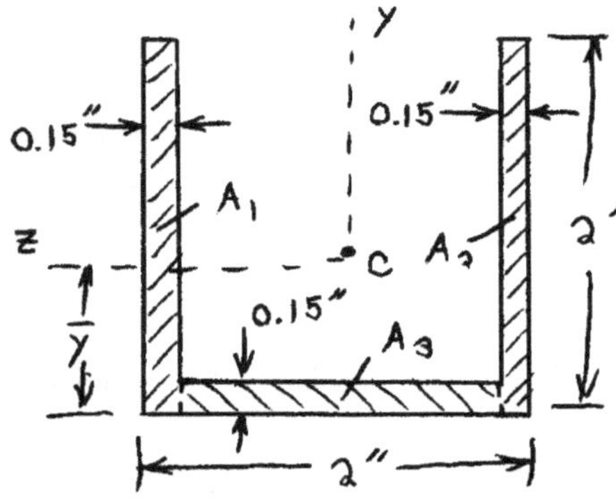

find : $\boxed{I_z = ?}$.

① break area into 3 rectangles:

$A_1,\ A_2,\ A_3$

② find $\bar{y}$ first:

	A_i (in²)	$\bar{y}_i$ (in)	$A_i\bar{y}_i$ (in³)
A_1	(0.15)(2) = 0.3	1	0.3
A_2	(0.15)(2) = 0.3	1	0.3
A_3	(0.15)(1.7) = 0.255	0.075	0.019125
Σ	0.855		0.61913

③ $\bar{y} = \dfrac{\Sigma A_i\bar{y}_i}{\Sigma A_i} = \dfrac{0.61913}{0.855} = 0.7241$ in

④ $I_{z_1} = I_{zc_1} + A_1 d_1^2 = \frac{1}{12}\cdot b_1 h_1^3 + A_1(\bar{y}_1 - \bar{y})^2$

⑤ $I_{z_1} = \frac{1}{12}(0.15)(2)^3 + (0.3)(1 - 0.7241)^2$

⑥ $I_{z_1} = 0.1228$ in⁴ $= I_{z_2}$

⑦ $I_{z_3} = I_{zc_3} + A_3 d_3^2 = \frac{1}{12} b_3 h_3^3 + A_3(\bar{y}_3 - \bar{y})^2$

⑧ $I_{z_3} = \frac{1}{12}(1.7)(0.15)^3 + (0.255)(0.075 - 0.7241)^2$

⑨ $I_{z_3} = 0.1079$ in⁴

⑩ $I_z = I_{z_1} + I_{z_2} + I_{z_3} = 2(0.1228) + 0.1079$

⑪ $\boxed{I_z = 0.3535 \text{ in}^4}$.

Problem C.29(b):

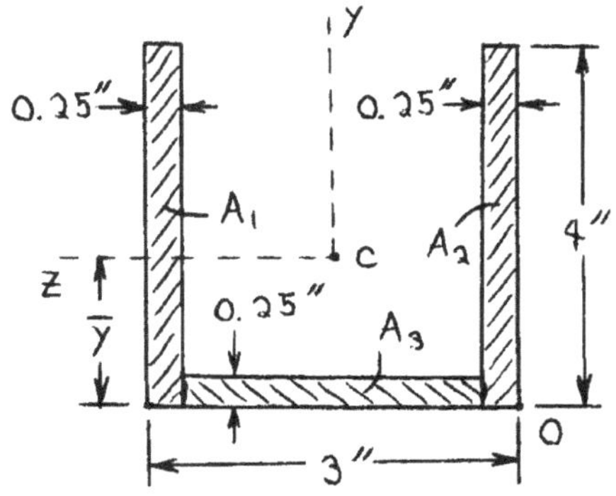

find: $\boxed{I_z = ?}$ (w.r.t. centroid)

① break area into 3 rectangles: A_1, A_2, A_3

② place O @ lower rt. corner

③

	A_i (in²)	$\bar{y}_i$ (in)	
A_1	$(0.25)(4) = 1$	$\frac{1}{2}(4) = 2$	2
A_2	$(0.25)(4) = 1$	$\frac{1}{2}(4) = 2$	2
A_3	$(2.5)(0.25) = 0.625$	$\frac{1}{2}(0.25) = 0.125$	0.078125
Σ	2.625		4.078125

④ $\bar{y} = \dfrac{\Sigma A_i \bar{y}_i}{\Sigma A_i} = \dfrac{4.078125}{2.625} = 1.554 \text{ in}$

⑤ now place coord. system @ centroid C

⑥ $I_{z_1} = I_{zc_1} + A_1 d_1^2 = \frac{1}{12} b_1 h_1^3 + A_1(\bar{y}_1 - \bar{y})^2$

⑦ $I_{z_1} = \frac{1}{12}(0.25)(4)^3 + (1)(2 - 1.554)^2$

⑧ $I_{z_1} = 1.333 + 0.1989 = 1.532 \text{ in}^4$

⑨ since A_2 is identical to A_1: $I_{z_2} = I_{z_1} = 1.532 \text{ in}^4$

⑩ $I_{z_3} = I_{zc_3} + A_3 d_3^2 = \frac{1}{12} b_3 h_3^3 + A_3(\bar{y}_3 - \bar{y})^2$

⑪ $I_{z_3} = \frac{1}{12}(2.5)(0.25)^3 + (0.625)(0.125 - 1.554)^2$

⑫ $I_{z_3} = 0.003255 + 1.276 = 1.279 \text{ in}^4$

⑬ $I_z = I_{z_1} + I_{z_2} + I_{z_3} = 2(1.532) + 1.279$

⑭ $\boxed{I_z = 4.343 \text{ in}^4}$

Problem C.34:

$$\begin{cases} \bar{x} = ? \\ \bar{y} = ? \end{cases}$$

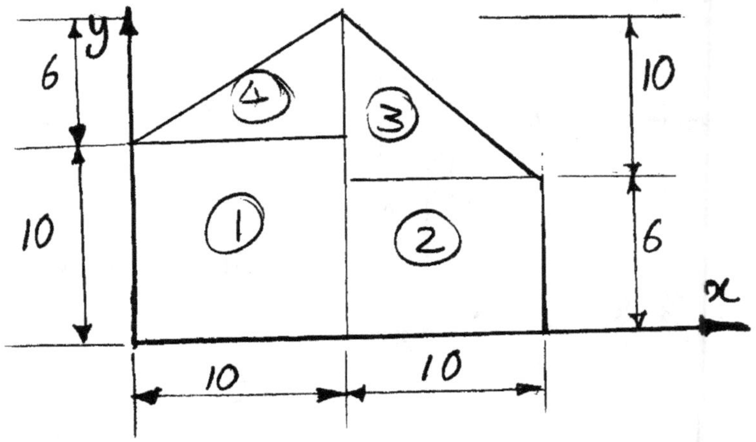

$\bar{x}_1 = 5$ $\quad \bar{y}_1 = 5$ $\quad A_1 = 100$

$\bar{x}_2 = 15$ $\quad \bar{y}_2 = 3$ $\quad A_2 = 60$

$\bar{x}_3 = 10 + \frac{10}{3} = 13.3$ $\quad \bar{y}_3 = 6 + \frac{10}{3} = 9.3$ $\quad A_3 = \frac{10 \times 10}{2} = 50$

$\bar{x}_4 = \frac{2}{3} \times 10 = 6.7$ $\quad \bar{y}_4 = 10 + \frac{6}{3} = 12$ $\quad A_4 = \frac{10 \times 6}{2} = 30$

$A_t = 240$

$$\bar{x} = \frac{\sum_{n=1}^{4} \bar{x}_n A_n}{A_t} = \frac{5 \times 100 + 15 \times 60 + 13.3 \times 50 + 6.7 \times 30}{240}$$

$$\therefore \boxed{\bar{x} = 9.4}$$

$$\bar{y} = \frac{\sum_{n=1}^{4} \bar{y}_n A_n}{A_t}$$

$$\therefore \boxed{\bar{y} = 6.3}$$

Problem C.34 - Alternate Method:

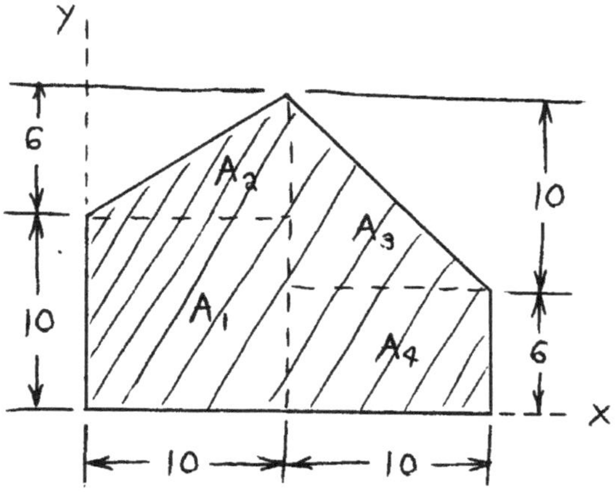

find: $\boxed{\bar{x}, \bar{y} = ?}$.

① break area into 4 parts: 2 rectangles + 2 triangles

②

	A_i	$\bar{x}_i$	$A_i \bar{x}_i$
A_1	$(10)^2 = 100$	5	500
A_2	$\frac{1}{2}(6)(10) = 30$	$\frac{2}{3}(10) = 6.667$	200
A_3	$\frac{1}{2}(10)^2 = 50$	$10 + \frac{1}{3}(10) = 13.333$	666.7
A_4	$(6)(10) = 60$	$10 + 5 = 15$	900
Σ	240		2,266.7

③ $\bar{x} = \dfrac{\Sigma A_i \bar{x}_i}{\Sigma A_i} = \dfrac{2,266.7}{240} \Rightarrow \boxed{\bar{x} = 9.445}$.

④

	A_i	$\bar{y}_i$	$A_i \bar{y}_i$
A_1	100	5	500
A_2	30	$10 + \frac{1}{3}(6) = 12$	360
A_3	50	$6 + \frac{1}{3}(10) = 9.333$	466.7
A_4	60	3	180
Σ	240		1,506.7

⑤ $\bar{y} = \dfrac{\Sigma A_i \bar{y}_i}{\Sigma A_i} = \dfrac{1,506.7}{240} \Rightarrow \boxed{\bar{y} = 6.278}$.

ANSWERS TO SELECT PROBLEMS

Chapter 1

1.2 103.0 N on Earth
39.02 N on Mars

1.3 $W_s = 11.43$ N

1.14 p = 152.8 psi

1.16 d = 2.985 in.

1.21(a) F= 10,680 N
1.21(f) F = 422.6 kN

1.22(a) F = 502.2 lb
1.22(c) F = 2.968 kip

1.27(a) $\sigma = 82.74$ MPa
1.27(d) $\sigma = 913.6$ MPa

1.28(a) $\sigma = 145.0$ ksi
1.28(d) $\sigma = 203.0$ psi

Chapter 2

2.16(a) $S_v = 8{,}660$ N, $\theta_{Sv} = 90°$
2.16(b) $D_v = 13{,}230$ N, $\theta_{Dv} = -40.9°$
2.16(c) $D_v = 13{,}230$ N, $\theta_{Dv} = 139.1°$

2.17(a) $S_v = 913.9$ N, $\theta_{Sv} = 70.04°$
2.17(b) $D_v = 3{,}093$ N, $\theta_{Dv} = -46.93°$
2.17(c) $D_v = 3{,}093$ N, $\theta_{Dv} = 133.1°$

2.18 $F_x = 15{,}105$ lb
$F_{y'} = -3{,}202$ lb

2.19 $F_{AB} = 1{,}949$ N; $F_{CB} = 1{,}647$ N

2.21 $S_v = 15.61$ kN; $\theta = 209.7°$

2.22 F = 8.640 kN; $\theta = 46.67°$

2.25 $F_1 = 7{,}072$ lb; $\theta = 49.0°$

2.26 $F_1 = 10{,}120$ lb; $\theta = 55.87°$

2.33 $M_O = 438.8$ N-m (CCW)

2.34 $M_O = 86.6$ N-m (CCW)

2.37(a) $\bar{r} = (7\bar{i} + 4\bar{j} + 6\bar{k})$ in.
r = 10.05 in.
$\alpha = 45.85°$, $\beta = 66.55°$, $\gamma = 53.34°$
2.37(b) $\bar{r} = (110\bar{i} + 65\bar{j} + 70\bar{k})$ mm
r = 145.7 mm
$\alpha = 40.98°$, $\beta = 63.50°$, $\gamma = 61.29°$

2.39(b) $F_x = 388.2$ lb, $F_y = 964.2$ lb
$F_z = 1082$ lb
$\bar{F} = (388.2\bar{i} + 964.2\bar{j} + 1082\bar{k})$ lb

2.39(d) $F_x = 16.07$ kip, $F_y = -19.15$ kip
$F_z = 0$
$\bar{F} = (16.07\bar{i} - 19.15\bar{j})$ kip

2.40 $\bar{F} = \bar{F}_1 + \bar{F}_2 = (26.25\bar{i} + 37.23\bar{j} - 15.9\bar{k})$ kN

2.41(a) $\theta = 87.35°$
2.41(b) $\theta = 18.28°$

2.43(a) $\bar{M}_O = (-6.4\bar{i} + 14.95\bar{j} - 7.85\bar{k})$ kN - m
2.43(b) $\bar{M}_O = (-14.5\bar{i} - 13.28\bar{j} + 3.5\bar{k})$ ft - kip

2.45 $x_B = -58.62$ ft; $y_B = -3.307$ ft

2.47 $\bar{F} = (5.057\bar{i} + 2.528\bar{j} - 20.23\bar{k})$ kN
See equation above for F_x, F_y, F_z.
$\bar{M}_O = (-80.91\bar{i} + 20.23\bar{j} - 17.70\bar{k})$ kN - m
See equation above for M_x M_y, M_z.
$M_O = 85.26$ kN-m; $\theta = 28.15°$
$\bar{u}_{M_O} = (-0.949\bar{i} + 0.237\bar{j} - 0.208\bar{k})$
$F_{OQ} = 18.52$ kN(Compression)

2.48 $\bar{F} = (741.8\bar{i} + 3{,}338\bar{j} - 6{,}677\bar{k})$ lb
See equation above for F_x, F_y, F_z.

2.48 Continued:

$\overline{M}_O = (-73.44\bar{i} + 33.38\bar{j} + 8.530\bar{k})$ ft - kip

See equation above for M_x M_y, M_z.

$M_O = 81.12$ ft-kip

$\bar{u}_{Mo} = (-0.9053\bar{i} + 0.4115\bar{j} + 0.1052\bar{k})$

$\theta = 36.09°$;

$F_{OQ} = 6.061$ kip (Compression)

Chapter 3

3.6 $F_{CB} = 220.9$ lb; $F_{CA} = 585.3$ lb

3.7 $F_{AC} = 11.97$ kN; $F_{BC} = 9.807$ kN

3.9 $T_{AC} = 1{,}189$ N; $T_{BC} = 1{,}158$ N

3.10 F = 1.879 ton
Pulley system is not effective.

3.11 F = 4.358 kN
Pulley system is effective.

3.15 $F = W(1 - (r_2/r_1))/2$

3.16 $F_{AB} = 421.7$ N; $F_{BD} = 730.4$ N;
$F_{CG} = 1461$ N; $F_{BC} = F_{CE} = 843.4$ N

3.18 $T_{AB} = 186.0$ N; $T_{BC} = 173.9$ N
$T_{CD} = 212.0$ N; $m_C = 7.413$ kg

3.19 $W_{max} = 6{,}200$ lb > 140 lb
Stuntman will remain dry.

3.21 $\delta_T = 3.851$ in.; $k_{eff} = 6.076$ lb/in.

3.22 $\delta_T = 48.57$ mm; $k_{eff} = 1.441$ N/mm

3.23 $\delta_T = 24.36$ mm; $k_{eff} = 47.00$ N/mm

3.26 $A_x = 0$; $A_y = 10$ kN; $B_y = 30$ kN

3.27 $A_x = 0$; $A_y = 1{,}406$ lb; $B_y = 2{,}344$ lb

3.28 $A_x = 0$; $A_y = 26.16$ kN; $B_y = 53.16$ kN

3.29 $A_x = 0$; $A_y = 11$ kip;
$M_{RA} = 121$ kip-ft (CCW)

3.30 $A_x = 0$; $A_y = 37.33$ kN;
$M_{RA} = 161.8$ kN-m (CCW)

3.31 $A_x = 625.8$ lb; $A_y = 491.8$ lb
$B_y = 559.0$ lb

3.33 $A_x = 0$; $A_y = B_y = 40$ kN

3.36 $W_{max} = 1.892$ ton

3.37 $R_x = 0$; $R_y = 38.25$ kN
$M_z = 860.6$ kN-m

3.38 $A_x = -13.5$ kip; $B_x = 7.5$ kip;
$A_y = B_y = 4.5$ kip

3.39 $P_A = 36$ kN; $P_B = 18$ kN

3.40 V = F/4; M = FL/8

3.42 $V = q_0L/6$; $M = q_0L^2/9$

3.44 V = – F; M = – (3/4) FL

3.46 V = – 4 kip; M = 76.8 kip-ft

3.48 P = 250 lb, V = 0; $M_z = -375$ in-lb

Chapter 4

4.1 $\sigma = 19.10$ ksi; $\varepsilon = 0.0006367$
$L_f = 22.014$ ft

4.4 $\sigma_{max} = 1.104$ MPa, $\sigma_{12} = 764.8$ kPa
$\sigma_{23} = 320.7$ kPa

4.7 $P_{design} = 11.77$ kN

4.9 $\sigma = 1744$ psi

4.13 4340 Hr — Gage # 4-Os
52100 A — Gage # 4-Os

4.18 $\sigma = 22$ MPa; $\delta = 0.1913$ mm

4.19 $\sigma_{design} = 16.88$ ksi; A = 1.185 in.^{2}

4.22 $\tau = 21.33$ ksi

4.25 $P_{max} = 1.651$ kip

4.27 $\sigma_\phi = (6.365)\sin^2 \phi$ MPa
$\tau_\phi = (6.365)\sin \phi \cos \phi$ MPa

4.28 $\sigma = 90.0$ MPa; $\theta = 161.6°$
F = 15.75 kN

4.32 $\delta = 2.235$ mm

4.35 $\sigma_1 = 30.30$ ksi; $\sigma_2 = -1.538$ ksi
$\delta = 0.01841$ in.

4.36 $\sigma_{nom} = 20$ ksi; $\sigma_{max} = 46.6$ ksi

4.37 $\sigma_{nom} = 25$ MPa; $\sigma_{max} = 43.75$ MPa

4.41 $P_{max} = 343.5$ kN

4.42 $t_{min} = 0.320$ in.

4.43 $\sigma_a = -174$ MPa; $\sigma_f = -20$ kPa;
$\delta = 4.833$ mm (shorter)

4.44 $\sigma_s = 33.03$ ksi; $\sigma_b = -8.256$ ksi
$\delta = 0.006192$ in. (shorter)

4.45 $\sigma_s = -7.761$ ksi; $\sigma_c = -1.164$ ksi;
$\delta = 0.02483$ in (shorter)

4.46 $\sigma_a = 19.04$ MPa; $\sigma_b = 761.7$ kPa;
$\delta_a = 0.1983$ mm (longer);
$\delta_b = 0.001731$ mm (longer)

4.47 $\sigma_c = -20.49$ MPa; $\sigma_s = -512.1$ MPa
$\delta_c = 0.4647$ mm (longer)
$\delta_s = 0.4647$ mm (shorter)

4.48 $\sigma_s = 1.784$ ksi; $\sigma_a = -467$ psi

Chapter 5

5.5 $S_y = 46.1$ ksi; $S_u = 62.2$ ksi
S_y (0.2%) = 56.7 ksi

5.6 $S_y = 36$ ksi; $S_u = 44$ ksi
S_y (0.2%) = 38.5 ksi

5.7 %e = 32.8%; %A = 31.38%

5.9 d = 0.4997 in.

5.12 $V = V_0 [1 + (1 - 2\nu)\varepsilon_a] = V_0$
when $\nu = ½$ and higher order strain terms are neglected.

5.13 $\varepsilon_y = 0.001267$

5.14 $\varepsilon_a = 0.003153$; $\varepsilon_t = -0.001009$

5.15 $\varepsilon_x = \varepsilon_y = 0.0004807$

5.18 $\sigma_a = 52.8$ MPa; $\sigma_h = 105.6$ MPa

5.19 $\sigma = 57.55$ ksi; $\sigma_T = 107.4$ ksi

5.23 $N = \infty$ infinite life in fatigue

5.26 $\sigma_{max} = 255$ MPa

5.33 E = 30,000 ksi; G = 11,490 ksi
$\nu = 0.3054$

5.34 $E = 9.888 \times 10^6$ psi, $\sigma_u = 59.08$ ksi
$\sigma_f = 48.89$ ksi; $\varepsilon_f = 0.2100$
σ_y (0.2%) = 40.21 ksi
%e = 21.00%; %A = 31.77%

5.35 E = 199.6 GPa, $\sigma_u = 300.0$ MPa
$\sigma_f = 150.0$ MPa; $\varepsilon_f = 0.1864$
σ_y (0.2%) = 199.6 MPa
%e = 18.64%; %A = 49.78%

Chapter 6

6.8 $P_{AC} = 26.25$ kip (T); $P_{BC} = 52.5$ kip (T)
$P_{CD} = 36.34$ kip (C); $P_{CE} = 42.5$ kip (T)

6.10 $P_{AC} = 67.48$ kip (T); $P_{BC} = 135.0$ kip (T)
$P_{CD} = 67.08$ kip (C); $P_{CE} = 97.48$ kip (T)

6.12 $A_{AB} = 6.261$ in.^{2}; $A_{AC} = 2.8$ in.^{2}
$A_{BD} = 2.8$ in.^{2}; $A_{BC} = 5.6$ in.^{2}

6.13 $A_{AB} = 1{,}933$ mm^2; $A_{AF} = 1{,}694$ mm^2;
$A_{BC} = 1{,}607$ mm^2; $A_{BF} = 285.7$ mm^2

6.17 $P_{DE} = 0$; $P_{DF} = P_{DB} = 30$ kip (C)

6.20 $P_{DC} = 100$ kN (T); $P_{DE} = 131.7$ kN (T)
$P_{DF} = 257.1$ kN (C)

6.23 $P_{AC} = 200(s/h)$;
$P_{AD} = -200[(s/h)^2 + 1]^{1/2}$

6.25 $P_{AC} = 3.907$ kN (T); $P_{CD} = 0.2504$ kN (C)
$P_{CE} = 4.206$ kN (T); $P_{DF} = 4.206$ kN (C)

6.26 $P_{CD} = 41.08$ kN (C); $P_{CJ} = 11.27$ kN (T)
$P_{KJ} = 34.53$ kN (T)

6.30 $\sigma_{BC} = 10.4$ ksi (T), $SF_{BC} = 4.038$

6.35 $A_{AB} = 4{,}340$ mm^2; $A_{BD} = 4{,}340$ mm^2
$A_{DE} = 5{,}482$ mm^2; $A_{EF} = 2{,}856$ mm^2
$A_{CF} = 2{,}284$ mm^2

Chapter 7

7.4 $P_{EB} = -8.944$ kips; $d_{EB} = 1.033$ in.
$P_{EC} = -4.472$ kips; $d_{EC} = 0.7305$ in.
$P_{ED} = 15.56$ kips; $d_{ED} = 1.363$ in.

7.13 $MOS_{AB} = 165.1\%$
$MOS_{AP} = MOS_{AQ} = 181.1\%$

7.14 $d_{AB} = 8.148$ mm; $d_{AC} = 7.374$ mm
$d_{AD} = 13.98$ mm

7.17 Gage # 2; steel wire

7.18 $P_{AD} = 165.9$ lb; $P_{BD} = 437.6$ lb
$P_{CD} = 441.1$ lb

7.19 Gage # 000s steel wire

7.20 W = 1007 lb; $A_x = -419.6$ lb
$A_y = 2{,}518$ lb; $A_z = -881.2$ lb
$P_{BC} = 1{,}730$ lb; $P_{DE} = 1{,}841$ lb

7.21 F= 104.2 kN; A_x =41.7 kN
$A_y = 0$; $A_z = 250$ kN
$B_x = -145.9$ kN; $B_z = 250$ kN

Chapter 8

8.4 F = 36.84 N

8.5 $G_y = 2{,}591$ N; MA = 8.637

8.12 $P_C = 30$ ton; MA 6

8.13 $F_C = 165$ lb; W = 8.4 in-lb
$\Delta d_C = 0.0509$ in.

8.14 $A_x = -4.5$ kip; $A_y = 7.3$ kip
$M_{RA} = 64.35$ kip-ft; $C_y = -1.443$ kip
$C_x = -12.75$ kip; $P_{BD} = -15.46$ kip

8.15 $B_y = 3.84$ kN; $C_x = 0$; $C_y = -9.6$ kN;
$E_x = 0$; $E_y = 0.96$ kN; $P_{AF} = 0$

8.17 $P_{AC} = 1{,}178$ N; $P_{BC} = -834.4$ N
$P_{CE} = 627.9$ N; D = 699.3 N

8.21 $C_x = 2.5$ kip; $C_y = 14.59$ kip; $A_x = 0$;
$A_y = 34$ kip; $M_{RA} = 50$ kip-ft (CCW);
T = 13.35 kip; F = 15.61 kip

Chapter 9

9.5(a) $\mu_{Ave} = 0.4156$; $\mu_{Range} = 0.0515$

9.5(b) $\mu_{Ave} = 0.4530$; $\mu_{Range} = 0.0656$

9.7 F = 1,845 N

9.9 $(F_{fW})_{max} = 57.95$ lb $< F_{fW} = 64.48$ lb
Worker slips and cannot move crate.

9.11 $(F_1)_{max} = 1094$ lb

9.12 $F_{fA} = 26.17$ lb < 57.93 lb $= (F_{fA})_{max}$
The ladder is stable.

9.13 F = 240 N; The crate begins to tip.

9.14 $(F_{fA})_{max} = 156.4$ N < 210 N $= F_{fA}$
Cylinder slips and remains in corner.

9.30 $F_{fB} = 21.25$ lb < 50 lb $= (F_{fB})_{max}$
$x_B = 0.4808$ ft < 1.5 ft
$F_{fA} = 21.25$ lb < 38.25 lb $= (F_{fA})_{max}$
Both blocks are in equilibrium.

9.31 $(W_1)_{max} = 1.667$ kN; $(W_1)_{min} = 0.3$ kN

Appendix C

C.17 $\bar{x} = 0$ $\bar{y} = \dfrac{4r}{3\pi}$

C.18 $\bar{x} = 0$ $\bar{y} = \dfrac{4b}{3\pi}$

C.23 $\bar{x} = 17.81$ mm $\bar{y} = 23.72$ mm

C.25 $I_x = (\pi/4)ab^3$; $I_y = (\pi/4)ba^3$

C.27 $\bar{y} = 0.7241$ in

C.28 $\bar{y} = 1.554$ in.

C.29(a) $I_z = 0.3535$ in^4

C.29(b) $I_z = 4.343$ $in.^4$

C.34 $\bar{x} = 9.445$ mm $\bar{y} = 6.278$ mm

www.ingramcontent.com/pod-product-compliance
Lightning Source LLC
Chambersburg PA
CBHW081240220326
41597CB00023BA/4221
* 9 7 8 1 9 3 5 6 7 3 3 4 7 *